NONLINEAR FUNCTIONAL ANALYSIS AND DIFFERENTIAL EQUATIONS

PURE AND APPLIED MATHEMATICS

A Program of Monographs, Textbooks, and Lecture Notes

LECTURE NOTES
IN PURE AND APPLIED MATHEMATICS

1. *N. Jacobson,* Exceptional Lie Algebras
2. *L.-Å. Lindahl and F. Poulsen,* Thin Sets in Harmonic Analysis
3. *I. Satake,* Classification Theory of Semi-Simple Algebraic Groups
4. *F. Hirzebruch, W. D. Newmann, and S. S. Koh,* Differentiable Manifolds and Quadratic Forms
5. *I. Chavel,* Riemannian Symmetric Spaces of Rank One
6. *R. B. Burckel,* Characterization of C(X) Among Its Subalgebras
7. *B. R. McDonald, A. R. Magid, and K. C. Smith,* Ring Theory: Proceedings of the Oklahoma Conference
8. *Yum-Tong Siu,* Techniques of Extension of Analytic Objects
9. *S. R. Caradus, W. E. Pfaffenberger, and Bertram Yood,* Calkin Algebras and Algebras of Operators on Banach Spaces
10. *Emilio O. Roxin, Pan-Tai Liu, and Robert L. Sternberg,* Differential Games and Control Theory
11. *Morris Orzech and Charles Small,* The Brauer Group of Commutative Rings
12. *S. Thomeier,* Topology and Its Applications
13. *Jorge M. López and Kenneth A. Ross,* Sidon Sets
14. *W. W. Comfort and S. Negrepontis,* Continuous Pseudometrics
15. *Kelly McKennon and Jack M. Robertson,* Locally Convex Spaces
16. *M. Carmeli and S. Malin,* Representations of the Rotation and Lorentz Groups: An Introduction
17. *George B. Seligman,* Rational Methods in Lie Algebras
18. *Djairo Guedes de Figueiredo,* Functional Analysis: Proceedings of the Brazilian Mathematical Society Symposium
19. *Lamberto Cesari, Rangachary Kannan, and Jerry D. Schuur,* Nonlinear Functional Analysis of Differential Equations: Proceedings of the Michigan State University Conference

Other volumes in preparation

Proceedings of the Michigan State University Conference

NONLINEAR FUNCTIONAL ANALYSIS AND DIFFERENTIAL EQUATIONS

Edited by

LAMBERTO CESARI
Department of Mathematics
University of Michigan
Ann Arbor, Michigan

RANGACHARY KANNAN
Department of Mathematical Sciences
University of Missouri
St. Louis, Missouri

JERRY D. SCHUUR
Department of Mathematics
Michigan State University
East Lansing, Michigan

MARCEL DEKKER, INC. New York and Basel

MARCEL DEKKER, INC.

270 Madison Avenue, New York, New York 10016

LIBRARY OF CONGRESS CATALOG CARD NUMBER: 76-1797

ISBN: 0-8247-6452-8

Current printing (last digit):
10 9 8 7 6 5 4 3 2 1

PRINTED IN THE UNITED STATES OF AMERICA

FOREWORD

In recent years there has been considerable research in nonlinear functional analysis and its applications to differential equations. Hence it seemed appropriate, in the spring of 1975, to plan a conference which would bring together workers in this field to exchange thoughts, lay out a view of the present state of the field, and explore areas of future research.

This conference was held June 9 to 12, 1975 at Michigan State University with R. Kannan, then a visitor at Michigan State, and J. D. Schuur as its directors. Its format was six one-hour lectures by the principal speaker, L. Cesari, who developed in depth some aspects of the general topic; nine one-hour lectures by other invited speakers, who reviewed different aspects of the general topic; and research reports by participating mathematicians. This volume contains the lectures of the principal speaker and of the one-hour speakers and constitutes therefore the proceedings of the conference.

For his topic, Professor Cesari chose "Functional Analysis, Nonlinear Differential Equations, and the Alternative Method." He emphasized points at present under investigation, in particular the use of Banach's fixed point theorem, a priori bounds and topological arguments, monotone operators, and Schauder's fixed point theorem.

The one-hour talks ranged from problems in bifurcation theory to measurability of solutions of differential equations to current aspects of degree theory.

In such a large and active area, a survey will necessarily be limited by the choice of topics and a cutting point in time. However, we hope that this volume will prove useful in furthering the aims of the conference.

The editors are grateful to the Michigan State University Mathematics Department and its chairman, Professor J.E. Adney, for financial support, advice, and encouragement.

We also appreciate the work of Marcel Dekker in producing this volume, and of Mary Reynolds in typing it.

Lamberto Cesari

Rangachary Kannan

Jerry D. Schuur

CONFERENCE PARTICIPANTS

Walter Becker, Department of Physics, Michigan State University

Tom Bowman, Department of Mathematics, University of Florida

Russell Caflisch, Department of Mathematics, Michigan State University

Alfonso Castro, Department of Mathematics, University of Cincinnati

Lamberto Cesari, Department of Mathematics, University of Michigan

Shannon Coffey, Department of Mathematics, University of Cincinnati

Peter Dunkelberg, Department of Mathematics, University of Florida

Dennis Dunninger, Department of Mathematics, Michigan State University

Eric Grosse, Department of Mathematics, Michigan State University

Clark Givens, Department of Mathematics, Michigan Technical University

Carl Ganser, Department of Mathematics, Michigan State University

Tom Haigh, Department of Mathematics, Rose Hulman Institute

David Hale, Department of Mathematics, Michigan State University

Dan Henry, Department of Mathematics, Northwestern University and University of Kentucky

William Hall, Department of Mathematics, University of Pittsburgh

Rangachary Kannan, Department of Mathematics, Michigan State University and University of Missouri at St. Louis

Gene Klaasen, Department of Mathematics, University of Tennessee

George Knightly, Department of Mathematics, University of Massachusetts

J. C. Kurtz, Department of Mathematics, Michigan State University

Alan Lazer, Department of Mathematics, University of Cincinnati

Chi Lo, Department of Mathematics, Michigan State University

John Locker, Department of Mathematics, Colorado State University

John Mallet-Paret, Department of Mathematics, Brown University

P.J. McKenna, Department of Mathematics, University of Michigan

David McSweeney, Department of Mathematics, University of Kansas

R. Kent Nagle, Department of Mathematics, University of Michigan

Aubrey Poore, Department of Mathematics, Colorado State University

Erich Rothe, Department of Mathematics, University of Michigan

Habib Salehi, Department of Mathematics, Michigan State University

Jerry Schuur, Department of Mathematics, Michigan State University

Howard Shaw, Department of Mathematics, University of Michigan

Harlan Stech, Department of Mathematics, Michigan State University

M.B. Suryanarayana, Department of Mathematics, Eastern Michigan University

Daniel Sweet, Department of Mathematics, University of Maryland

Russell Thompson, Department of Mathematics, Northern Illinois University

Curtis Travis, Department of Mathematics, University of Tennessee

Nancy Waller, Department of Mathematics, Michigan State University

Richard Weinacht, Department of Mathematics, University of Delaware

P.K. Wong, Department of Mathematics, Michigan State University

Dave Zachmann, Department of Mathematics, Colorado State University

CONTENTS

FUNCTIONAL ANALYSIS, NONLINEAR DIFFERENTIAL EQUATIONS, AND THE ALTERNATIVE METHOD

Lamberto Cesari

University of Michigan

Section 1. The Alternative Method

1. Introduction. In this presentation of the alternative, or bifurcation method emphasis is given to certain points of recent research, in particular in connection with the use of Banach's fixed point theorem (nos. 6-19), a priori bounds and topological arguments (nos. 20-27), monotone operators (nos. 28-33), and Schauder's fixed point theorem (nos. 34-42).

We are interested in solving an operator equation $Ax = 0$ in a space X. We shall think of A as the difference $A = E - N$ of a linear operator E, not necessarily bounded, and of an operator N, not necessarily linear, and thus the equation takes the form

$$Ex = Nx, \qquad x \in X.$$

If E^{-1} exists, or equivalently the null space of E is trivial, then this equation can be written as $x - E^{-1}Nx = 0$, or $(I + KN)x = 0$. This is a Hammerstein equation and very extensive work has been done in this direction. For references to the wide literature one can see [147].

Whenever KN is compact, concepts and methods of topological degree theory in function spaces are relevant, and we should refer to the work of Leray-Schauder [167], Rothe [184], and Nagumo [177, 178].

We are particularly interested in the case where E has a nontrivial null space, the case which is often mentioned as the "problem at resonance." Thus we shall not exclude that E may have a nonzero null space. This will be actually the most interesting case.

In the line of the bifurcation process of Poincaré [180], Lyapunov [171], and Schmidt [187] we should decompose the equation $Ex = Nx$ into a system of two equations, possibly in different spaces. Much work has been done in this direction, for which we refer to [191] and [192].

The injection of ideas of functional analysis in the last years has made the process a remarkably fine tool of analysis, particularly in the difficult "problem at resonance" in the usual terminology.

The general theory which has ensued, with all its variants and ramifications through the works of many authors, is often referred to as the bifurcation theory, or the alternative method ([6], [7], [65], and recent expositions, e.g., [21-23], [63]). It is being used in theoretical existence analysis of the solutions, in methods of successive approximations of the solutions, in estimating the error of approximate solutions, in problems of perturbations, and in problems of bifurcation of eigenvalues.

2. The Alternative Scheme. To be specific, let us consider an equation of the form

$$Ex = Nx, \tag{2.1}$$

where $E:\mathcal{D}(E) \to Y$ is a linear operator, and $N:\mathcal{D}(N) \to Y$ is an operator, not necessarily linear, both E and N have domains $\mathcal{D}(E)$, $\mathcal{D}(N)$ in a Banach space X and ranges $\mathcal{R}(E)$, $\mathcal{R}(N)$ in a Banach space Y, $\mathcal{D}(E) \cap \mathcal{D}(N) \neq \emptyset$.

Let us assume for a moment that the following holds (we shall see in no. 3 that these assumptions are quite natural, and they are also easily verified in the so-called selfadjoint case). Let us assume that there are projection operators $P:X \to X$, $Q:Y \to Y$ (that is, linear, bounded, idempotent, or $PP = P$, $QQ = Q$), and a linear operator H, which we shall consider as a partial inverse of E satisfying

$$H(I - Q)Ex = (I - P)x \quad \text{for all } x \in \mathcal{D}(E), \qquad (k_1)$$

$$QEx = EPx \quad \text{for all } x \in \mathcal{D}(E), \qquad (k_2)$$

$$EH(I - Q)Nx = (I-Q)Nx \quad \text{for all } x \in \mathcal{D}(E) \cap \mathcal{D}(N). \qquad (k_3)$$

Under these assumptions, the equation $Ex = Nx$ is equivalent to the system of two equations

$$x = Px + H(I - Q)Nx, \qquad (2.2)$$

$$Q(Ex - Nx) = 0. \qquad (2.3)$$

Indeed, if (2.1) is satisfied, then by applying Q to equation (2.1) we obtain (2.3). By applying $H(I - Q)$ to (2.1) we obtain $H(I - Q)Ex = H(I - Q)Nx$, and by using (k_1) we obtain (2.2). Conversely, if (2.2) and (2.3) are satisfied, then, by applying E to (2.2) we have $Ex - EPx = EH(I - Q)Nx$, and by using (k_2), (k_3), also $Ex - QEx = (I - Q)Nx$, or $Ex - Nx = Q(Ex - Nx) = 0$.

Equations (2.2) and (2.3) are usually denoted as the auxiliary and the bifurcation equations, respectively.

Let $X_0 = PX$, $X_1 = (I - P)X$, $Y_0 = QY$, $Y_1 = (I - Q)Y$, so that X and Y have the decompositions $X = X_0 + X_1$, $Y = Y_0 + Y_1$ (direct sums), $X_0 \cap X_1 = \{0\}$, $Y_0 \cap Y_1 = \{0\}$. The linear subspace of X of all elements $x \in X$ for which $Ex = 0$ is often denoted as the null space of E, or kernel of E, and denoted by $\ker E$. When there are elements $x \neq 0$ in X with $Ex = 0$, that is, $\ker E$ is not trivial, then we say that we have a resonance case. We shall always assume that $\ker E \subset X_0$. Also, we shall assume that $Y_1 \subset \mathcal{R}(E)$, $\mathcal{D}(H) = Y_1$, $\mathcal{R}(H) = \mathcal{D}(E) \cap X_1$. Thus $EHy = y$ for all $y \in Y_1$, $HEx = x$ for all $x \in \mathcal{D}(E) \cap X_1$. Often, we may assume $\ker E = X_0$, $Y_1 = \mathcal{R}(E)$. In this particular situation the bifurcation equation reduces to $QNx = 0$.

If we write $x^* = Px$, then the auxiliary equation can always be written in the form of a fixed point statement:

$$x = Tx, \qquad Tx = x^* + H(I - Q)Nx, \tag{2.4}$$

where T is easily seen to map the fiber $P^{-1}x^*$ into itself. The auxiliary equation can also be written in the form of a Hammerstein equation:

$$x + KNx = x^*, \qquad K = -H(I - Q). \tag{2.5}$$

Finally the auxiliary equation, written in the equivalent form

$$Ex - Nx = Q(Ex - Nx),$$

is often called the relaxed equation, or relaxed problem.

There are certain rather general situations (N locally Lipschitzian, N monotone, etc.) under which the operators P, Q, H can be so chosen that the ensuing map T is a contraction map on suitable sets, or has other suitable functional analysis properties, so that the auxiliary equation is uniquely solvable for every $x^* \in X_0$, and we may write its solution in the form $x = \mathcal{J}(x^*)$ with $Px = P\mathcal{J}(x^*) = x^*$, $x^* \in X_0$, or

$$x = [I - H(I - Q)N]^{-1}x^*.$$

By replacing $x = \mathcal{J}(x^*)$ in the bifurcation equation we have reduced the original problem $Ex = Nx$ to the sole equation:

$$Q(E - N)\mathcal{J}(x^*) = 0, \quad \text{or} \tag{2.6}$$

$$Gx^* = Q(E - N)[I - H(I - Q)N]^{-1}x^* = 0, \tag{2.6)'$$

or other analogous forms, all being equations in X_0. This is often a finite dimensional space. We say that the original problem $Ex = Nx$ has been reduced to an alternative one.

On the other hand, the auxiliary and bifurcation equations can be studied together (see below).

Concerning the auxiliary equation, the following simple remarks are of interest. Let F denote the map $F = H(I - Q)N$ so that $T = P + F$.

Let c, d be real numbers, $0 < c < d$, and let $W = [x^* \in X_0 \mid \|x^*\| \leq c]$, and for every $x^* \in W$ let $V^* = V(x^*) = [x \in X \mid Px = x^*, \|x\| \leq d]$. Let W^* be the set of all $x \in X$ with $\|Px\| \leq c$, $\|x\| \leq d$. We assume that $W^* \subset \mathcal{D}(N)$.

(2.i) If $x \in V^*$ implies $\|Fx\| \leq \delta$, $c + \delta \leq d$, then $T: V^* \to V^*$.

Indeed, for $x \in V^*$ and $y = Tx = Px + Fx$, we have $Py = PPx + PFx = Px = x^*$, and

$$\|y\| \leq \|x^*\| + \|Fx\| \leq c + \delta \leq d.$$

(2.ii) Under the conditions of (2.i), if F is a contraction on V^*, then T also is a contraction on V^*.

Indeed, for $x_1, x_2 \in V^*$ we have $Tx_i = Px_i + Fx_i$, $Px_i = x^*$, $i = 1,2$, and $Tx_1 - Tx_2 = Fx_1 - Fx_2$, and then $\|Tx_1 - Tx_2\| \leq k\|x_1 - x_2\|$ for some $0 \leq k < 1$.

(2.iii) Under the conditions of (2.i), $T|V^*$ has at least one fixed element in V^* if $T|V^*$ is compact.

Indeed, V^* is convex and closed, and thus Schauder's fixed point theorem applies.

(2.iv) Under the conditions of (2.ii), $T|V^*$ has a unique fixed point.

Indeed, V^* is a metric complete space (since X is a Banach space), and Banach's fixed point theorem applies.

(2.v) If F is a contraction in W^*, then the unique fixed point of $T|V^*$ is a continuous function of x^*, say $x = Tx = \mathcal{T}x^* \in V(x^*)$.

Indeed, for any two points $x_1^*, x_2^* \in W$ and $x_i = Tx_i \in V(x_i^*)$, $i = 1,2$, we have $x_i = x_i^* + Fx_i$, $x_i \in W^*$, $i = 1,2$, and

$$\|x_1 - x_2\| \leq \|x_1^* - x_2^*\| + \|Fx_1 - Fx_2\| \leq \|x_1^* - x_2^*\| + k\|x_1 - x_2\|,$$

$$\|\mathcal{T}x_1^* - \mathcal{T}x_2^*\| = \|x_1 - x_2\| \leq (1 - k)^{-1}\|x_1^* - x_2^*\|.$$

Let W' denote the interior of W, and G any open subset of X containing W^*. If Nx has a continuous Fréchet derivative $D(x)$, $x \in G$, then T, defined by $T(x^*, x) = x^* + H(I - Q)Nx$, has continuous Fréchet derivatives

I and $H(I - Q)D(x)$, with respect to x^* and x respectively. Moreover if F is a contraction on W^*, then $\mathcal{J}:W' \to W^*$ possesses a continuous Fréchet derivative $L(x^*)$ for $x^* \in W'$. This is a consequence of well-known results in functional analysis (see, e.g., Hale [64], p.7). Analogously, if $N(x^*,\mu)$ depends on x and also on a parameter μ varying in an open set S of some Banach space M, if N possesses continuous Fréchet derivatives with respect to x and μ respectively (continuous in $G \times S$), and if F is a uniform contraction in W^* for μ in S, then $\mathcal{J}(x^*,\mu)$ also possesses continuous Fréchet derivatives with respect to x^* and μ.

Concerning the operator F as a contraction, the following remark can be made. Let us assume for a moment that besides (k_{123}) the following holds: (k_4) There are positive nondecreasing functions $\alpha(\rho)$, $\beta(\rho)$ such that for all x, x' with $\|x\|$, $\|x'\| \leq \rho$ we have

$$\|H(I - Q)Nx\| \leq \beta(\rho),$$

$$\|H(I - Q)Nx - H(I - Q)Nx'\| \leq \alpha(\rho)\|x - x'\|.$$

Thus, if $\|H(I - Q)\| \leq K$, and $\|Nx - Nx'\| \leq L\|x - x'\|$ for $\|x\|$, $\|x'\| \leq \rho$, then we can take $\alpha(\rho) = LK$.

As we shall see in no. 3 for selfadjoint problems $(X = Y$, $P = Q)$, and in no. 5 for nonselfadjoint problems, it is always possible, under mild hypotheses, to choose P, Q, H so as to make $\|H(I - Q)\|$ as small as we want, and thus to make $F = H(I - Q)N$ a contraction. Cesari in [6], [7] mentioned the above scheme for selfadjoint problems, with many applications, and the remark that F can be made into a contraction.

These considerations were soon extended to the nonselfadjoint case by Locker [95], Hale, Bancroft and Sweet [65], and Hale [64]. (Also, see Cesari [23],[27],[28], and McKenna [109]).

As mentioned, the auxiliary and bifurcation equations (2.2), (2.3) can be studied together. For instance, system (2.2), (2.3) can be reduced to the fixed point problem for the transformation $T': (x, x^*) \to (\bar{x}, \bar{x}^*)$ defined by

$$T': \begin{aligned} \bar{x} &= T_1'(x, x^*) = x^* + H(I - Q)Nx, \\ \bar{x}^* &= T_2'(x, x^*) = x^* + Q(E - N)x, \end{aligned} \qquad (2.7)$$

where $x, \bar{x} \in X$, $x^*, \bar{x}^* \in X_0$. The last term of the second equation can be replaced by $-Q(E - N)x$, or by $\pm Q(E - N)\bar{x}$. (Landesman and Lazer [92], Williams [125], Cesari [26,27,28]) (cf. Section 7).

Also, if we take $x = x^* + y$, $x^* \in X_0 = PX$, $y \in X_1 = (I - P)X$, then the system of auxiliary and bifurcation equations (2.2), (2.3) becomes

$$y = H(I - Q)N(x^* + y), \qquad (2.8)$$

$$Q(E - N)(x^* + y) = 0. \qquad (2.9)$$

In turn, this system can be reduced to the fixed point problem relative to the map $T'': (x^*, y) \to (\bar{x}^*, \bar{y})$ defined by

$$T'': \bar{y} = H(I - Q)N(x^* + y), \quad \bar{x}^* = x^* + Q(E - N)(x^* + y),$$

where $x = x^* + y$, $\bar{x} = \bar{x}^* + \bar{y}$, $x^*, \bar{x}^* \in X_0$, $y, \bar{y} \in X_1$, and where alternatively we may replace the last term by $-Q(E - N)(x^* + y)$. (Mawhin [106], Cesari and Kannan [36], McKenna [109]) (cf. no. 4c).

Remark 1. The following main ideas have been repeatedly used in the study of auxiliary and bifurcation equations, and we indicate points of the present exposition where these ideas are used.

(a) Schauder's fixed point theorem (Cesari [6], 1963; Landesman and Lazer [92], 1970; Cesari and Kannan [37], Cesari [26,27,28]) (cf. no. 4 and Section 7).

(b) Banach's fixed point theorem for contraction maps (Cesari [7], 1964, and much subsequent work) (cf. nos. 3, 4, 5 in this Section, and Sections 2, 3, 4).

(c) Invariance properties of topological degree (Cesari [6], 1963; Knobloch [88,89], 1964; Cronin [39,40]; in particular, Leray-Schauder topological degree argument, Mawhin [102, 106,108], 1969, Kannan and McKenna [85], McKenna [109]; and Borsuk-Ulam theorem, Cesari and Kannan [35,36] (cf. nos. 24,34).

(d) Brouwer's fixed point theorem, particularly the Golomb-Cinquini-Miranda equivalent form (Cesari [10], 1960; Knobloch [88,89], 1964) (cf. no. 4).

(e) Theory of monotone and maximal monotone operators in Banach spaces (Gustafson and Sather [44,45], 1972; Cesari and Kannan [34], 1974) (cf. Section 6).

(f) Schauder's principle of invariance of domain (Kannan [79], 1974) (cf. Section 6).

(g) Implicit function theorem and Newton's "polygon" method in Banach spaces (Sather [119], 1970).

(h) Decomposition of the main operators into a product of operators (square root of a positive operator, Gustafson and Sather [44,45], 1972; $E = TT^*$ with suitable boundary conditions on T, T^*, Kannan and Locker [82,83], 1974) (cf. Section 5).

We mention here that Harris, Sibuya, and Weinberg [72] used the scheme of the present no. and contraction maps for a direct proof of the classical theorems on singular points of linear differential equations in the complex field (Cauchy, Frobenius, Perron, Lettenmeyer).

Remark 2. In association with the maps T' and T'', and for $X = Y$, $P = Q$, X_0 finite dimensional, Williams ([124], 1968) considered the map W defined by

$$Wz = Pz + H(I - P)Nz - P(E - N)z, \quad z \in X,$$

and noted that the fixed points of W are exactly the solutions of the equation $Ex - Nx = 0$. Williams worked in situations where $T = P + H(I - P)N$ is a contraction on the fibers $P^{-1}x^*$ of P, but T is not necessarily compact. However, Williams noted that, in such a situation, the composite map W^* defined by $W^*z = W\mathcal{T}z$ has the same fixed points as W, that is, the solutions of $Ex - Nx = 0$; but W^* now is compact, and a Leray-Schauder degree $d_{LS}[I - W^*, \Omega, 0]$ can be defined. Williams proved that the Leray-Schauder degree just mentioned is equal to the Brouwer degree of the finite dimensional map $T_2\mathcal{T}$, or briefly T_2, of the sole bifurcation equation, say,

$$d_{LS}[I - W', \Omega, 0] = d_B[T_2, \Omega_0, 0].$$

Deep seated work in the comparison of Leray-Schauder degree for T' and T'' and Brouwer degree for T_2 has just been announced by Rothe.

Later Mawhin ([106], 1972, and subsequent work) has reconsidered the map W for nonselfadjoint problems, with

$X_0 = \ker E$, $Y = \operatorname{coker} E$. If $S: Y_0 \to X_0$ is any continuous map, linear or nonlinear, with the property that $S^{-1}(0) = 0$, then for the map defined by

$$Wz = Pz + H(I - Q)Nz - SQN,$$

the fixed points of W are exactly the solutions of the original problem $Ex - Nx = 0$. In the problems considered by Mawhin, W is always a compact map, because the total map T', or T'', is already compact. Then the Leray-Schauder degree $d_{LS}(I - W, \Omega, 0)$ can be defined, and it can be estimated under suitable a priori bounds. This is Mawhin's coincidence degree. This was used, for instance, in the study of solutions in the large of Liénard systems with forcing terms. Cesari and Kannan [35,36] obtained improved results by direct use of maps T', the Borsuk-Ulam lemma, and suitable a priori bounds (cf. Section 5). The coincidence degree argument is well within the frame and structure of the alternative method.

3. The Selfadjoint Case. General Reduction to an Alternative Problem By Contraction Maps. Let us assume here that X is a real separable Hilbert space with inner product (x,y) and norm $\|x\| = (x,x)^{1/2}$. (In applications, E may be a linear differential operator on some domain $G \subset E^{\nu}$, $\nu \geq 1$, with given homogeneous linear boundary conditions on the boundary ∂G of G.) Here we need to know that the associated linear problem $Ex + \lambda x = 0$ has countably many real eigenvalues λ_i and eigenvectors φ_i, $\lambda_i \leq \lambda_{i+1}$, $E\varphi_i + \lambda_i \varphi_i = 0$, $i = 1,2,\ldots$, with $\lambda_i \to +\infty$ as $i \to \infty$, and $\{\varphi_i, i = 1,2,\ldots\}$ is a complete orthonormal system in X.

(Then, in the general situation of no. 2, we would have $X = L_2(G)$.)

Thus, every element $x \in X$ has Fourier series $x = \sum_1^\infty c_i\varphi_i$, $c_i = (x, \varphi_i)$, which is convergent in X. If $y = \sum_1^\infty d_i\varphi_i$ is any other element, then $(x,y) = \sum_1^\infty c_i d_i$, $\|x\| = (\sum_1^\infty c_i^2)^{1/2}$. Since $\lambda_i \leq \lambda_{i+1}$ and $\lambda_i \to +\infty$, there is some integer m so that $\lambda_{m+1} > 0$, and we shall define $P:X \to X$ by taking $Px = \sum_1^m c_i\varphi_i$, $c_i = (x,\varphi_i)$, $i = 1,\ldots,m$. Then, the range $X_0 = PX$ of P is the m-dimensional space $X_0 = \{\varphi_1,\ldots,\varphi_m\}$ spanned by $\varphi_1,\ldots,\varphi_m$. If X_1 is the complementary space, then any element $x \in X_1$ has Fourier series $x = \sum_{m+1}^\infty c_i\varphi_i$, and we can define $H:X_1 \to X_1$ by taking

$$Hx = -\sum_{m+1}^\infty c_i\lambda_i^{-1}\varphi_i, \quad x \in X_1. \tag{3.1}$$

This is possible since we chose m so that $\lambda_{m+1} > 0$ and thus $\lambda_i \geq \lambda_{m+1} > 0$ for all $i \geq m + 1$. Then, we also have

$$\|Hx\| = (\sum_{m+1}^\infty c_i^2\lambda_i^{-2})^{1/2} \leq \lambda_{m+1}^{-1}(\sum_{m+1}^\infty c_i^2)^{1/2} = \lambda_{m+1}^{-1}\|x\|, \tag{3.2}$$

or $\|Hx\| \leq k\|x\|$, $k = \lambda_{m+1}^{-1}$, for all $x \in X_1$. We see that H is a linear bounded operator, and that H can always be made into a contraction by taking m so that $\lambda_{m+1} > 1$.

It is apparent that $P:X \to X$ is a linear projection operator, $PP = P$, $\|P\| = 1$, and if I denotes the identity operator in X, then also $I - P$ is a projection operator, $\|I - P\| = 1$, and $X_1 = (I - P)X$. If $\mathcal{D}(E)$ denotes the linear subspace of X consisting of all $x = \sum_1^\infty c_i\varphi_i$, $c_i = (x,\varphi_i)$, with $\sum_1^\infty c_i^2\lambda_i^2 < \infty$, then $E:\mathcal{D}(E) \to X$, and

P, E, H have the properties (k_{123}) of no. 2 with $X = Y$, $P = Q$, that is

$$H(I - P)E = I - P, \tag{h_1}$$

$$PE = EP, \tag{h_2}$$

$$EH(I - P) = I - P. \tag{h_3}$$

The operator H has other interesting properties we shall need later. First

$$\begin{aligned}(-Hx,x) &= \sum_{m+1}^{\infty} c_i^2\lambda_i^{-1} \geq \lambda_{m+1}\left(\sum_{m+1}^{\infty} c_i^2\lambda_i^{-2}\right) \\ &= \lambda_{m+1}\|Hx\|^2 \geq 0.\end{aligned} \tag{3.3}$$

Moreover, H as defined by (3.1), is compact in the Hilbert space X (see, e.g., [185], p.234).

As in no. 2 we consider now an equation of the form

$$Ex = Nx \tag{3.4}$$

where $N:X \to X$ is any operator in X. From no. 2 we derive that this equation is equivalent to the system of auxiliary and bifurcation equations

$$x = x^* + H(I - P)Nx, \qquad Px = x^* \in X_0, \tag{3.5}$$

$$P(Ex - Nx) = 0, \tag{3.6}$$

and the auxiliary equation is of the form $x = Tx$ with T defined by $Tx = x^* + H(I - P)Nx$.

(3.i) Theorem (Cesari [7]). Under the hypotheses above, if $N:X \to X$ is uniformly Lipschitzian of given constant L, that is $\|Nx_1 - Nx_2\| \leq L\|x_1 - x_2\|$ for all $x_1, x_2 \in X$, then by taking m so that $\lambda_{m+1} > L$ the mapping T is a contraction map of $x^* + X_1$ into itself, and the auxiliary equation (3.5) is uniquely solvable for every $x^* \in X_0$.

<u>Proof</u>. Here we have $Tx = x^* + H(I - P)Nx$, and thus T maps X into $x^* + X_1$. Then certainly T maps $x^* + X_1$ into itself as mentioned in no. 2. For any two elements $x_1, x_2 \in x^* + X_1$ we also have

$$\|Tx_1 - Tx_2\| = \|H(I - P)(Nx_1 - Nx_2)\| \leq \lambda_{m+1}^{-1}\|Nx_1 - Nx_2\|$$
$$\leq (L\lambda_{m+1}^{-1})\|x_1 - x_2\|,$$

where $k = L\lambda_{m+1}^{-1} < 1$ because of the choice of m. By Banach's fixed point theorem, there is one and only one element $x = \mathcal{T}(x^*) \in x^* + X_1$ such that $x = Tx$, and $\mathcal{T}(x^*)$ satisfies the auxiliary equation (3.5). This proves (3.i).

Thus, for every $x^* \in X_0$, the auxiliary equation has a unique solution $x = Tx = \mathcal{T}x^*$, as mentioned in no. 2, and the bifurcation equation takes the form $P(E - N)\mathcal{T}x^* = 0$, an equation in X_0, a finite dimensional space (the alternative problem) (no. 2).

If x^* is written in the form $x^* = \gamma_1\varphi_1 + \cdots + \gamma_m\varphi_m$, then we have

$$P(E - N)\mathcal{T}x^* = c_1(x^*)\varphi_1 + \cdots + c_m(x^*)\varphi_m,$$

and the bifurcation equation $P(E - N)\mathcal{T}x^* = 0$ takes the form

$$c_1(x^*) = 0, \ldots, c_m(x^*) = 0, \tag{3.7}$$

a system of m equations in the m unknowns $\gamma_1, \ldots, \gamma_m$.

Note that if, instead of solving the system of auxiliary and bifurcation equations, we simply solve the bifurcation equation, namely $P(E - N)z = 0$ with $z \in X_0$, then we obtain what are usually called Galerkin's approximations, with respect to the eigenfunctions $\varphi_1, \ldots, \varphi_m$, to the solutions $x \in X$ of the original problem $Ex = Nx$. Note that if we write $z \in X_0$ in the form $z = \gamma_1\varphi_1 +\cdots+ \gamma_m\varphi_m$, $\gamma = (\gamma_1,\ldots,\gamma_m)$, then we have

$$P(E - N)z = \Gamma_1(z)\varphi_1 +\cdots+ \Gamma_m(z)\varphi_m,$$

with $\Gamma_i(z) = (P(E - N)z,\varphi_i)$, $i = 1,\ldots,m$, and equation $P(E - N)z = 0$, $z \in X_0$, reduces to the system of the m Galerkin equations in the m unknowns $\gamma_1,\ldots,\gamma_m$,

$$\Gamma_1(z) = 0,\ldots,\Gamma_m(z) = 0.$$

We shall see in no. 18 that by the alternative method a way arises to prove the existence of an exact solution $x \in X$ in some neighborhood of the approximation $z \in X_0$ and to numerically estimate the error $\|x - z\|$.

4. An Analysis of Auxiliary and Bifurcation Equations for the Problem of Periodic Solutions. (a) Two examples of analysis of the auxiliary equation.

To be specific, we shall consider the problem of the periodic solutions of period $T = 2\pi/\omega$, $\omega > 0$, of the differential system $x' = f(t,x)$, or

$$x_i' = f_i(t,x_1,\ldots,x_n), \quad i = 1,\ldots,n, \tag{4.1}$$

where the functions f_i are also periodic in t of period $2\pi/\omega$ for every x.

We may take $Ex = x'$, $(Nx)(t) = f(t,x(t))$, and we may restrict ourselves to $[0,T]$ and associate to E the boundary condition (or periodicity condition) $x(0) = x(T)$. Here we assume that the functions f are continuous in x, and that for given constants $M,R > 0$ we have

$$|f_i(t,x)| \le M, \qquad |f_i(t,x) - f_i(t,y)| \le M|x - y| \tag{4.2}$$

for all t and for all x,y with $|x|,|y| \le R$. Here $|x|$

denotes the norm $|x| = \max|x_j|$ for $x \in R^n$. For every $x \in (L_2[0,T])^n$, each component x_i has Fourier series

$$x_i(t) = a_{0i} + \sum_{s=1}^{\infty} (a_{si} \cos s\omega t + b_{si} \sin s\omega t),$$

and for the square norm $\|x\| = \max\|x_i\|$, we have

$$\|x_i\| = (T^{-1}\int_0^T x_i^2(t)dt)^{1/2} = [a_{0i}^2 + 2^{-1}\sum_{s=1}^{\infty} (a_{si}^2 + b_{si}^2)]^{1/2}.$$

Then, $\ker E$ is a space of constants $c = (c_1,\ldots,c_n)$, or $\ker E = R^n$, and for any $m \geq 1$ we take $Px = (Px_1,\ldots,Px_n)$, $P = Q$,

$$Px_i = a_{0i} + \sum_{s=1}^{m} (a_{si} \cos s\omega t + b_{si} \sin s\omega t). \tag{4.3}$$

Finally, for $Px = 0$, or $a_{si} = 0$, $s = 0,\ldots,m$, $b_{si} = 0$, $s = 1,\ldots,m$, $i = 1,\ldots,n$, we take

$$(Hx_i)(t) = \sum_{s=m+1}^{\infty} (s\omega)^{-1}(-b_{si} \cos s\omega t + a_{si} \sin s\omega t),$$

and

$$\|Hx\|_i = [2^{-1}\sum_{s=m+1}^{\infty} (s\omega)^{-2}(a_{si}^2 + b_{si}^2)]^{1/2}$$
$$\leq \omega^{-1}(m + 1)^{-1}\|x_i\|,$$

or $\|Hx\| \leq \omega^{-1}(m + 1)^{-1}\|x\|$ for all $x \in (L_2[0,T])^n$. For $x,y \in (L_2[0,T])^n$ with $|x_i(t)|, |y_i(t)| \leq R$ we also have

$$\|H(I - P)Nx - H(I - P)Ny\| \leq \omega^{-1}(m + 1)^{-1}M\|x - y\|.$$

Thus, $H(I - P)N$ can be made into a contraction by taking m sufficiently large.

On the other hand, if $\|x\|_\infty = \sup|x(t)|$, then we have

$$|H(I - P)x_i(t)| =$$
$$= |\sum_{s=m+1}^{\infty} (s\omega)^{-1}(-b_{si} \cos s\omega t + a_{si} \sin s\omega t)|$$
$$\leq [\sum (s\omega)^{-2}]^{1/2}[\sum (-b_{si} \cos s\omega t + a_{si} \sin s\omega t)^2]^{1/2}$$

$$\leq 2^{1/2}\omega^{-1}[\Sigma\, s^{-2}]^{1/2}[2^{-1}\Sigma\,(a_{si}^2 + b_{si}^2)]^{1/2}$$

$$\leq 2^{1/2}\omega^{-1}\sigma(m)\,\|(I - P)x_i\|,$$

or $\|H(I - P)x\|_\infty \leq 2^{1/2}\omega^{-1}\sigma(m)\,\|x\|$, where $\sigma^2(m)$ is the sum of the series $(m + 1)^{-2} + (m + 2)^{-2} + \cdots$, and $\sigma(m) \to 0$ as $m \to \infty$ (Cesari [6,17]).

Now the auxiliary equation for our problem becomes

$$x = x^* + H(I - P)Nx,$$

or $x = Tx$, where T is the second member of this equation, and where $x^* = Px$, or x_c^*, is expressed as a set of n trigonometrical polynomials (4.3), depending on $(2n + 1)m$ coefficients, say

$$c = (a_{0i}, a_{si}, b_{si},\quad s = 1, \ldots, m,\ i = 1, \ldots, n) \tag{4.4}$$

Let R_0, R_1 be positive constants with $R_0 + R_1 \leq R$. Let us assume m sufficiently large so that $r = \omega^{-1}(m + 1)^{-1}M < 1$ and also $\rho = \rho_m = 2^{1/2}\omega^{-1}\sigma(m) \leq R_1M^{-1}$. For every x_c^* with $\|x_c^*\|_\infty \leq R_0$, let K_c denote the set of all $x = x_c^* + y$ with $Py = 0$, $\|y\|_\infty \leq R_1$. Then $\|x\|_\infty \leq R_0 + R_1 \leq R$, and because of (4.2) we have $\|Nx\|_\infty \leq M$, $\|Nx\| \leq M$, and then, for $z = H(I - P)Nx$, we have also $\|z\|_\infty \leq \rho M \leq R_1$, and then $Tx = x_c^* + z \in K_c$, or $T{:}K_c \to K_c$. On the other hand, for any two elements $x, y \in K_c$ we also have $\|Tx - Ty\| = \|H(I - P)(Nx - Ny)\| \leq r\|x - y\|$ with $r < 1$, that is, $T{:}K_c \to K_c$ is a contraction in the square norm.

Alternatively, we note that $\sigma(m) \to 0$ as fast as $m^{-1/2}$ and thus $\rho_m \to 0$ as $m^{-1/2}$. On the other hand, if x is any bounded measurable element, then $\|(I - P)x\|_\infty \leq k_m\|x\|_\infty$ where k_m is a constant which depends on m, and for which, in Fourier series theory, estimates are known of the form

$k_m \leq A + B \log m$, A, B constants (cf. Knobloch [88,89]).
Thus,

$$\|H(I - P)x\|_\infty \leq \rho_m \|(I - P)x\| \leq \rho_m \|(I - P)x\|_\infty$$
$$\leq \rho_m k_m \|x\|_\infty,$$

and $\rho_m k_m \to 0$ as $m \to \infty$, as fast as $m^{-1/2} \log m$.

If R_0, R_1 are positive constants with $R_0 + R_1 \leq R$, let us choose m sufficiently large so that $r' = \rho_m k_m M < \min[1, R_1]$. Then, as before, for any x_c^* with $\|x_c^*\|_\infty \leq R_0$, we consider the class K_c of all elements of the form $x_c^* + y$, $Py = 0$, $\|y\|_\infty \leq R_1$, and we note that, for $x \in K_c$, then $\|Nx\|_\infty \leq M$, and then, for $z = H(I - P)Nx$ we have also $\|z\|_\infty \leq \rho_m k_m M \leq R_1$ and then $Tx = x_c^* + z \in K_c$, or $T: K_c \to K_c$. On the other hand, for any two elements $x, y \in K_c$ we also have

$$\|Tx - Ty\|_\infty = \|H(I - P)(Nx - Ny)\|_\infty \leq r'\|x - y\|_\infty,$$

with $r' < 1$, that is, $T: K_c \to K_c$ is a contraction in the sup norm.

By either remark above we see that we can choose m such that the auxiliary equation has a solution $x_c(t) = x_c^*(t) + y(t)$, $\|x_c^*\|_\infty \leq R_0$, $Py = 0$, $\|y\|_\infty \leq R_1$.

By the considerations above, $m = 2$, and the analysis of the bifurcation equation by topological degree, Cesari proved in [6], [17], that the equation

$$x'' + x^3 = \sin t$$

has some 2π-periodic solution $x(t)$.

(b) <u>A discussion of the bifurcation equation</u>. To discuss the bifurcation equation we shall need a well known lemma:

(4.ii) If $F(z) = (F_1,\ldots,F_\mu)$, $z = (z^1,\ldots,z^\mu)$, is any continuous vector-valued function in the interval I of R^μ of all $z = (z^1,\ldots,z^\mu)$ with $-R_i \le z^i \le R_i$, $i = 1,\ldots,\mu$, and $F_i(z)$ has constant opposite signs on the faces $z^i = \pm R_i$, $i = 1,\ldots,\mu$, (that is, $F_i \ge 0$ for $z \in I$, $z^i = R_i$, and $F_i \le 0$ for $z \in I$, $z^i = -R_i$, or the same interchanging signs), then there is at least one point $z \in I$ with $F(z) = 0$.

This statement was noted by Poincaré (1884), Golomb (1935), Cinquini (1935), and was shown by Miranda (1941) to be equivalent to Brouwer's fixed point theorem.

For problem (4.1), the bifurcation equation reduces to a system of $(2m + 1)n$ equations

$$L_{is1}(c) = 0, \quad L_{is2}(c) = 0, \quad L_{io}(c) = 0, \quad s = 1,\ldots,m,$$
$$i = 1,\ldots,n,$$

with

$$L_{is1}(c) = a_{si}s\omega + 2T^{-1}\int_0^T f_i(t,x_c(t))\sin s\omega t\,dt,$$
$$L_{is2}(c) = b_{si}s\omega - 2T^{-1}\int_0^T f_i(t,x_c(t))\cos s\omega t\,dt,$$
$$L_{io}(c) = \int_0^T f_i(t,x_c(t))\,dt.$$

We assume that (α) there is a constant M such that relations (4.2) hold for <u>all</u> t, x, y; and (β) there is a constant $R_0 > 0$ such that $f_i \ge 0$ for $x_i \ge R_0$ and $f_i \le 0$ for $x_i \le -R_0$ (or the same interchanging signs). For $\mu = (2m + 1)n$, we consider the space R^μ of all constant vectors c as in (4.4), and we note that, for $a_{si} = \pm R$, $b_{si} = \pm R$, with $R > 0$ sufficiently large, then $L_{is1}(c)$ and

$L_{is2}(c)$ have constant sign, namely the sign of a_{si}, independently of the values of all other constants entering c.

Moreover, having so fixed $R > 0$, and all constants a_{si}, b_{si} having values in absolute values $\leq R$, $s = 1,\ldots,m$, $i = 1,\ldots,n$, we see that each component x_i of $x_c(t) = x_c^*(t) + y_c(t)$, $Py_c = 0$, $\|y_c\|_\infty \leq R_2$, has constant term $2^{-1}a_{0i}$ and we can take $a_{0i} = \pm R'$ with $R' > 0$ so large that $x_i(t)$ has constant sign, the same as a_{0i}, and actually has values $\geq R_0$ for $a_{0i} = +R'$, and values $\leq -R_0$ for $a_{0i} = -R'$. In other words $Lc = (L_{is1}, L_{is2}, L_{i0}$, $s = 1,\ldots,m$, $i = 1,\ldots,n)$ satisfies all conditions of Lemma (4.ii) in the interval I of R^μ defined by $|a_{is}| \leq R$, $|b_{is}| \leq R$, $s = 1,\ldots,m$, $i = 1,\ldots,n$, $|a_{i0}| \leq R'$, $|b_{i0}| \leq R'$, $i = 1,\ldots,n$. There is, therefore at least one point $c \in I$ with $Lc = 0$, and the bifurcation equation is solvable.

By the considerations of (a) and (b) we see that, under the stated hypotheses (α), (β), the differential system (4.1) has at least one T-periodic solution $x(t) = (x_1,\ldots,x_n)$ (Knobloch [88,89]).

In any particular situation in which an a priori bound for the solutions is known, the functions f_i may be modified outside a large ball so as to make them satisfy conditions (α), (β) in the whole space. On the basis of these considerations, Knobloch [88,89] indeed obtained qualitative conditions for the existence of periodic solutions of the scalar equation $x'' = f(t,x,x')$, and related comparison theorems [90] for the same nonlinear equation. The remarks in (c) below, based on Schauder's fixed point theorem, show that the Lipschitz hypothesis is not needed, that the auxiliary and bifurcation equations can be also treated together, and that m need not be large, namely we shall take $m = 0$ in (c).

(c) An example of analysis of auxiliary and bifurcation equations without Lipschitz hypothesis. Continuing our example of the periodic solutions of (4.1), we only assume here that f is continuous, and that (α) $|f_i| \leq M$ for some constant M and all t,x, and that (β) for some constant $R_0 \geq 0$ we have $f_i \geq 0$ for $x^i \geq R_0$ and $f_i \leq 0$ for $x^i \leq -R_0$, $i = 1,\ldots,n$, independently of the values of the remaining coordinates.

We take here $m = 0$, so that P is the mean value operator $T^{-1}\int_0^T x(t)dt$, $X_0 = PX$ is a space of constant vectors, or $X_0 = R^n$, and then all elements $y \in X_1 = (I - P)X$ are T-periodic functions of mean value zero. Here $X_0 = \ker E$; hence the bifurcation equation reduces to $PNx = 0$. The system of auxiliary and bifurcation equations can be reduced to the search for the fixed points of the transformation $T_0: (x^*, y) \to (\bar{x}^*, \bar{y})$ with T_0 defined by

$$T_0: \bar{y} = H(I - P)N(x^* + y), \qquad \bar{x}^* = x^* - PN(x^* + y). \tag{4.5}$$

We restrict x^*, y to the set C of all pairs x^*, y with $x^* \in X_0$, $y \in X_1$, with $\|x^*\|_\infty \leq R_1$, $\|y\|_\infty \leq R_2$, and R_1, $\bar{R}_1$, R_2 are arbitrary constants for which we only require

$$R_2 \geq \rho_0 k_0 M, \qquad \bar{R}_1 \geq R_2 + R_0, \qquad \bar{R}_1 \geq M, \qquad R_1 \geq \bar{R}_1 + M.$$

Now, for $(x^*, y) \in C$, $x = x^* + y$, we have

$$\|\bar{y}\|_\infty = \|H(I - P)Nx\|_\infty \leq \rho_0 k_0 M \leq R_2.$$

Relation $\bar{x}^* = x^* - PN(x^* + y)$, written in component form, becomes

$$2^{-1}\bar{a}_{Oi} = 2^{-1}a_{Oi} - T^{-1}\int_O^T f_i(t, x^*(t) + y(t))dt,$$
$$i = 1, \ldots, n.$$

For $(x^*, y) \in C$ with $\|x_i^*\| \leq \bar{R}_1$, or $|2^{-1}a_{Oi}| \leq \bar{R}_1$, we have

$$|2^{-1}\bar{a}_{Oi}| \leq |2^{-1}a_{Oi}| + T^{-1}\int_O^T |f_i|dt \leq \bar{R}_1 + M \leq R_1.$$

For $(x^*, y) \in C$ with $\bar{R}_1 \leq 2^{-1}a_{Oi} \leq R_1$, then $x_i^*(t) = 2^{-1}a_{Oi} + y(t) \geq \bar{R}_1 - R_2 \geq R_O$, hence $f_i(t, x^*(t) + y(t)) \geq 0$, $PN(x^* + y) \geq 0$, and finally

$$0 \leq \bar{R}_1 - M \leq 2^{-1}\bar{a}_{Oi} \leq R_1.$$

For $(x^*, y) \in C$ with $-R_1 \leq 2^{-1}a_{Oi} \leq -\bar{R}_1$ we have analogously $-R_1 \leq 2^{-1}\bar{a}_{Oi} \leq 0$. We see that in any case $T: C \to C$.

Here T is a compact map. To see this we note that H is the operation of integration, on functions y of mean value zero and taking for Hy the only primitive of mean value zero. Hence uniformly bounded functions are mapped by H into equibounded and equilipschitzian functions. Thus, H is compact. Now for any bounded sequence $[(x_k^*, y_k)]$ of elements of C, certainly $N(x_k^* + y_k)$ is bounded (by M), and $\bar{y}_k = H(I - Q)N(x_k^* + y_k)$ is a sequence in X_1 from which we can extract a convergent subsequence. Then x_k^* and $PN(x_k^* + y_k)$ are bounded sequences of elements of a finite dimensional space. Thus, there is some subsequence $(\bar{x}_{k_s}^*, \bar{y}_{k_s})$ which is convergent, and T is compact.

Since C is a closed bounded convex set and $T: C \to C$ is compact, by Schauder's fixed point theorem there is a fixed element $(x^*, y) = T(x^*, y)$ and the system of auxiliary and bifurcation equations is solvable.

Note that hypothesis (β) can be expressed by saying that $(Nx, x^*) \geq 0$ for $x^* = Px$, $\|x^*\| \geq R$ for some constant R. We shall encounter this condition again in Sections 6 and 7.

5. <u>General Reduction to an Alternative Problem by Contraction Maps in the Nonselfadjoint Case</u>. We have seen in no. 3 how selfadjoint problems can be reduced to an alternative one by contraction maps. Let us show here that essentially the same holds for nonselfadjoint problems.

(a) Let us assume that E, P, Q are given as in no. 2, so that $\mathcal{R}(P) \subset \mathcal{D}(E)$, $\ker E \subset X_0 = PX$, $(I - Q)\mathcal{R}(E) = \mathcal{R}(E) \cap Y_1$, $QEx = EPx$. Let E' be the map E restricted to $D' = \mathcal{D}(E') = X_1 \cap \mathcal{D}(E)$, and let $R' = \mathcal{R}(E')$, so that $E':D' \to R'$ is one-one and onto. Then necessarily E' has an inverse $H:R' \to D'$ which is linear and onto. As a consequence of closed graph theorem we also have:

(5.i) If the linear map E has closed graph and closed range, then H is a bounded linear map.

Concerning estimates on the norm $\|H\|$ of H the following simple remark is relevant.

(5.ii) We have $\|H\| \leq k$ if and only if $\|(I - P)x\| \leq k\|Ex\|$ for all $x \in \mathcal{D}(E)$ [63].

Indeed, for $y \in R'$ and $x' \in D'$ with $y = E'x'$ we have $Hy = HE'x' = (I - P)x$. Then, $\|(I - P)x\| \leq k\|Ex\|$ if and only if $\|Hy\| \leq k\|y\|$. Thus, we need not compute H to estimate its norm $\|H\|$. The latter can be computed directly from E and P.

(b) Let X and Y be Banach spaces, $E:\mathcal{D}(E) \to Y$ a linear operator from a linear subspace $\mathcal{D}(E)$ of X with closed graph and closed range, and $\mathcal{D}(E)$ everywhere dense in X. Let us make the assumption that there are projection operators $P_0:X \to X$, $Q_0:Y \to Y$ such that $\ker E = P_0X = X_{00}$, $\mathcal{R}(E) = (I - Q_0)Y = Y_{01}$, and let $X = X_{00} + X_{01}$, $Y = Y_{00} + Y_{01}$ (direct sums). Let $H_0:Y_{01} \to X_{01} \cap \mathcal{D}(E)$ denote the inverse operator of $E' = E|_{X_{01} \cap \mathcal{D}(E)}:X_{01} \cap \mathcal{D}(E) \to Y_{01}$. Since E' is linear, one-one and onto, H_0 exists and is a linear map. Having assumed that E has closed graph and closed range, then H_0 is a linear, bounded operator.

Note that now, for every $x \in \mathcal{D}(E)$ we certainly have

$Q_0Ex = 0$, $EP_0x = 0$, $E(I - P_0) = E$,

$H_0(I - Q_0)Ex = H_0Ex = H_0E(I - P_0)x = (I - P_0)x$,

that is, (k_1) and (k_2) hold. Also, for every $y \in Y$ and with $\bar{y} = (I - Q_0)y$, we have

$EH_0(I - Q_0)y = EH_0\bar{y} = \bar{y} = (I - Q_0)y$,

that is, (k_3) holds. Consequently, as in no. 2, we know that equation $Ex = Nx$ is equivalent to the system of two equations (2.2), (2.3), which now has the simpler form

$$x = P_0x + H_0(I - Q_0)x, \quad Q_0Nx = 0. \tag{5.1}$$

(5.iii) (a lemma) If $S:X \to X$ is any projector operator with $SX \subset X_{01} \cap \mathcal{D}(E)$, $SP_0 = 0$, $P_0S = 0$, then $P = P_0 + S$ is also a projection operator, and system (5.1) is equivalent to

$$x = Px + (I - P)H_0(I - Q_0)Nx, \tag{5.2}$$

$$Sx = SH_0(I - Q_0)Nx, \quad Q_0Nx = 0. \tag{5.3}$$

Proof. Here $SP_0 = 0$, $(I - P)P_0 = (I - P_0 - S)P_0 = 0$, and by applying $I - P$ and S to the first of relations (5.1), then the same relations (5.1) yield (5.2), (5.3). Relation (5.2) is the new auxiliary equation; relations (5.3) together are the bifurcation equation. It is clear that, under the same assumptions above, relations (5.2), (5.3) imply $Ex = Nx$. Thus, equation $Ex = Nx$ and system (5.2), (5.3) are equivalent.

(5.iv) (a lemma) Suppose X is a Hilbert space, Y a Banach space, and there are projections $P_0 : X \to X$, $Q_0 : Y \to Y$ with $P_0 X = \ker E = X_{00}$, $(I - Q_0)Y = \mathcal{R}(E) = Y_{01}$, and thus we have the decompositions $X = X_{00} + X_{01}$, $X_{00} \cap X_{01} = \{0\}$, $Y = Y_{00} + Y_{01}$, $Y_{00} \cap Y_{01} = \{0\}$, $X_{01} = (I - P_0)X$, $Y_{01} = (I - Q_0)Y$. Suppose that the set $V = [x \in (I - P_0)\mathcal{D}(E) \mid \|Ex\| \leq 1]$ is compact. Then, for any $\sigma > 0$ there is a projection $S : X \to X$ with finite dimensional range $\Sigma = \mathcal{R}(S) \subset X_{01} \cap \mathcal{D}(E)$, $SP_0 = 0$, $P_0 S = 0$, such that for $P = P_0 + S$ we have $\|(I - P)H_0(I - Q_0)\| \leq \sigma\|I - Q_0\|$ [63].

Proof. Let B be the unit ball in Y_{01}. Then $H_0(B) = V$ and then H_0 is a compact map (and hence bounded). Since V is compact, for any $\sigma > 0$ there is a finite system $[x_j]$ of points $x_j \in V$, $j = 1, \ldots, n$, $n = n(\sigma)$, such that, for any $z \in B$ there is at least one x_j with $\|H_0 z - x_j\| \leq \sigma$. Let S' denote the orthogonal projection $S' : X_{01} \to \Sigma$, whose range Σ is the finite dimensional space spanned by $x_1, \ldots, x_n$. The kernel of S' is then the orthogonal complement $\Sigma^{\perp}$ of Σ in X_{01}. Then

$S = S'(I - P_0)$ is also a projection $S:X \to X$ with range Σ, and $P_0 S = 0$, $SP_0 = 0$. Thus, $P = P_0 + S$ is a projection operator with range $\Sigma \times X_{00}$, S' and $I - S'$ are orthogonal projections, and $\|S'\| = 1$, $\|I - S'\| = 1$. For any x_j we certainly have $S'x_j = x_j$, $(I - S')x_j = 0$. For any $z \in B$ there is now an x_j such that $\|H_0 z - x_j\| \leq \sigma$, and hence

$$\|(I - S')H_0 z\| = \|(I - S')(H_0 z - x_j)\|$$
$$\leq \|I - S'\|\|H_0 z - x_j\| \leq \sigma.$$

Thus, for any $y \in Y_{01}$ we have $y/\|y\| \in B$, and by the linearity of $(I - S')H_0$ also

$$\|(I - S')H_0 y\| = (I - S')H_0(\|y\|(y/\|y\|))$$
$$= \|y\|\|(I - S')H_0(y/\|y\|)\| \leq \sigma\|y\|.$$

Now note that $S = S'(I - P_0)$ is also a projection operator $S:X \to X$ with range Σ, and $P_0 S = SP_0 = 0$. Thus, $P = P_0 + S$ is also a projection $P:X \to X$ with range $X_0 = PX = X_{00} + \Sigma$ (direct sum). Also

$$I - P = I - P_0 - S = I - P_0 - S'(I - P_0)$$
$$= (I - S')(I - P_0),$$

and we have a new decomposition $X = X_0 + X_1$, $X_0 \cap X_1 = \{0\}$, $X_0 = PX$, $X_1 = (I - P)X \subset X_{01}$. Thus, for any $y \in Y$, let $\bar{y} = (I - Q_0)y \in Y_{01}$, $\bar{x} = H_0\bar{y}$, and then

$$\|(I - P)H_0(I - Q_0)y\| = \|(I - S')(I - P_0)H_0\bar{y}\|$$
$$= \|(I - S')H_0\bar{y}\| \leq \sigma\|\bar{y}\|$$
$$\leq \sigma\|I - Q_0\|\|y\| \tag{5.4}$$

for every $y \in Y$. This proves lemma (5.iv).

(5.v) <u>Theorem</u>. Under the assumptions of (5.iv), given any number k, there are projections $P:X \to X$, $Q:Y \to Y$ such that, if $PX = X_0$, $(I - P)X = X_1$, $QY = Y_0$, $(I - Q)Y = Y_1$, then $X_0 \supset \ker E = X_{00}$, $Y_1 \subset \mathcal{R}(E) = Y_{01}$, and if $H = H_0|_{Y_1}$, then relations (k_{123}) hold for E, P, Q, H, and in addition $\|H(I - Q)\| \leq k$ (Hale [63]).

<u>Proof</u>. Take $\sigma > 0$ so that $\sigma\|I - Q_0\| \leq k$, and let us define Σ, S', S, $P = P_0 + S$ as in (5.iv). Then P is a projection with $X_1 = \mathcal{R}(I - P) = \mathcal{R}(I - P_0 - S)$, and we have the decompositions $X = X_{00} + X_{01} = X_{00} + \Sigma + X_1$ (direct sum). We take now

$$\Sigma' = \mathcal{R}(E|_{\Sigma}), \qquad Y_1 = \mathcal{R}(E|_{\mathcal{D}(E) \cap X_1}).$$

Let $R': Y_{01} \to Y_{01}$ be the operator defined by taking $R'\bar{y} = ES'H\bar{y}$ for all $\bar{y} \in Y_{01}$. Then the range of R' is Σ'. Also, let $R: Y \to Y$ be the operator defined by $Ry = R'(I - Q_0)y$ for all $y \in Y$, and whose range again is Σ'. Since $RQ_0 = 0$, $Q_0R = 0$, the operator $Q = Q_0 + R$ is a projection, with $QY = Y_{00} + \Sigma'$, $(I - Q)Y = Y_1$, and we have the decompositions $Y_{01} = \Sigma' + Y_1$, $Y = Y_{00} + \Sigma' + Y_1$ (direct sums). Now, for any $x \in \mathcal{D}(E)$, we have $x = x_0 + x' + x_1$, $P_0x = x_0 \in \ker E$, $Sx = x'$, $(I - P)x = x_1$, and then, for $y' = Ex' \in \Sigma'$, $y_1 = Ex_1 \in Y_1$, also

$$\begin{aligned} H_0(I - Q)Ex &= H_0(I - Q_0 - R)E(x_0 + x' + x_1) \\ &= H_0(I - Q_0 - R)(y' + y_1) = H_0y_1 = x_1. \end{aligned}$$

Thus, $H_0(I - Q)Ex = (I - P)x$ for all $x \in \mathcal{D}(E)$ and (k_1) is proved. Analogously

$$QEx = (Q_0 + R)E(x_0 + x' + x_1) = (Q_0 + R)(y' + y_1) = y',$$

$$EPx = E(P_0 + S)(x_0 + x' + x_1) = E(x_0 + x') = y';$$

thus $QEx = EPx$ for all $x \in \mathcal{D}(E)$ and (k_2) is proved.

Finally, for any $y \in Y$ we have $y = y_0 + y' + y_1$, $Q_0 y = y_0 \in Y_{00}$, $Ry = y' \in \Sigma'$, $(I - Q)y = y_1 \in Y_1$, and also

$$(I - Q)y = (I - Q_0 - R)(y_0 + y' + y_1) = y_1 \in Y_1 \subset Y_{10} = \mathcal{R}(E).$$

In other words, $(I - Q)y \in \mathcal{R}(E)$ for all $y \in Y$, and we know that EH_0 is the identity on $\mathcal{R}(E)$. Thus, $EH_0(I - Q)y = (I - Q)y$ for all $y \in Y$, and (k_3) is proved.

Moreover, for every $y \in Y$ we have

$$\begin{aligned}(I - P)H_0(I - Q_0)y &= (I - P_0 - S)H_0(I - Q_0)(y_0 + y' + y_1)\\ &= (I - P_0 - S)H_0(y' + y_1)\\ &= (I - P_0 - S)(x' + x_1) = x_1,\end{aligned}$$

$$H_0(I - Q)y = H_0(I - Q_0 - R)(y_0 + y' + y_1) = H_0 y_1 = x_1,$$

and thus $(I - P)H_0(I - Q_0)y = H_0(I - Q)y$ for all $y \in Y$. By (5.4) we have

$$\|H_0(I - Q)\| = \|(I - P)H_0(I - Q_0)\| \le \sigma\|I - Q_0\| \le k.$$

Statement (5.v) is thereby proved.

If we split equation $Ex = Nx$ by using the operators P, Q, and $H = H_0|_{Y_1}$ of (5.v), we obtain as usual the equivalent system

$$x = Px + H(I - Q)Nx, \qquad \text{(auxiliary equation)},$$

$$Q(Ex - Nx) = 0, \qquad \text{(bifurcation equation)}.$$

Statement (5.v) shows that $H(I - Q)$ can be made into a contraction map of constant k as small as we want. Thus, if N is Lipschitzian of constant L in some ball $W = [x \,|\, \|x\| \le \rho]$ in X, and we take k so that $kL < 1$, we see that $H(I - Q)N$ is a Lipschitz operator of constant $\le kL < 1$, that is, a contraction map in W. In other words, the auxiliary equation is theoretically uniquely solvable by Banach's fixed point theorem, and problem $Ex = Nx$ is theoretically reducible to the sole bifurcation equation, or determining problem (also equation (2.3)), in the space $X_0 = PX = X_{00} + \Sigma$. If $X_{00} = \ker E$ is finite dimensional, so is X_0. In other words, we can always suitably enlarge $\ker E$ into a space $\ker E + \Sigma$, Σ finite dimensional, so as to make $H(I - Q)N$ a contraction map, and equation $Ex = Nx$ is reducible to an alternative problem.

Section 2. Problems of Perturbation for Ordinary Differential Equations

6. Boundary Value Problems for Ordinary Differential Equations. We present here a local treatment of homogeneous linear boundary value problems for ordinary differential systems based on the considerations of no. 2, namely we consider problems of the type

$$\begin{aligned} &x' - A(t)x = f(t,x(t),x'(t)), \quad a \le t \le b, \\ &B_1x(a) + B_2x(b) = 0, \end{aligned} \tag{6.1}$$

where $x = \mathrm{col}(x_1,\dots,x_n)$, $f = \mathrm{col}(f_1,\dots,f_n)$, A is an $n \times n$ matrix and B_1, B_2 are $m \times n$ matrices, $0 \le m \le n$. This treatment, for which we refer to Nagle [110] for details, is more general than previous ones on the subject because we allow the underlying linear problem to be nonselfadjoint, because $f(t,x,x')$ may depend on the derivatives x', and because f is only measurable in t, though bounded and Lipschitzian in x,x' for x,x' bounded. The corresponding boundary value problems for ordinary differential equations of order n are included, that is, problems of the form

$$y^{(n)} + a_1(t)y^{(n-1)} + \cdots + a_n(t)y$$
$$= f_0(t, y(t), \ldots, y^{(n)}(t)),$$
$$B_1 z(a) + B_2 z(b) = 0, \quad z = \mathrm{col}(y, y', \ldots, y^{(n-1)}).$$

In nos. 8-12 we consider problems of the perturbation type, where f is replaced by εg (f_0 by εg_0), ε a small parameter. This perturbation analysis is intended to include previous work on perturbation problems for periodic solutions of ordinary differential equations and systems, though we only mention briefly some of this latter work (Hale [51-60], Gambill [41-43], Hale and Gambill [66], Cesari [10,11], Cesari and Bailey [30], Bailey and Gambill [1]). We refer to Cesari's book [9], and to those of Coddington and Levinson [140] and Cole [141], for the needed background in ordinary differential equations. As usual in perturbation problems, the theorems state (e.g. (9.i)) that solutions certainly exist for $|\varepsilon|$ sufficiently small. In a number of examples Cesari [6,7,17] has shown, by the present alternative technique, that $|\varepsilon|$ can be allowed to be "fairly large". Some examples are presented in Section 4. The same has occurred for partial differential equations (Cesari [18], Shaw [120]). We present some examples in Sections 3 and 4. Thus, we consider here boundary value problems (6.1) for systems of n first order ordinary differential equations, where $x = \mathrm{col}(x_1, \ldots, x_n)$ are the n unknowns, $A(t) = [a_{ij}(t)]$ is an $n \times n$ matrix with bounded measurable entries, and B_1, B_2 constant $m \times n$ matrices. Thus, if E denotes the linear operator defined by $Ex = x' - A(t)x$, with boundary conditions $B_1x(a) + B_2x(b) = 0$, and N is the not necessarily linear operator defined by $Nx = f(t,x(t),x'(t))$, then (6.1) takes the usual form $Ex = Nx$.

The problem of periodic solutions $x(t)$, $-\infty < t < +\infty$, of some period T, or $x(t+T) = x(t)$, is certainly contained in (6.1) by taking $a = 0$, $b = T$, $B_1 = I$, $B_2 = -I$, $n = m$, I the identity matrix in E^n, and by assuming either $f(t + T,x,x') = f(t,x,x')$ (the periodic case), or $f(x,x')$ independent of t (the autonomous case). (Problems of harmonics and subharmonics for forced oscillations, as in [66], can also be included.) (Cf. no. 11).

In (6.1) the underlying linear problem is

$$x' - A(t)x = 0, \quad a \leq t \leq b, \quad B_1x(a) + B_2x(b) = 0, \quad (6.2)$$

or $Ex = 0$, including the boundary conditions in E as stated.

As usual, a solution $x(t) = \mathrm{col}(x_1,\ldots,x_n)$, $a \leq t \leq b$, is an n-vector with absolutely continuous (AC) entries in $[a,b]$, satisfying the differential system a.e. in $[a,b]$.

Let $|\ |$ denote the Euclidean norm in E^n. In (6.2), $f(t,x,x') = \mathrm{col}(f_1,\ldots,f_n)$ is an $n \times 1$ vector function defined on $[a,b] \times E^{2n}$, whose entries are measurable in t for every (x,x'), and continuous in (x,x') for every t. Moreover we assume here that for any given pair of constants R_1, R_2 there are two other constants L, M such that

$$|f(t,x,x')| \leq M,$$

$$|f(t,x,x') - f(t,y,y')| \leq L[|x - y| + |x' - y'|]$$

for all $a \leq t \leq b$, $|x|,|y| \leq R_1$, $|x'|,|y'| \leq R_2$.

For any n-vector $z(t) = (z_1,...,z_n)$, $a \leq t \leq b$, we denote by $\|z\|_0$ the usual Ess Sup norm, or $\|z\|_0$ = Ess $\text{Sup}_{a \leq t \leq b} |z(t)|$, and by $\|z\|_1$ the L_1-norm $\|z\|_1 = (b - a)^{-1} \int_a^b |z(t)| dt$. Let X denote the (Sobolev) space of all AC vector functions $x(t) = (x_1,...,x_n)$, $a \leq t \leq b$, for which we may as well take the norm $\|x\|_1^1 = \|x\|_0 + \|x'\|_1$ (since this norm is equivalent to the Sobolev norm $\|x\|_1 + \|x'\|_1$). Let Y be the space of all vector functions $y(t) = \text{col}(y_1,...,y_n)$ with L_1-integrable entries. Let $\|y\|_1$ be the norm for Y.

7. <u>The Linear Problem. Basic Statements</u>. We refer to Coddington and Levinson [140] and Cole [141] for boundary value problems for linear differential equations and systems. Here we mention only a few definitions and statements given in Cole ([141], p.139).

Let $\tilde{A}$ denote the transpose of any given matrix A. Then the linear problem adjoint to (6.1) is

$$d\tilde{y}/dt + \tilde{y}A(t) = 0, \quad \tilde{y}(a) = \tilde{\alpha}B_1, \quad \tilde{y}(b) = -\tilde{\alpha}B_2, \tag{7.1}$$

or

$$dy/dt + \tilde{A}(t)y = 0, \quad y(a) = \tilde{B}_1\alpha, \quad y(b) = -\tilde{B}_2\alpha, \tag{7.1$'$}$$

where $y(t) = \text{col}(y_1,...,y_n)$, or $\tilde{y}(t) = \text{row}(y_1,...,y_n)$, and where $\alpha = \text{col}(\alpha_1,...,\alpha_m)$ denotes any arbitrary real $m \times 1$ vector (parametric form of the adjoint problem). Then, the linear operator E^* adjoint to E is defined by relations (7.1), or (7.1)$'$, and as usual ker E = coker E^*, coker E = ker E^*. We denote by p and q the dimensions of ker E and coker E, respectively. Let X(t) denote any fundamental

$n \times n$ matrix for the homogeneous problem $x' - A(t)x = 0$, and let $h(t)$, $a \leq t \leq b$, denote any element of $(L_1[a,b])^n = Y$. Then the unique solution $x(t)$, with initial data $x(a) = X(a)c$, $c \in R^n$, of the nonhomogeneous system $x' - A(t)x = h(t)$ is given by

$$x(t) = X(t)\int_a^t X^{-1}(s)h(s)ds + X(t)c. \tag{7.2}$$

Thus x satisfies the boundary problem

$$x' - A(t)x = h(t), \quad a \leq t \leq b, \quad B_1x(a) + B_2x(b) = 0,$$
$$h \in Y,$$

if and only if

$$Dc = -B_2X(b)\int_a^b X^{-1}(s)h(s)ds, \tag{7.3}$$

where D is the $m \times n$ matrix $D = B_1X(a) + B_2X(b)$. On the other hand, the n rows of X^{-1} are n linearly independent solutions of the homogeneous adjoint system (7.1) since $(d/dt)X^{-1} = -X^{-1}X'X^{-1} = -X^{-1}A$. In other words, every solution of (7.1) has the form $\tilde{z} = \tilde{c}X^{-1}$, and thus $\tilde{z}$ satisfies the adjoint boundary data (7.1) if and only if

$$\tilde{c}X^{-1}(a) = \tilde{\alpha}B_1, \quad \tilde{c}X^{-1}(b) = -\tilde{\alpha}B_2.$$

If we multiply these two relations at the right by $X(a)$ and $X(b)$ respectively, we obtain $0 = \tilde{\alpha}D$. In other words, the adjoint boundary value problem (7.1) has a nontrivial solution $\tilde{z}$ if and only if there is $\tilde{\alpha}$ such that $\tilde{\alpha}D = 0$, and then $\tilde{z} = \tilde{c}X^{-1}$ with $\tilde{c} = \tilde{\alpha}B_1X(a) = -\tilde{\alpha}B_2X(b)$. Equivalently, any solution $\tilde{z}$ of the adjoint boundary value problem (7.1) has the form $\tilde{z} = \tilde{\alpha}B_1X(a)X^{-1} = -\tilde{\alpha}B_2X(b)X^{-1}$ for $\tilde{\alpha}D = 0$.

(7.i) (Fredholm's alternative theorem). The homogeneous boundary value problem

$$x' - A(t)x = h(t), \quad a \le t \le b, \quad B_1x(a) + B_2x(b) = 0,$$
$$h \in (L_1[a,b])^n = Y, \tag{7.4}$$

has a solution $x(t)$, $a \le t \le b$, $z \in X$, if and only if $\int_a^b \tilde{z}(s)h(s)ds = 0$ for every solution $z(t)$, $a \le t \le b$, $z \in X$, of the adjoint boundary value problem (7.1).

Proof. As we have noted, the given boundary value problem (6.1) has a solution x if and only if relation (7.2) holds with $x(a) = X(a)c$, $c \in R^n$, that is, if and only if there exists $c \in R^n$ with $Dc = \beta$, where β is given by the second member of (7.3). Here D is an $m \times n$ matrix and we know by linear algebra that the range of D is the annihilator of the null space of D [160, p.103]. In other words, β is in the range of D if and only if $\tilde{\beta}\gamma = 0$ for every γ such that $\tilde{D}\gamma = 0$, or $\tilde{\gamma}\beta = 0$, that is

$$\int_a^b (-\tilde{\gamma}) B_2 X(b) X^{-1}(s)h(s)ds = 0 \tag{7.5}$$

for every γ with $\tilde{\gamma}D = 0$. But we have just mentioned that any solution $\tilde{z}$ of the adjoint boundary value (7.1) is of the form $\tilde{z} = -\tilde{\gamma}B_2X(b)X^{-1}$ with $\tilde{\gamma}D = 0$. Thus (7.5) takes the form $\int_a^b \tilde{z}(s)h(s)ds = 0$, and (7.i) is proved.

8. The Projection Operators P and Q and the Partial Inverse Operator H.

Let U be an $n \times p$ matrix whose p columns form a basis for the solutions to the given boundary value problem (6.1), and let V be a $q \times n$ matrix whose q rows form a basis for the solutions of the adjoint boundary value problem (7.1). Let c and d denote the $p \times p$ and $q \times q$ matrices, whose entries are numbers, defined by

$$c = \int_a^b \tilde{U}(s)U(s)ds, \quad d = \int_a^b V(s)\tilde{V}(s)ds. \tag{8.1}$$

These matrices are nonsingular. Indeed, if say c were singular, then there would be a linear combination of its column which is identically zero, say $ck = 0$ with $k = \mathrm{col}(k_1, \ldots, k_p) \neq 0$. Hence $\int_a^b \tilde{U}(s)[U(s)k]ds = 0$, and also $\int_a^b [\tilde{k}\tilde{U}(s)][\tilde{U}(s)k]ds = 0$, or $\int_a^b |\varphi(s)|^2 ds = 0$ where $\varphi = U(s)k$. Thus $\varphi = 0$ in $[a,b]$, a contradiction. An analogous proof holds for d.

We define now the projection operators $P:X \to X$ and $Q:Y \to Y$ by taking

$$(Px)(t) = U(t)\alpha \quad \text{where} \quad \alpha = c^{-1}\int_a^b \tilde{U}(s)x(s)ds,$$
$$x \in X, \tag{8.2}$$
$$(Qy)(t) = \tilde{V}(t)\beta \quad \text{where} \quad \beta = d^{-1}\int_a^b V(s)y(s)ds,$$
$$y \in Y. \tag{8.3}$$

Whenever $p = 0$ we take $P = 0$, whenever $q = 0$ we take $Q = 0$. It is evident that the range of P is the subspace of X spanned by the p columns of U, that is, the null space of E, or $\mathcal{R}(P) = \ker E$. Analogously, $\mathcal{R}(Q) = \ker E^* = \mathrm{coker}\, E$.

It remains to prove that $P:X \to X$, $Q:Y \to Y$ are bounded (linear) operators. Let A_0 denote the largest of the $\| \|_1$-norms of the n^2 entries of $A(t)$. Let $U_0 = \|U\|_0$, $U_1 = \|U'\|_1$ denote the largest of the $\| \|_0$- and $\| \|_1$-norms of the p columns u_i of $U(t)$ and u_i' of $U'(t)$ respectively, $i = 1,\ldots,p$. Each u_i is AC in $[a,b]$ and $u_i' = A(t)u_i$. Thus, U_0 is finite and $U_1 \leq n^2 A_0 U_0$. Let $V_0 = \|V\|_0$, $V_1 = \|V\|_1$ denote the largest of the $\| \|_0$- and $\| \|_1$-norms of the q rows v_i, $i = 1,\ldots,q$, of $V(t)$. Then, for $z = Px$, $x,z \in X$, we have $z(t) = U(t)\alpha$ where α is given by (8.2). Hence, $z'(t) = U'(t)\alpha$, and $\|z\|_0 \leq U_0|\alpha|$, $\|z'\|_1 \leq U_1|\alpha|$, $|\alpha| \leq (b - a)\|c^{-1}\|U_0\|x\|_0$. Analogously, if $z = Qy$, $y, z \in Y$, we have $z(t) = \tilde{V}(t)\beta$. Hence $\|z\|_1 \leq V_1|\beta|$, $|\beta| \leq (b - a)\|d^{-1}\|V_0\|y\|_1$, and finally

$$\begin{aligned} &\|Px\|_1^1 = \|z_0\|_0 + \|z'\|_1 \leq K\|x\|_0, \\ &K_0 = (b - a)\|c^{-1}\|U_0(U_0 + U_1), \end{aligned} \tag{8.4}$$

$$\begin{aligned} &\|Qy\|_1 = \|z\|_1 \leq K_1\|y\|_1, \\ &K_1 = (b - a)\|d^{-1}\|V_1V_0. \end{aligned} \tag{8.5}$$

Thus, $P:X \to X$ is a bounded linear operator with respect to the norm $\| \|_1^1$ in X, and $Q:Y \to Y$ is a bounded linear operator with respect to the norm $\| \|_1$ in Y.

Moreover, for every $y \in Y$, the element $z = Qy$ is bounded, and we have $\|z\|_0 \leq V_0|\beta|$, and

$$\|Qy\|_0 \leq K\|y\|_1, \quad K = (b - a)\|d^{-1}\|V_0^2. \tag{8.6}$$

We have proved above that P and Q are projection operators in X and Y respectively, and we take as usual $X_0 = PX$, $X_1 = (I - P)X$, $Y_0 = QY$, $Y_1 = (I - Q)Y$.

(8.i) Theorem. If $h \in (L_1[a,b])^n$, then boundary value problem (7.4) has a solution $x \in X$ if and only if $Qh = 0$. If $Qh = 0$, then boundary value problem (7.4) has a unique solution Hh with $PHh = 0$ and all the other solutions are of the form $Hh + Ua$, a an arbitrary row p-vector. Furthermore, $H:Y_1 \to X_1$ is a bounded linear map.

Proof. The first part is only a restatement of (7.i). If $Qh = 0$, then (7.4) has at least a solution $\bar{x}(t)$, $a \leq t \leq b$, $\bar{x} \in X$. For any two solutions x_1, x_2 of (7.4), the difference $x_1 - x_2$ is a solution of the corresponding homogeneous problem (6.2), or $x_1 - x_2 \in \ker E$. Hence, any solution x of (7.4) is of the form $x = \bar{x} + Ua$. Thus, we have to determine the p-vector a in such a way that $Px = P\bar{x} + PUa = 0$, or

$$U(t)[c^{-1}\int_a^b \tilde{U}(s)(U(s)a)ds] = -U(t)[c^{-1}\int_a^b \tilde{U}(s)\bar{x}(s)ds],$$

and finally

$$a = -c^{-1}\int_a^b \tilde{U}(s)\bar{x}(s)ds. \tag{8.7}$$

This proves the existence and uniqueness of a solution $x = Hh$ of (7.4) satisfying $Px = 0$. For any $h \in Y$ we know from (7.2), (7.3) that

$$x(t) = X(t)\int_a^t X^{-1}(s)h(s)ds + X(t)\gamma, \quad a \leq t \leq b, \tag{8.8}$$

is a solution of $x' = A(t)x + h$ provided

$$D\gamma = \beta, \quad \beta = -B_2X(b)\int_a^b X^{-1}(s)h(s)ds,$$

$$D = B_1X(a) + B_2X(b).$$

Here D is an $m \times n$ matrix, and β is in the range of D, or $\beta \in \Gamma = \mathcal{R}(D)$, where Γ is a linear subspace of R^m. We know that there is an $n \times m$ matrix D^* such that $\gamma = D^*\beta$ satisfies $D\gamma = \beta$. In other words, D^* is a partial inverse of D. If $\bar{x}$ denotes the solution (8.8) with $\gamma = D^*\beta$, then the solution $x = Hh$ with $Px = 0$ is given by $\bar{x} + Ua$ with a given by (8.7). In other words, if $x(t) = Hh$, we have

$$\begin{aligned} x(t) &= \bar{x}(t) + U(t)a \\ &= X(t)\int_a^t X^{-1}(s)h(s)ds + X(t)\gamma + U(t)a, \end{aligned} \tag{8.9}$$

$$\gamma = -D^*B_2X(b)\int_a^b X^{-1}(s)h(s)ds,$$

$$a = -c^{-1}\int_a^b \tilde{U}(s)\bar{x}(s)ds.$$

The components of the matrices $X(t)$, $X^{-1}(t)$ are all bounded functions in $[a,b]$, and B_2, D^* are constant matrices. Thus, using the constant U_0 defined above, and for suitable new constants $M_0, M_1, \ldots,$ we have

$$|\gamma| \leq M_0(b - a)^{-1}\int_a^b |h(s)|ds = M_0\|h\|_1,$$

$$|\bar{x}(t)| \leq M_1(b - a)\int_a^b |h(s)|ds + M_2|\gamma|$$

$$= (M_1 + M_2M_0)\|h\|_1 = \bar{M}_1\|h\|_1,$$

$$|a| \leq \|C^{-1}\|U_0\int_a^b |\bar{x}(t)|dt \leq \|C^{-1}\|U_0\bar{M}_1\|h\|_1 = \bar{M}_2\|h\|_1,$$

$$|x(t)| \leq \bar{M}_1\|h\|_1 + U_0\bar{M}_2\|h\|_1 = \bar{M}_3\|h\|_1. \qquad (8.10)$$

From (8.9) and (8.10) we have now

$$x'(t) = A(t)\bar{x}(t) + U'(t)a + h(t),$$

$$|x'(t)| \leq |h(t)| + M_4\bar{M}_1\|h\|_1 + \|U'\|_0|a|$$

$$= |h(t)| + (M_4\bar{M}_1 + \|U'\|_0\bar{M}_2)\|h\|_1$$

$$= |h(t)| + \bar{M}_4\|h\|_1. \qquad (8.11)$$

Thus, for $x(t) = Hh$, we have the relations

$$\|Hh\|_1^1 = \|x\|_0 + \|x'\|_1 \leq \bar{M}_3\|h\|_1 + (1 + M_4)\|h\|_1$$

$$= M_5\|h\|_1. \qquad (8.12)$$

Thus $H:Y_1 \to X_1$ is a bounded linear map with respect to the norms $\|\ \|_1$ in Y and $\|\ \|_1^1$ in X, and (8.i) is proved.

(8.ii) For X, Y, E, P, Q, H as defined above, relations (k_{123}) of no. 2 hold.

<u>Proof</u>. First, for $x \in X$ we have $Ex = x' - Ax = y$; hence, by (8.i), $Qy = 0$, or $y \in Y_1$, and $QEx = 0$, while $Px \in X_0$ and $EPx = 0$, and (k_2) is proved. For $y \in Y_1$ we certainly have $EHy = y$, hence $EH(I - Q)z = (I - Q)z$ for every $z \in Y$ and (k_3) is proved. Now, if $x \in \mathcal{D}(E)$, then $B_1x(a) + B_2x(b) = 0$, and $Ex = x' - Ax = z \in Y$. By (8.i), then $Qz = 0$, hence $z \in Y_1$, and $x_0 = H(I - Q)Ex \in X_1$ is the unique solution of

$$x_0' - A(t)x_0 = z, \quad B_1x_0(a) + B_2x_0(b) = 0, \quad Px_0 = 0.$$

In other words, $x - x_0 = u \in X_0$, hence $Px - Px_0 = Pu$, or $Px = u$, and $x_0 = x - u = (I - P)x$. We have proved that for $x \in \mathcal{D}(E)$ we have $H(I - P)Ex = (I - P)x$, and this is (k_1). Relations (k_{123}) have been proved.

As a consequence of no. 2, boundary value problem (6.1), or $Ex = Nx$, is now equivalent to the system of auxiliary and bifurcation equations

$$x = Px + H(I - Q)Nx, \qquad Q(E - N)x = 0,$$

and since $QEx = 0$, also

$$x = x^* + H(I - Q)Nx, \quad x^* = Px, \tag{8.13}$$

$$QNx = 0. \tag{8.14}$$

9. <u>The Perturbation Problem</u>. Let us assume now that f in (6.1) is of the form εg, and that the matrices A, B_1, B_2 depend on ε; in other words, let us consider systems (6.1) of the form

$$\begin{aligned} &x' - A(t,\varepsilon)x = \varepsilon g(t,x,x',\varepsilon), \quad a \le t \le b, \\ &B_1(\varepsilon)x(a) + B_2(\varepsilon)x(b) = 0. \end{aligned} \tag{9.1}$$

Here ε is a real parameter, $-\varepsilon_0 \le \varepsilon \le \varepsilon_0$, for some $\varepsilon_0 > 0$, g is defined in $[a,b] \times R^{2n} \times [-\varepsilon_0, \varepsilon_0]$, and for any given pairs of constants $R_1, R_2 > 0$ there are also two other constants L, M such that

$$\begin{aligned} &|g(t,x,x',\varepsilon)| \le M, \\ &|g(t,x,x',\varepsilon) - g(t,y,y',\varepsilon| \\ &\quad \le L[|x - y| + |x' - y'|], \end{aligned} \tag{9.2}$$

for all $t \in [a,b]$, $\varepsilon \in [-\varepsilon_0, \varepsilon_0]$, and all $x,y,x',y' \in R^n$ with $|x|, |y| \le R_1$, $|x'|, |y'| \le R_2$. We shall also assume that $A(t,\varepsilon)$ is an $n \times n$ matrix whose entries are bounded functions on $[a,b] \times [-\varepsilon_0, \varepsilon_0]$, each measurable in t for every ε, and continuous in ε for every t. Also we assume that $B_1(\varepsilon)$, $B_2(\varepsilon)$ are $m \times n$ matrices whose entries are continuous functions of ε in $[-\varepsilon_0, \varepsilon_0]$.

We assume here that the homogeneous boundary value problem (6.2), which now depends on ε, has a well determined number p, $0 \le p \le n$, of independent solutions, the same p for all ε sufficiently small, say $|\varepsilon| \le \varepsilon_0$, and that we can construct an $n \times p$ matrix $U(t,\varepsilon)$ whose p columns form a base for ker E, and whose entries are AC functions of t in $[a,b]$ and continuous functions of ε in $[-\varepsilon_0, \varepsilon_0]$. We make analogous assumptions on the adjoint boundary value problem and the $q \times n$ matrix $V(t,\varepsilon)$. Since c and d are nonsingular, then necessarily $|\det c| \ge \mu > 0$, $|\det d| \ge \mu > 0$ for some $\mu > 0$ and all $\varepsilon \in [-\varepsilon_0, \varepsilon_0]$, and $\|c^{-1}\|$, $\|d^{-1}\|$ are less than some given constant, independent of ε.

The following statement shows that the auxiliary equation always has a solution for ε small.

(9.i) Given $R_1, R_2 > 0$, so that (9.2) holds for some L, M as stated, then there are $\rho > 0$, $\varepsilon_1 > 0$ such that, for any constant p-vector α and number ε with $|\alpha| \le \rho$, $|\varepsilon| \le \varepsilon_1$, there is a unique vector function $x = x(t,\alpha,\varepsilon)$, $a \le t \le b$, $x \in X$, $|\alpha| \le \rho$, $|\varepsilon| \le \varepsilon_1$, such that

$$x = U\alpha + \varepsilon H(I - Q)g(\cdot, x, x', \varepsilon), \tag{9.3}$$

and $x \in X$ depends continuously on α and ε. Moreover, if there is an $\alpha = \alpha(\varepsilon)$ with $|\alpha(\varepsilon)| \le \rho$, $-\varepsilon_1 < \varepsilon < \varepsilon_1$, such that

$$G(\cdot,\alpha(\varepsilon),\varepsilon) \equiv Qg(\cdot,x(\cdot,\alpha(\varepsilon),\varepsilon),x'(\cdot,\alpha(\varepsilon),\varepsilon),\varepsilon) \equiv 0, \tag{9.4}$$

then $x(\cdot,\alpha(\varepsilon),\varepsilon)$ is a solution of the given boundary value problem (9.1).

<u>Proof</u>. From no. 8 we know that $Q:Y \to Y$ is a bounded linear map. Then so is $I - Q:Y \to Y$, and $K_1 = \|I - Q\|_1$ is finite. We have then in no. 8 that $H:Y_1 \to X_1$ is a bounded linear map. If U_0, U_1 are the constants defined in no. 8, $\bar{M}_3$, $\bar{M}_4$ the constants in (8.10), (8.11), and K_1, K the constants in (8.5), (8.6), let $\bar{K}_1 = 1 + K_1$, $\bar{K} = 1 + K + \bar{M}_4(1 + K_1)$, and take

$$\begin{aligned}\rho &= \min[2^{-1}U_0^{-1}R_1,\ 2^{-1}\|U'\|_0^{-1}R_2],\\ \varepsilon_1 &= \min[\varepsilon_0, 2^{-1}\bar{M}_3^{-1}\bar{K}_1^{-1}M^{-1}R_1, 2^{-1}\bar{K}^{-1}M^{-1}R_2, 2^{-1}M_5^{-1}\bar{K}_1^{-1}L^{-1}].\end{aligned} \tag{9.5}$$

Let S denote the subset of all $x \in X$, $x(t) = (x_1,\ldots,x_n)$, with $|x(t)| \le R_1$, $a \le t \le b$, $|x'(t)| \le R_2$, $t \in [a,b]$ (a.e.), satisfying $B_1x(a) + B_2x(b) = 0$, $Px = U\alpha$, $|\alpha| \le \rho$. For any $|\alpha| \le \rho$ let S_α denote the subset of all $x \in S$ with $Px = U\alpha$. Every set S_α is closed in the topology of X. For $x \in S_\alpha$, $y = Tx$, where T is the same transformation of no. 2, or $T = P + H(I - Q)N$, we have

$$\begin{aligned}y(t) &= Tx = U(t)\alpha + \varepsilon H(I - Q)g(t,x(t),x'(t),\varepsilon),\\ y'(t) &= U'(t)\alpha + \varepsilon(d/dt)H(I - Q)g,\end{aligned} \tag{9.6}$$

where $|x(t)| \le R_1$, $|x'(t)| \le R_2$, hence $|g| \le M$, $\|g\|_1 \le M$. For $|\alpha| \le \rho$, $|\varepsilon| \le \varepsilon_1$, and by using relations (8.5), (8.6), (8.10), (8.11) and (9.5) we have

$$\begin{aligned}|y(t)| &\le |U(t)\alpha| + |\varepsilon|\bar{M}_3\|(I - Q)g\|_1\\ &\le U_0|\alpha| + |\varepsilon|\bar{M}_3(1 + K_1)\|g\|_1 \le R_1/2 + R_1/2 = R_1,\end{aligned}$$

$$|y'(t)| \le |U'(t)\alpha| + |\varepsilon|[|(I - Q)g| + \bar{M}_4\|(I - Q)g\|_1]$$
$$\le \|U'\|_0|\alpha| + |\varepsilon|[(1 + K) + \bar{M}_4(1 + K_1)]\|g\|_1$$
$$\le R_2/2 + R_2/2 = R_2.$$

Thus, $y = Tx \in S$, and also $y \in S_\alpha$ since $Py = U\alpha$. Moreover, if $x_1, x_2 \in S_\alpha$ and $y_i = Tx_i$, $i = 1,2$, we also have from (8.12), (9.2), (9.5), and for $|\alpha| \le \rho$, $|\varepsilon| \le \varepsilon_1$,

$$y_1(t) - y_2(t) = \varepsilon H(I - Q)[g(t,x_1(t),x_1'(t),\varepsilon) - g(t,x_2(t),x_2'(t),\varepsilon)],$$

$$\|y_1 - y_2\|_1^1 \le \varepsilon M_5(1 + K_1)L\|x_1 - x_2\|_1^1 \le 2^{-1}\|x_1 - x_2\|_1^1.$$

Thus, $T:S_\alpha \to S_\alpha$ is a contraction, and possesses a unique fixed point $x = x(t,\alpha,\varepsilon)$, $a \le t \le b$, $|\alpha| \le \rho$, $|\varepsilon| \le \varepsilon_1$, that is, $x = Tx$, and (9.3) holds, that is, we have solved the auxiliary equation.

We shall now consider the bifurcation equation $QNx = 0$, or $Qf = 0$, or $\varepsilon Qg = 0$. By dividing through by ε we have the bifurcation equation $Qg = 0$, or explicitly in the form (9.4).

If $\alpha(\varepsilon)$ exists as stated in (9.i) so that (9.4) holds, then also the bifurcation equation is satisfied, and the original problem $Ex = Nx$ is solved.

Now we assume $f = \varepsilon g(t,x,x',\varepsilon) = \varepsilon\ \mathrm{col}(g_1,\dots,g_n)$ with g as before satisfying (9.2) in $[a,b] \times R^{2n} \times [-\varepsilon_0,\varepsilon]$, and we assume also that $g(t,x,x',\varepsilon)$ is continuously differentiable with respect to x,x',ε for a.a. $t \in [a,b]$. In other words, the partial derivatives $g_{ix_j}, g_{ix_j'}, g_{i\varepsilon}$, $i,j = 1,\dots,n$, exist for $t \in [a,b]$ (a.e.) and are continuous with respect to x,x',ε, (and are measurable in t for every

$x, x', \varepsilon)$. Obviously $|g_{ix_j}|, |g_{ix'_j}| \leq L$, and we assume that $|g_{i\varepsilon}| \leq h(t)$ for some fixed function $h \in L_1[a,b]$. Then, the solution $x = x(t,\alpha,\varepsilon)$, $x \in X$, $|\alpha| < \rho$, $|\varepsilon| < \varepsilon_1$, $\|x\|_0 < R_1$, $\|x'\|_0 < R_2$ of the auxiliary equation is continuously differentiable in α and ε. This can be derived from the general statements mentioned in no. 2 (Hale [64], p.7) and suitable variants. (The Fréchet derivatives at x are computed here relative to elements $z \in S$.)

The bifurcation equation $QN = 0$, or briefly $Qf = 0$, is satisfied if and only if f is orthogonal to all elements of Y_0, or $\int_a^b V(s) f ds = 0$. Thus, for $f = \varepsilon g$, and division by ε, we obtain the bifurcation equation in the form

$$K(\alpha, \varepsilon) = \int_a^b V(s) g(s, x(s;\alpha,\varepsilon), x'(s;\alpha,\varepsilon), \varepsilon) ds = 0. \qquad (9.7)$$

Here $K = (K_1, \ldots, K_q)$ is a q-vector valued function of α, ε for $|\alpha| \leq \rho$, $|\varepsilon| \leq \varepsilon_1$. Under the hypotheses on g stated above, K is continuously differentiable in α, ε. By the implicit function theorem of calculus we have now:

(9.ii) If $p \geq q$, and if there is an $\bar{\alpha} \in R^p$, $|\bar{\alpha}| < \rho$, such that $K(\bar{\alpha}_1, 0) = 0$, and such that the $q \times p$ Jacobian matrix $\partial K/\partial\alpha$, or $[\partial K_i(\bar{\alpha}, 0)/\partial\alpha_j$, $i = 1, \ldots, q$, $j = 1, \ldots, p]$ has maximal rank q, then there if $\bar{\varepsilon}$, $0 < \bar{\varepsilon} \leq \varepsilon_1 \leq \varepsilon_0$, such that, for every $|\varepsilon| \leq \bar{\varepsilon}$, there is an $\alpha = \alpha(\varepsilon)$, $\alpha(0) = \bar{\alpha}$, $|\alpha(\varepsilon)| \leq \rho$, with $K(\alpha(\varepsilon), \varepsilon) = 0$, and then $x = x(t;\alpha(\varepsilon),\varepsilon)$ is a solution of the given boundary value problem (9.1). Here $\alpha(\varepsilon)$, and thus $x = x(t;\alpha(\varepsilon),\varepsilon)$, $x \in X$, are continuously differentiable functions of ε in $[-\bar{\varepsilon}, \bar{\varepsilon}]$.

10. Examples.

1. Periodic solutions of period 2π of the second order equation

$$x'' = \varepsilon[-e^{x''} + x^2\sin^2 t] \qquad (10.1)$$

For $y_1 = x$, $y_2 = x'$, $y = \mathrm{col}(y_1, y_2)$ we have the differential system

$$y' = Ay + \varepsilon g(t, y, y'),$$

$$y(0) - y(2\pi) = 0,$$

where $n = 2$, $A = [0,1,0,0]$, $B_1 = I$, $B_2 = -I$, and $g = \mathrm{col}(g_1, g_2)$, $g_1 = 0$, $g_2 = -e^{-y_2'} + y_1^2 \sin^2 t$. Here we have $p = q = 1$, $U(t) = \mathrm{col}(1,0)$, $V(t) = \mathrm{row}(0,1)$. Then for ε small the auxiliary equation has solution $x(t,\alpha,\varepsilon) = \alpha + O(\varepsilon)$, α scalar, and

$$K(\alpha, \varepsilon) = \int_0^{2\pi} [-e^{O(\varepsilon)} + (\alpha^2 + O(\varepsilon))\sin^2 t]dt.$$

So $K(\alpha, 0) = \pi\alpha^2 - 2\pi$, $\partial K(\alpha,0)/\partial\alpha = 2\pi\alpha$, and thus problem (10.1) has a 2π-periodic solution of the form $x(t) = \alpha + O(\varepsilon)$ with $\alpha = \pm\sqrt{2}$.

2. Periodic solutions of period $2\pi/\sigma$ (harmonics) of the van der Pol equation

$$x'' + \sigma^2 x = \varepsilon[(1 - x^2)x' + a\sigma^{-1}x'' + b\sigma\cos(\sigma t + \alpha)]. \qquad (10.2)$$

For $y_1 = x$, $y_2 = x'$, $y = \mathrm{col}(y_1, y_2)$ we have the system

$$y' = Ay + \varepsilon g(t, y, y'),$$

$$y(0) - y(\tfrac{2\pi}{\sigma}) = 0,$$

where $n = 2$, $A = [0,1;-\sigma^2,0]$, $B_1 = I$, $B_2 = -I$, and $g = \mathrm{col}(g_1, g_2)$, $g_1 = 0$, $g_2 = (1 - y_1^2)y_2 + a\sigma^{-1}y_2' + b\sigma\cos(\sigma t + \alpha)$. Here we have $p = q = 2$, and we may take

$$U(t) = (\sin \sigma t, \cos \sigma t;\ \sigma \cos \sigma t, -\sigma \sin \sigma t),$$

$$V(t) = (\sigma \sin(\sigma t + \theta_0), \cos(\sigma t + \theta_0);\ \sigma \cos(\sigma t + \theta_0),$$
$$- \sin(\sigma t + \theta_0)),$$

where θ_0 is for the moment arbitrary.

Then the solution of the auxiliary equation has the form

$$x(t) = c_1\sigma^{-1}\sin(\sigma(t) + c_2\sigma^{-1}\cos(\sigma t) + O(\varepsilon)$$
$$= \lambda(\varepsilon)\sigma^{-1}\sin(\sigma t + \theta(\varepsilon)) + O(\varepsilon),$$

where $\lambda(\varepsilon) = \lambda_0 + O(\varepsilon)$, $\theta(\varepsilon) = \theta_0 + O(\varepsilon)$. Here we have $K = \mathrm{col}(K_1, K_2)$, with

$$K_1(\lambda, \theta, \varepsilon) = \int_0 \{[1 - \lambda^2\sigma^{-2}\sin^2(\sigma t + \theta)]\lambda \cos(\sigma t + \theta)$$
$$+ b\sigma \cos(\sigma t + \alpha)$$
$$- \alpha\lambda \sin(\sigma t + \theta)\}\cos(\sigma t + \theta_0)dt;$$

$$K_2(\lambda, \theta, \varepsilon) = \int_0 \{[1 - \lambda^2\sigma^{-2}\sin^2(\sigma t + \theta)]\lambda \cos(\sigma t + \theta)$$
$$+ b\sigma \cos(\sigma t + \alpha)$$
$$- \alpha\lambda \sin(\sigma t + \theta)\}\sin(\sigma t + \theta_0)dt,$$

$$K_1(\lambda, \theta, 0) = -4^{-1}\sigma^{-3}\pi[\lambda_0^3 - 4\lambda_0\sigma^2 - 4b\sigma^3\cos(\alpha - \theta)],$$

$$K_2(\lambda, \theta, 0) = -\sigma^{-1}\pi[a\lambda_0 + b\sigma \sin(\alpha - \theta)].$$

Thus $K = 0$ has a solution for $a \ll b$, $b \ll 1$, of the form $\lambda_0^2 \cong 4\sigma^2$, $\theta_0 \simeq \alpha$. For these values of λ_0 and θ_0 the Jacobian determinant becomes a nonzero term times $3\lambda_0^2 + 4\sigma^2 - 4a\sigma \tan(\alpha - \theta_0) \simeq 16\sigma^2 \neq 0$. Thus, for $a \ll b$, $b \ll 1$, problem (10.2) has a periodic solution of period $2\pi/\sigma$ of the form $x(t) = \lambda(\varepsilon)\sigma^{-1} \sin(\sigma t + \theta(\varepsilon)) + O(\varepsilon)$, with $\lambda(\varepsilon) \simeq 2\sigma + O(\varepsilon)$, $\theta(\varepsilon) \simeq \alpha + O(\varepsilon)$.

3. Solutions of the boundary value problem

$$x'' + x = \varepsilon(x^3 + \sin t), \quad x'(0) = x'(2\pi).$$

For $y_1 = x$, $y_2 = x'$, $y = \mathrm{col}(y_1, y_2)$ we have the problem

$$y' = Ay + \varepsilon g(t,y,y'), \quad y_2(0) = y_2(2\pi),$$

where $n = 2$, $A = [0,1;-1,0]$, $g = \mathrm{col}(g_1, g_2)$, $g_1 = 0$, $g_2 = x^3 + \sin t$, $B_1 = \mathrm{row}[0,1]$, $B_2 = \mathrm{row}[0,-1]$. Here $p = 2$, $q = 1$, and we may take

$$U(t) = [\sin t, \cos t; \cos t, -\sin t],$$

$$V(t) = (\sin t, \cos t).$$

The solution of the auxiliary equation is of the form $x = \alpha_1 \sin t + \alpha_2 \cos t + O(\varepsilon)$ and the bifurcation equation becomes

$$K(\alpha, \varepsilon) = \int_0^{2\pi} [(\alpha_1 \sin t + \alpha_2 \cos t + O(\varepsilon))^3 + \sin t] \cos t \, dt = 0,$$

$$K(\alpha,0) = (3\pi/4)(\alpha_1^2\alpha_2 + \alpha_2^3) = 0,$$

$$\partial K(\alpha,0)/\partial\alpha = \mathrm{row}[(3\pi/2)\alpha_1\alpha_2, (3\pi/4)(\alpha_1^2 + 3\alpha_2^2)].$$

If $\alpha_1 \neq 0$, $\alpha_2 = 0$, then $K(\alpha,0) = 0$ and $\partial K(\alpha,0)/\partial\alpha$ has rank 1. Hence, problem (10.3) has a solution of the form $x = \alpha_1 \sin t + O(\varepsilon)$.

4. Periodic solutions of period 2π (harmonics) of the equation

$$x^{(n)} = -a_1(\varepsilon)x^{(n-1)} - \cdots - a_n(\varepsilon)x + \varepsilon g_0(t,x,x',\ldots,x^{(n-1)};\varepsilon), \qquad (10.4)$$

where $a_1(0) = \cdots = a_n(0) = 0$. We assume that $a_1'(0),\ldots,a_n'(0)$ exist. For $y = x$, $y_2 = x', \ldots, y_n = x^{(n-1)}$, $y = \mathrm{col}(y_1,\ldots,y_n)$, we have the problem

$$y' - Ay = \varepsilon g, \quad B_1 y(0) + B_2 y(2\pi) = 0,$$

$$A = [0, 1, 0, \ldots, 0; 0, 0, 1, \ldots, 0; \ldots; 0, 0, 0, \ldots, 0],$$

$$B_1 = I, \quad B_2 = -I,$$

where I is the $n \times n$ identity matrix, $m = n$, $g = \mathrm{col}(g_1,\ldots,g_n)$, $g_1 = \cdots = g_{n-1} = 0$, and

$$g_n = -\varepsilon^{-1}a_n(\varepsilon)y_1 - \cdots - \varepsilon^{-1}a_1(\varepsilon)y_n + g_0(t,y,y_1,\ldots,y_n).$$

Here $p = 1$, $q = 1$, and we can take for U and V the $n \times 1$ and $1 \times n$ matrices

$$U = \mathrm{col}(1,0,\ldots,0), \quad V = \mathrm{row}(0,\ldots,0,1).$$

Then, the solution of the auxiliary equation has the form

$$y_1 = \lambda + O(\varepsilon), \quad y_2 = O(\varepsilon),\ldots,y_n = O(\varepsilon),$$

and the bifurcation equation takes the form

$$K(\lambda,\varepsilon) = \int_0^{2\pi} V(s)g_0(t,\lambda + O(\varepsilon),O(\varepsilon),\ldots,O(\varepsilon);\varepsilon)ds$$

$$= \int_0^{2\pi} [-\varepsilon^{-1}a_n(\varepsilon)(\lambda + O(\varepsilon),O(\varepsilon),\ldots,O(\varepsilon);\varepsilon)ds,$$

$$K(\lambda,0) = -2\pi a_n'(0)\lambda + \int_0^{2\pi} g_0(t,\lambda,0,\ldots,0;0)dt.$$

Thus, the bifurcation equation becomes

$$(2\pi)^{-1}\int_0^{2\pi} g_0(t,\lambda,0,\ldots,0)dt = a_n'(0)\lambda. \tag{10.5}$$

If this equation has a solution λ_0 and $\partial K(\lambda_0,0)/\partial\lambda \neq 0$, then problem (10.4) has a 2π-periodic solution of the form

$$x(t) = \lambda + O(\varepsilon), \quad \lambda = \lambda_0 + O(\varepsilon).$$

For instance, for the differential equation

$$x'' = -\varepsilon x + \varepsilon x^3\sin^2 t, \tag{10.6}$$

the bifurcation equation (10.5) has the form

$$(2\pi)^{-1}\int_0^{2\pi} \lambda^3\sin^2 t\,dt = \lambda,$$

or $\lambda^3 - 2\lambda = 0$ with root $\lambda = 2^{1/2}$, and the derivative $3\lambda^2 - 2$ is $\neq 0$ at that point. Thus, (10.6) has a 2π-periodic solution of the form $x(t) = 2^{1/2} + O(\varepsilon)$.

11. Some Results for Periodic Solutions of Certain Typical Systems. The following results concerning systems of equations of typical forms are of relevance in applications. They cover both the problem of harmonic solutions, and of subharmonic, or analogous solutions.

Consider the system of equations

$$\begin{aligned} &x_j'' + 2\alpha_j x_j' + \beta_j x_j = \varepsilon g_j(t,x,x',\varepsilon), \quad j = 1,\ldots,n, \\ &x_j' + \gamma_j x_j = \varepsilon g_j(t,x,x',\varepsilon), \quad j = n+1,\ldots,N, \end{aligned} \tag{11.1}$$

where $x = (x_1,\ldots,x_N)$, $x' = (x_1',\ldots,x_n')$, $0 \le n \le N$. Assume that α_j, β_j, γ_j are real constants, or continuous functions of ε, and all g_j are real continuous functions of ε, uniformly Lipschitzian in x, x', periodic in t of a fixed period $T = 2\pi/\omega$, bounded and measurable in $[0,T]$ for $|x_j| \le R_1$, $j = 1,\ldots,N$, $|x_j'| \le R_2$, $j = 1,\ldots,n$, $|\varepsilon| \le \varepsilon_0$, $-\infty < t < +\infty$, ω, ε_0, R_1, $R_2 > 0$. Assume that, for some ℓ, ν, μ, $0 \le \ell \le \nu \le n \le \mu \le N$, we have (i) $\beta_j = \sigma_j^2$, $\alpha_j(0) = 0$, $\sigma_j(0) = \tau_j = k_j\omega/m_j$, k_j, m_j positive relatively prime integers, for all $j = 1,\ldots,\ell$; (ii) $\alpha_j(0) = 0$, $\beta_j(0) = 0$, $j = \ell + 1,\ldots,\nu$ (iii) $\beta_i = \sigma_i^2$ and either $\alpha_j(0) = 0$, $\sigma_j(0) > 0$, $\sigma_j(0) \ne k\omega/m$, $m = m_1,\ldots,m_\ell$, $k = 1,2,\ldots$, or $\alpha_j(0) \ne 0$, $\sigma_j(0) > 0$, $\sigma_j^2 - \alpha_j^2 > 0$ for all $j = \nu + 1,\ldots,n$; (iv) $\beta_j(0) \ne 0$ for all $j = n + 1,\ldots,\mu$; (v) $\gamma_j(0) = 0$ for all $j = \mu + 1,\ldots,N$. We shall assume that the derivatives $\alpha_j'(0)$, $\beta_j'(0)$, $\gamma_j'(0)$, $\sigma_j'(0)$ exist, and the functions P_{jo}, Q_{jo}, R_{jo}, S_{jo} below have first order continuous partial derivatives in

$\lambda_1,\ldots,\lambda_\ell$, $\theta_1,\ldots,\theta_\ell$, $\delta_1,\ldots,\delta_{\nu-\ell}$, $\eta_1,\ldots,\eta_{N-\mu}$. If the $2\ell + (\nu - \ell) + (N - \mu)$ finite equations in these variables

$$\begin{aligned} P_{jo}(\lambda,\theta,\delta,\eta) &= \lambda_j\alpha_j'(0), \quad j = 1,\ldots,\ell, \\ Q_{jo}(\lambda,\theta,\delta,\eta) &= \lambda_j\sigma_j'(0), \quad j = 1,\ldots,\ell, \\ R_{jo}(\lambda,\theta,\delta,\eta) &= \delta_j\beta_j'(0), \quad j = \ell + 1,\ldots,\nu, \\ S_{jo}(\lambda,\theta,\delta,\eta) &= \eta_j\gamma_j'(0), \quad j = 1,\ldots,N - \mu, \end{aligned}$$

have a solution λ_{jo}, θ_{jo}, δ_{jo}, η_{jo} with Jacobian different from zero at λ_{jo}, θ_{jo}, δ_{jo}, η_{jo}, then system (11.1) has a periodic solution of period $T = 2\pi m/\omega$ of the form

$$x_j(t,\varepsilon) = \lambda_j \tau_j^{-1} \sin(\tau_j t + \theta_j) + O(\varepsilon), \quad j = 1,\dots,\ell,$$

$$x_j(t,\varepsilon) = \delta_{j-\ell} + O(\varepsilon), \quad j = \ell + 1,\dots,\nu,$$

$$x_j(t,\varepsilon) = O(\varepsilon), \quad j = \nu + 1,\dots,\mu,$$

$$x_j(t,\varepsilon) = \eta_{j-\mu} + O(\varepsilon), \quad j = \mu + 1,\dots,N,$$

where $\lambda_j = \lambda_{jo} + O(\varepsilon)$, $\theta_j = \theta_{jo} + O(\varepsilon)$, $j = 1,\dots,\ell$, $\delta_j = \delta_{jo} + O(\varepsilon)$, $j = 1,\dots,\nu-\ell$, $\eta_j = \eta_{jo} + O(\varepsilon)$, $j = \mu+1,\dots,N$, for all $|\varepsilon| \le \varepsilon_1$ and some $0 < \varepsilon_1 \le \varepsilon_0$. The functions P_{jo}, Q_{jo}, R_{jo}, S_{jo} are given by

$$P_{jo} = T_1^{-1}\int_0^{T_1} g_{jo} \cos(\tau_j t + \theta_j)dt, \quad j = 1,\dots,\ell,$$

$$Q_{jo} = T_1^{-1}\int_0^{T_1} g_{jo} \sin(\tau_j t + \theta_j)dt, \quad j = 1,\dots,\ell,$$

$$R_{jo} = T_1^{-1}\int_0^{T_1} g_{jo}\, dt, \quad j = \ell + 1,\dots,\nu,$$

$$S_{jo} = T_1^{-1}\int_0^{T_1} g_{jo}\, dt, \quad j = \mu + 1,\dots,N,$$

where g_{jo} denotes g_j when the arguments ε, x_j, x_j' are replaced by $\varepsilon = 0$, and by $x_j(t,0)$, $x_j'(t,0)$. This theorem was proved by alternative arguments by Cesari and Hale [32] and Cesari [10]. For versions of this theorem, and a great many examples discussed in detail by alternative arguments, we refer to Hale and Gambill [66]. The parallel discussion of the autonomous case cannot be given here, and we refer to Cesari's book [9], and the papers mentioned above.

12. <u>Symmetries and Families of Solutions</u>. In this no. we are now going to study nonlinear boundary value problems (9.1) which satisfy Hale's symmetry condition, or property (S)

[56,64]. Under this condition, the number of components in the bifurcation equation can be reduced (in some cases the bifurcation equation is identically satisfied), and the existence of families of solutions to the given boundary value problem can be proved.

A differential system $x' = g(t,x)$, where x and g are n—vectors, is said to have property (S) if there is a symmetric, constant, $n \times n$ matrix S such that

$$S^2 = I, \quad \text{and} \quad Sg(-t,Sx) = -g(t,x), \quad -a \le t \le a, \quad x \in R^n.$$

If $x(t)$ is a solution to a system $x' = g(t,x)$ which has property (S), then $y(t) = Sx(-t)$ is also a solution for the system $x' = g(t,x)$. Indeed, $y' = (d/dt)Sx(-t) = -Sx'(-t) = -Sg(-t,x(-t)) = -Sg(-t,S(Sx(-t)) = g(t,Sx(-t)) = g(t,y)$. Throughout this no. we will consider only boundary value problems

$$x' = A(t)x + f(t,x,\varepsilon), \quad t \in [-a,a], \tag{12.1}$$

$$B_1x(-a) + B_2x(a) = 0. \tag{12.2}$$

We make on x, A, f, B_1, B_2, ε the same assumptions as in no. 9. The operators P, Q, H and matrices U, V, c, d are those of no. 8.

(12.i) (a lemma). Let S be a symmetric constant $n \times n$ matrix with $S^2 = I$. Let $A(t)$ be an $n \times n$ matrix whose entries are bounded measurable functions over the interval $[-a,a]$, and let $h(t) = (h_1,\dots,h_n)$ be an element of $(L_1[-a,a])^n$. Let us assume that (a) $SA(-t) = A(t)S$; (b) $Sh(-t) = -h(t)$; (c) if $z \in X$ satisfies (12.1), then $Sz(-t)$ also satisfies (12.1); (d) if $z \in X$ satisfies the adjoint boundary conditions (7.1), that is $\tilde{z}(-a) = \tilde{\alpha}B_1$,

$\tilde{z}(a) = -\tilde{\alpha}B_2$ for some $\alpha \in R^m$, then so does $Sz(-t)$. Then (α) $SU(-t)\alpha = U(t)\alpha$ for $t \in [-a,a]$ if $SU(0)\ \alpha = U(0)\alpha$ for some $\alpha \in R^p$; (β) $SQh(-t) = -Qh(t)$; (γ) $S[H(I - Q)h](-t) = [H(I - Q)h](t)$ [110].

Proof. First note that A satisfies (a) and thus system $x' = A(t)x$ has property (S). Hence, $SU(-t)\alpha$ is a solution of $x' = A(t)x$ as is $U(t)\alpha$. If $SU(0)\alpha = U(0)\alpha$ for some $\alpha \in R^p$, then by uniqueness we must have $SU(-t)\alpha = U(t)\alpha$ for $t \in [-a,a]$. This proves (α). Now the p columns of the $n \times p$ matrix $U(t)$ form a basis for the solutions of $x' = A(t)x$ satisfying (12.2). Then the p columns of $SU(-t)$ also form a basis for the same solutions since (c) assures that the columns of $SU(-t)$ satisfy $x' = A(t)x$. Therefore, there is a constant nonsingular $p \times p$ matrix G such that $SU(-t) = U(t)G$, $t \in [-a,a]$. Similarly, there is a nonsingular $q \times q$ matrix G_0 such that $V(-t)S = G_0V(t)$, $t \in [-a,a]$. Now let $h \in (L_1[a,b])^n$ be such that (b) holds. From the definition of d we derive that $G_0 d\tilde{G}_0 = d$. Moreover, since S is symmetric,

$$\begin{aligned} SQh(-t) &= S\tilde{V}(-t)d^{-1}\int_{-a}^{a} V(u)h(u)du \\ &= -\tilde{V}(t)\tilde{G}_0 d^{-1}G_0\int_{-a}^{a} V(u)h(u)du. \end{aligned}$$

Also, $V(-t)S = G_0V(t)$ and then $V(-t) = V(-t)S^2 = G_0V(t)S = G_0^2V(-t)$, and $G_0^2 = I$. From here we derive that $\tilde{G}_0d^{-1}G_0 = d^{-1}$, and finally $SQh(-t) = -Qh(t)$, and (β) is proved. From (b) and (β) we see that $S(I - Q)h(-t) = -(I - Q)h(t)$. Thus, to prove ($\gamma$) it suffices to prove ($\gamma$) with $(I - Q)h$ replaced by any h with $Qh = 0$. Now, Hh is the unique solution of the boundary value problem

$$x' = A(t)x + h(t), \quad t \in [-a,a], \quad Px = 0, \tag{12.3}$$

$$B_1x(-a) + B_2x(a) = 0. \tag{12.4}$$

From (a) and (b) we derive that system (12.3) has property (S). Hence, $SHh(-t)$ also is a solution of (12.3-4) and by (c), $SHh(-t)$ satisfies (12.3-4). Since Hh is the unique solution satisfying (12.3), (12.4), it remains to show that $P(S\ h)(-t) = 0$. Indeed,

$$\begin{aligned} P[SKh(\cdots)(t) &= U(t)C^{-1}\int_{-a}^{a} \tilde{U}(s)SHh(-s)ds \\ &= U(t)C^{-1}\int_{-a}^{a} \tilde{G}\tilde{U}(-s)Hh(-s)ds \\ &= U(t)C^{-1}\tilde{G}\int_{-a}^{a} \tilde{U}(s)Hh(u)du = 0. \end{aligned}$$

This proves (γ) and completes the proof of (12.i).

We consider now the nonlinear problem

$$x' = A(t)x + \varepsilon g(t,x,\varepsilon), \quad t \in [-a,a], \tag{12.5}$$

$$B_1x(-a) + B_2x(a) = 0. \tag{12.6}$$

(12.ii) Theorem (Nagle [110]). Under the conditions of (9.i) for system (12.5-6), let $x_0(t) = x_0(t) = x_0(t;\alpha,\varepsilon)$ be the solution of the auxiliary equation. Let us assume that system (12.5-6) has property (S) for $|\varepsilon| \leq \varepsilon_0$, and that assumptions (c) and (d) of (12.i) hold. If $SU(0)\alpha = U(0)\alpha$, then we also have $(Sx_0)(-t) = x_0(t)$, $-a \leq t \leq a$. Moreover, if $K(\alpha,\varepsilon)$ is the vector function defined by (9.7), then

$$(I + G_0)K(\alpha,\varepsilon) = 0, \tag{12.7}$$

where G_0 is the nonsingular $q \times q$ matrix such that $V(-t)S = G_0V(t)$, $-a \leq t \leq a$.

Proof. Here $SU(0)\alpha = U(0)\alpha$, and let S_α be defined as in no. 9. Let S_α^* denote the subset of all $y \in S_\alpha$ with

$Sy(-t) = y(t)$. We consider now the transformation T defined by (9.6), or $y = Tx$, which now becomes

$$y = T(\alpha, \epsilon)x = U(t)\alpha + \epsilon H(I - Q)g(t, x(t), \epsilon).$$

Since (12.5) has property S, we have

$$Sg(-t, x(-t), \epsilon) = Sg(-t, Sx(t), \epsilon) = -g(t, x(t), \epsilon),$$
$$-a \leq t \leq a.$$

Hence, $H(I - Q)g(t, x(t), \epsilon)$ is in S_α^*. Moreover, since $SU(0)\alpha = U(0)\alpha$, then $U(t)\alpha$ also is in S_α^*, and $y = Tx \in S_\alpha^*$. Thus, $T: S_\alpha^* \to S_\alpha^*$ (for $|\epsilon|$ sufficiently small), and the fixed point $x_0 = Tx_0$ of T is in S_α^*. Thus, $x_0(t; \alpha, \epsilon) \in S_\alpha^*$; or $(Sx_0)(-t) = x_0$. Now

$$\begin{aligned} K(\alpha, \epsilon) &= \int_{-a}^{a} V(s)g(s, x_0(s), \epsilon)\,ds \\ &= \int_{-a}^{a} V(s)g(s, (Sx_0)(-s), \epsilon)\,ds \\ &= -\int_{-a}^{a} V(s)Sg(-s, x_0(-s), \epsilon)\,ds \\ &= -G_0\int_{-a}^{a} V(-s)g(-s, x_0(-s), \epsilon)\,ds \\ &= -G_0\int_{-a}^{a} V(s)g(s, x_0(s), \epsilon)\,ds = -G_0K(\alpha, \epsilon). \end{aligned}$$

This proves (12.7) and (12.ii) is thereby proved.

<u>Examples</u>.

1. Let us consider the boundary value problem in $[-\pi, \pi]$:

$$x_1' = x_2 + \epsilon g_1, \quad x_2' = -x_1 + \epsilon g_2, \quad x_3' = \epsilon g_3, \tag{12.8}$$
$$x_1(-\pi) - x_1(\pi) = 0, \quad x_2(-\pi) + x_2(\pi) = 0,$$
$$x_3(-\pi) - x_3(\pi) = 0. \tag{12.9}$$

For $x = \operatorname{col}(x_1, x_2, x_3)$, $g = \operatorname{col}(g_1, g_2, g_3)$ we have the problem $x' = Ax + \epsilon g$, $B_1x(-\pi) + B_2x(\pi) = 0$, with $A = [0,1,0;-1,0,0;0,0,0]$, $B_1 = \operatorname{diag}(1,1,1)$, $B_2 = \operatorname{diag}(-1,1,-1)$.

Here $g_i(t,x)$, or $g_i(t,x_1,x_2,x_3)$, $i = 1,2,3$, are assumed to have the symmetry properties

$$\begin{aligned} g_1(-t,x_1,-x_2,x_3) &= -g_1(t,x), \\ g_2(-t,x_1,-x_2,x_3) &= g_2(t,x), \\ g_3(-t,x_1,-x_2,x_3) &= -g_3(t,x). \end{aligned} \tag{12.10}$$

In this problem, $p = 2$, $q = 2$. Let $S = \text{diag}(1,-1,1)$, so that S is a symmetric constant matrix with $S^2 = I$. It can be immediately verified that

$$SA(-t)S = -A(t), \quad Sg(-t,Sx) = -g(t,x)$$

so that (a), (b) are satisfied. Furthermore, if z satisfies (12.9), then

$$B_1Sz(-(-\pi)) + B_2Sz(-\pi) = B_1z(-\pi) + B_2z(\pi) = 0$$

and thus (c) is satisfied. The adjoint boundary conditions are the same as (12.9), and thus (d) also is satisfied. For the 3×2 matrix U and the 2×3 matrix V we may take here

$$U(t) = [\cos t, 0; -\sin t, 0; 0, 1],$$

$$V(t) = [\cos t, -\sin t, 0; 0, 0, 1],$$

and now we can determine the 2×2 constant matrix G_0 such that $V(-t)S = G_0V(t)$. Since $V(-t)S = V(t)$, we see that $G_0 = I$ is the 2×2 identity matrix. By Theorem (12.ii), $K(\alpha, \varepsilon) = 0$ for all α and ε sufficiently small. Finally, if α is any 2-vector, then $SU(0)\alpha = U(0)\alpha$, as we can verify. Thus, problem (12.8-9) has a 2-parameter family of solutions, of the form

$$x_1 = \lambda \sin t + O(\varepsilon), \quad x_2 = \lambda \cos t + O(\varepsilon),$$

$$x_3 = \eta + O(\varepsilon).$$

2. Let us consider the boundary value problem in $[-a,a]$:

$$x_1' = x_2 + \varepsilon g_1, \quad x_2' = x_3 + \varepsilon g_2, \quad x_3' = \varepsilon g_3, \tag{12.11}$$

$$x_1(-a) - x_1(a) = 0, \quad x_2(-a) + x_2(a) = 0,$$
$$x_3(-a) - x_3(a) = 0. \tag{12.12}$$

For $x = \text{col}(x_1,x_2,x_3)$, $g = \text{col}(g_1,g_2,g_3)$ we have the problem

$$x' = Ax + \varepsilon g, \quad B_1x(-\pi) + B_2x(\pi) = 0,$$

with $A = [0,1,0;0,0,1;0,0,0]$, $B_1 = \text{diag}(1,1,1)$, $B_2 = \text{diag}(-1,1,-1)$, and where g_1, g_2, g_3 satisfy the same symmetry conditions (12.10) as in Example 1. Again, we have $p = q = 2$, and we can take for U and V the 3×2 and 2×3 matrices

$$U(t) = [1,t^2;0,2t;0,2],$$
$$V(t) = [0,0,1;2,-2t,t^2].$$

Then, for $S = \text{diag}(1,-1,1)$, the symmetry relations are satisfied, $G_0 = I$ is the 2×2 identity matrix, and $SU(0)\alpha = U(0)\alpha$ for every 2-vector α. Then, again $(I + G_0)K(\alpha,\varepsilon) = 0$ implies that $K(\alpha,\varepsilon) = 0$ for all $|\alpha|$, ε sufficiently small. Thus, (12.11-12) has a 2-parameter family of solutions of the form

$$x_1 = \lambda_1 + \lambda_2 t^2 + O(\varepsilon), \quad x_2 = 2\lambda_2 t + O(\varepsilon),$$
$$x_3 = 2\lambda_2 + O(\varepsilon).$$

Section 3. Problems of Perturbation for Partial Differential Equations

13. Boundary Value Problems for Partial Differential Equations.

We present here a local treatment of homogeneous linear boundary value problems for nonlinear partial differential equations, based on the alternative method, and for which we refer to Shaw [120] for details. We consider problems of the type

$$\begin{aligned} &Ex = f(t,x,Dx,\ldots,D^m x), \quad t \in G \subset E^\nu, \\ &Bx = 0, \quad t \in \partial G, \quad t = (t_1,\ldots,t_\nu), \end{aligned} \tag{13.1}$$

where E is a linear elliptic operator of order m in a region G of the ν-dimensional space E^ν, with linear homogeneous boundary conditions B on ∂G. Of particular interest is the case where there are nonzero functions $x_0(t)$, $t \in G$, with $Ex_0 = 0$, $Bx_0 = 0$, that is, the fundamental operator pair (E,B) has nontrivial null space. A particular case is, of course, the Neumann problem with $E = \Delta$ and boundary condition $\partial x/\partial n = 0$ on ∂G, whose null space is made up of all constant functions on G. In (13.1) the operator E has the form

$$Ex = \sum_{|\alpha|\leq m} a_\alpha(t) D^\alpha x, \qquad D^\alpha = \partial^{|\alpha|}/\partial t_1^{\alpha_1},\ldots,\partial t_\nu^{\alpha_\nu},$$

with smooth real coefficients satisfying a uniformly elliptic condition

$$c^{-1}|\xi|^m \leq \sum_{|\alpha|=m} a_\alpha(t)\xi^\alpha \leq c|\xi|^m$$

for some constant $c > 0$ and all $\xi = (\xi_1,\ldots,\xi_\nu)$ real. Here the pair (E,B) need not be selfadjoint.

Here G is a bounded connected open subset of E^ν with piecewise smooth boundary. In (13.1), $f(t,x,Dx,\ldots,D^m x)$ denotes a smooth real-valued function defined on $(G \cup \partial G) \times R^N$, where $N = 1 + \nu + \nu(\nu + 1)/2 + \cdots$ is the number of all possible distinct partial derivatives $x, Dx, \ldots, D^m x$ of orders $\leq m$ in R^ν. We allow in f derivatives of order $\leq m$, that is, up to and including the order m of the operator E.

14. <u>The Linear Problem. Sobolev Spaces. Basic Statements</u>. We list here a number of definitions, and some properties of the solutions of the linear problem $Ex = h(t)$, $t \in G$, $Bx = 0$, $t \in \partial G$. These properties are well known under very mild assumptions on G, E, B (e.g., smoothness hypotheses on ∂G, a Garding inequality holds, the coefficients $a_{\alpha\beta}$ of E are smooth on G, B is represented by linear operators on ∂G with smooth coefficients). (We refer here to the expositions of Agmon, Lions, Magenes, et al.)

Let (x,y) and $\|x\| = \|x\|_0 = (x,x)^{1/2}$ denote the inner product and norm in the real Hilbert space $L_2(G) = H_0(G)$. Let $H_s(G)$, $s \geq 0$, s an integer, be the (Sobolev) space of all real-valued functions $h(t)$, $t = (t_1,\ldots,t_\nu) \in G \subset E^\nu$, $h \in L_2(G)$, whose distribution partial derivatives $D_t^\alpha h$, or $D^\alpha h$, are actual functions $D^\alpha h(t)$, $t \in G$, $D^\alpha h \in L_2(G)$, for any ν-multi-index $\alpha = (\alpha_1,\ldots,\alpha_\nu)$, $0 \leq |\alpha| = \alpha_1 + \cdots + \alpha_\nu \leq s$. Then, H_s is also a real Hilbert space with inner product $(x,y)_s = \sum_{|\alpha| \leq s} (D^\alpha x, D^\alpha y)$, and corresponding norm $\|x\|_s = (x,x)_s^{1/2}$.

As usual, let $C^\infty(G)$ be the linear space of all functions $x(t)$ of class C^∞ on $\bar{G}$, and let $C_0^\infty(G)$ be the subspace of all $x \in C^\infty(G)$ with compact support in G. Then H_s coincides with the completion of $C^\infty(G)$ in the norm $\| \|_s$, and we denote by $H_s^0(G)$ the completion of $C_0^\infty(G)$ with respect to the same norm. Thus, for any two integers $r \geq 0$, $s \geq 1$, we have $H_{r+s} \subset H_r$, and for $x \in H_{r+s}$ we obviously have $\|x\|_{r+s} \geq \|x\|_r$. The injection map $j:H_{r+s} \to H_r$ is a linear, bounded, and compact operation (Rellich's theorem).

For any $r \geq 0$, $x \in H_r$, and any $\alpha = (\alpha_1,\ldots,\alpha_\nu)$, $0 \leq |\alpha| \leq r$, the partial derivative $D^\alpha x$ belongs to $H_{r-|\alpha|}$. In particular, for $|\alpha| = r$, $D^\alpha x \in H_0 = L_2(G)$.

For $r > \nu/2$ all functions $x \in H_r$ are continuous on $\bar{G}$, hence bounded, and there is a constant $C > 0$ such that

$$\sup_{t\in G} |x(t)| \leq C\|x\|_r, \quad r > \nu/2, \tag{14.1}$$

where C is a constant which may depend on r, ν and G, but not on $x \in H_r$ (Sobolev's imbedding theorem).

A first corollary of this theorem is that, for $r > \nu/2$, $x \in H_r$, and multi-indices α with $r - |\alpha| > \nu/2$, all derivatives $D^\alpha x$ are continuous on $\bar{G}$, and

$$\sup_{t\in G} |D_t^\alpha x(t)| \leq C\|x\|_r, \quad r - |\alpha| > \nu/2.$$

In other words, for $r > \nu/2$, $m \geq 0$, $x \in H_{m+r}$, all derivatives $D^\alpha x$ with $|\alpha| \leq m$, are certainly continuous in $\bar{G}$, and bounded with $\sup_{t\in G} |D^\alpha x(t)| \leq C\|x\|_{r+m}$ for a constant C which depends only on r, m and of course ν and G. Note that, if $x,y \in H_{r+m}$, then $x - y \in H_{r+m}$, and then

$$\sup_{t\in G} |D^\alpha x(t) - D^\alpha y(t)| \leq C\|x - y\|_{r+m}, \quad 0 \leq |\alpha| \leq m. \tag{14.2}$$

Finally, for any $r > \nu/2$, any system of ν-multi-indices $\beta_1,\ldots,\beta_h$, with $|\beta_i| > 0$, and $|\beta_1| + \cdots + |\beta_h| \leq r$, the map

$$(x_1,\ldots,x_h) \to D^{\beta_1}x_1 D^{\beta_2}x_2 \ldots D^{\beta_h}x_h$$

is continuous, as a map from $H_r \times \cdots \times H_r$ (h times) into $L_2(G) = H_0$ (Schauder's theorem).

Concerning the operators E, B we assume that the null space $X_0 = \ker(E,B)$ of (E,B) is finite dimensional, that is, the space X_0 of all $x \in H^m$ with $Ex = 0$, $Bx = 0$, and let $p = \dim X_0$. Actually, we assume that these elements are of class $C^\infty(G)$, and we choose a basis $u_1,\ldots,u_p$ for X_0 which is orthonormal in $L_2(G)$. Analogously, we assume that the space $Y_0 = \operatorname{coker}(E,B)$ is also finite dimensional, $Y_0 \subset C^\infty(G)$, $\dim Y_0 = q$, and we choose a basis $v_1,\ldots,v_q$ for Y_0 which is orthonormal in $L_2(G)$. Thus, for any $h \in L_2(G)$ the problem $Ex = h$, $Bx = 0$, $x \in H^m$, has a solution if and only if $(h,v_i) = 0$, $i = 1,\ldots,q$. If $(h,v_i) = 0$, $i = 1,\ldots,q$, and $\bar{x}$ is any solution of $Ex = h$, $Bx = 0$, $x \in H^m$, then any other solution is of the form $x = \bar{x} + \sum_{i=1}^{m} \xi_i u_i$, ξ_i arbitrary real numbers. We note here that for any integer $r \geq 0$, $E:H^{r+m} \to H^r$ is a linear operator, and we assume that for some constant $C > 0$ and every $x \in H^{r+m}$ we have

$$\|x\|_{r+m} \leq C(\|Ex\|_r + \|x\|_{r+m-1}), \tag{14.3}$$

where C may depend on r but not on $x \in H_{r+m}$ (coerciveness property).

15. <u>The Operators P, Q and H.</u> We first define the two projection operators P and Q onto the kernel and cokernel of (E,B) as the usual projections $P:H_{r+m} \to X_0$, $Q:H_r \to Y_0$ in $L_2(G)$, or

$$Px = \sum_{i=1}^{p} (x,u_i)u_i, \quad x \in H_{r+m}, \tag{15.1}$$

$$Qy = \sum_{i=1}^{q} (y,v_i)v_i, \quad y \in H_r. \tag{15.2}$$

Note that while P and Q are bounded operators on H_{r+m} and H_r respectively, their operator norms may be larger than one, since they are defined by orthogonal projection in the sense of $L_2 = H_0$, not in the sense of H_{r+m} or H_r. Note that for $h \in H^r(G)$ the problem $Ex = h$, $Bx = 0$, $x \in H^{m+r}$, has solutions if and only if $Qh = 0$. If $Qh = 0$ and $\bar{x} \in H^{r+m}$ is any solution of the problem, then any other solution is of the form $\bar{x} + \sum_{i=1}^{p} \xi_i u_i$.

We intend to solve the equation $Ex = Nx$ in G, $Bu = 0$ on ∂G. Let $Px = x^* = cu = c_1u_1 + \cdots + c_pu_p$, so that $x = y + cu$, where $Py = 0$. Our equation can now be written in the equivalent form

$$Ey = N(y + cu) \quad \text{in } G, \quad Py = 0 \quad \text{in } G, \quad By = 0 \quad \text{on } \partial G.$$

We now split this problem into an auxiliary (or relaxed) equation, and a bifurcation equation:

$$Ey = (I - Q)N(y + cu) \quad \text{in } G; \quad Py = 0 \quad \text{in } G,$$

$$By = 0 \quad \text{on } \partial G; \tag{15.3}$$

$$QN(y + cu) = 0 \quad \text{in } G. \tag{15.4}$$

It is evident that the solution of this system is equivalent to the solution of the original problem $Ex = Nx$.

Note that, if $Y = H^r$, $Y_0 = QY = \text{coker}(E,B)$, and then $Y_1 = (I - Q)Y = \mathcal{R}(E,B)$, $Y_1 = \ker Q$. Analogously, if $X = H^{r+m}$, $X_0 = PX = \ker(E,B)$, and then $X_1 = (I - P)X = \ker P$. Moreover E, restricted to X_1, maps X_1 onto Y_1 and $E: X_1 \to Y_1$ is one-one. Thus, the inverse H of $E|X_1$ exists. We can characterize H by noting that, for $h \in \mathcal{R}(E,B) = \ker Q = Y_1$, we know that $Ex = h$, $Bx = 0$, $x \in H^{r+m}$, is solvable. The solution x is defined up to the addition of an arbitrary element of $\ker(E,B) = X_0$. Thus, we choose for Hh the unique element x with $Px = 0$. In other words, if $x_0 \in H^{r+m}$ denotes any solution of $Ex = h$, $Bx = 0$, then we take $Hh = x_0 - Px_0$. It is easily seen that Hh is independent of the choice of x_0, and that $PHh = 0$ for $h \in Y_1$. Thus, EH is the identity on Y_1, and HE is the identity on X_1. By force of (14.3) we have now

$$\|Hh\|_{r+m} \leq C(\|h\|_r + \|Hh\|_{r+m-1}). \tag{15.5}$$

From here we can derive that

$$\|Hh\|_{r+m} \leq C_0\|h\|_r \tag{15.6}$$

for another constant C_0, by the following standard argument of functional analysis. If no such inequality (15.6) holds, then there is a sequence h_i, $i = 1,2,\ldots$, of elements of Y_1 with $\|Hh_i\|_{r+m} = 1$ for all i, but $\|h_i\|_r \leq 1/i$. Since the sequence Hh_i is bounded in H_{r+m}, then a subsequence, say still [i] for the sake of simplicity, must converge weakly in H_{r+m} by weak compactness of the unit ball in any Hilbert space (see e.g., Yosida [207], p.126). By Rellich's theorem on the compact embedding of H_{r+m} into H_{r+m-1} this subsequence must converge strongly in H_{r+m-1}. Let us say,

Hh_i converges strongly to x in H_{r+m-1}. We also know that $EHh_i = h_i$, $i = 1,2,\ldots,$ converges strongly to the zero element in H_r. Note that E is not defined on all of H_{r+m-1}, but E has a closed graph when restricted on the dense domain H_{r+m} of H_{r+m-1}. It follows that x is in the domain of E and that $Ex = 0$. Thus, x is in the kernel of (E,B), or $x \in X_0$, and $x = Px$. Since $PHh_i = 0$, we have $Px = 0$, and then $x = 0$. Thus, Hh_i converges to zero strongly in H_{r+m-1}, that is, $\|Hh_i\|_{r+m-1} \to 0$, which contradicts (15.5). Relation (15.6) is thereby proved for a constant C_0 which may depend on r but not on h. Relation (15.6) shows that $H:Y_1 \to X_1$ is a bounded linear operator.

Let us consider now the equation

$$y = H(I - Q)N(y + cu), \tag{15.7}$$

and note that this is equivalent to the original formulation of the auxiliary equation (15.3). To solve $Ex = Nx$, we first consider the auxiliary equation (15.7) and solve it as a fixed point equation in y for every $c \in R^p$. We shall then seek such c that will satisfy the bifurcation equation

$$QN(y + cu) = 0, \tag{15.8}$$

where $y = y(c)$ is any solution of (15.7) corresponding to c.

16. Problems of Perturbation. The Auxiliary Equation. We consider now problems of perturbation, that is, problems of the form

$$Ex = \varepsilon f(t,x,Dx,\ldots,D^m x), \tag{16.1}$$

$$Bx = 0, \tag{16.2}$$

where ϵ is a small parameter. The auxiliary equation is here

$$y = \epsilon H(I - Q)f(t, \mathcal{D}(y + cu)) \tag{16.3}$$

where now we denote by $\mathcal{D}y$ the N-vector $(y, Dy, \ldots, D^m y)$. Thus, we seek fixed points for the transformation $T: y \to \epsilon H(I - Q)f$. We take now any integer $r > \nu/2$ and any two constants $R_0, R_1 > 0$. Let K_0 denote the set of all $y \in H_{r+m}$ with $Py = 0$, $\|y\|_{r+m} \leq R_0$. Let K_1 denote the set of all $c \in R^p$ with $|c| \leq R_1$.

(16.i) There is an $\epsilon_1 > 0$ such that, for every $|\epsilon| \leq \epsilon_1$ and $|c| \leq R_1$, the transformation T maps K_0 into itself and is a contraction there. Hence, there is one and only one fixed element $x = Tx$ in K_0, and this element $x = x(c)$ is the unique solution in K_0 of the auxiliary equation (16.3).

Proof. Since $u_1, \ldots, u_p$ are functions of class C^∞ on G, we see that for $c \in R^p$, $|c| \leq R_1$, all functions $y + cu$ are in H_{r+m} as y, and that $\|y + cu\|_{r+m} \leq R_3$ for some constant $R_3 > 0$. Since $r > \nu/2$, all partial derivatives $D^\alpha(y + cu)$, $0 \leq |\alpha| \leq m$, are continuous and bounded in $\bar{G}$ with

$$|D^\alpha(y + cu)(t)| \leq C\|y + cu\|_{r+m} \leq CR_3, \quad t \in \bar{G},$$
$$|\alpha| \leq m. \tag{16.4}$$

If τ denotes the $N + \nu$ vector $\tau = (t_1, \ldots, t_\nu, z_1, z_2, \ldots, z_N)$, then any partial derivative of $f(t, z) = f(t_1, \ldots, t_\nu, z_1, \ldots, z_N)$ will be denoted by $D_\tau^\gamma f$ where γ is a multi-index $(\gamma_1, \ldots, \gamma_{N+\nu})$. Since f is of class C^∞, so are its partial derivatives $D_\tau^\gamma f$ for $0 \leq |\gamma| \leq r$, and thus f and all these $D_\tau^\gamma f$, $|\gamma| \leq r$, are continuous, bounded, and therefore Lipschitzian for $t \in \bar{G}$ and $|z_i| \leq CR_3$, $i = 1,2,\dots,N$, or

$$|D_\tau^\gamma f(t,z)| \leq M_0, \quad |D_\tau^\gamma f(t,z) - D_\tau^\gamma f(t,\bar{z})| \leq L|z - \bar{z}| \tag{16.5}$$

for $t \in \bar{G}$, $|z_i|, |\bar{z}_i| \leq CR_3$, $i = 1,\dots,N$, and any $|\gamma| \leq r$. If y_1, y_2 are any two elements of K_0, then $y_1, y_2 \in H_{r+m}$, and so $y_1 - y_2 \in H_{r+m}$, so that

$$\sup_{t\in\bar{G}} |D^\alpha y_1(t) - D^\alpha y_2(t)| \leq C\|y_1 - y_2\|_{r+m},$$
$$0 \leq |\alpha| \leq m,$$

while we know already $|D^\alpha(y_i + cu)(t)| \leq CR_3$, $t \in \bar{G}$, $|\alpha| \leq m$, $i = 1,2$. Then

$$\begin{aligned}\|f(t,\mathcal{D}(y_1(t) + cu(t)) - f(t,\mathcal{D}(y_2(t) + cu(t))\| \\ \leq L \sum_{|\alpha|\leq m} |D^\alpha y_1(t) - D^\alpha y_2(t)| \\ \leq LNC\|y_1 - y_2\|_{r+m},\end{aligned} \tag{16.6}$$

and hence

$$\|f(t,\mathcal{D}(y_1 + cu)) - f(t,\mathcal{D}(y_2 + cu)\|_0$$
$$\leq LNC|G|^{1/2}\|y_1 - y_2\|_{r+m},$$

where $|G|$ denotes the measure of G. To obtain an analogous relation for the norm $\|\ \|_r$, we need first consider the derivatives $D_t^\alpha f(t,\mathcal{D}(y(t) + cu(t))$, $t \in G$, with $|\alpha| \leq r$. By the chain rule of calculus we see that each of these derivatives is a well determined finite sum of products $U(t)V(t)$, $t \in G$, where U and V have the form

$$U(t) = D_\tau^\gamma f(t,\mathcal{D}(y + cu)),$$
$$V(t) = \prod_{i=1}^{\mu} D^{\alpha_i+s_i}(y + cu),$$

where γ is a $(N+\nu)$-multi-index with $|\gamma| \leq r$, where μ is an integer, $0 \leq \mu \leq r$, where α_i, s_i are ν-multi-indices,

$|\alpha_i| \leq |\alpha| \leq m$, $|s_i| \geq 1$, $|s_1| + \cdots + |s_\mu| \leq |\alpha| \leq r$. We have seen above that each factor U is bounded by M_0, and repeating the argument for (16.6) we have, for all $t \in G$ and $|\gamma| \leq |r|$,

$$|D_\tau^\gamma f(t, \mathcal{B}(y_1 + cu)) - D_\tau^\gamma f(t, \mathcal{B}(y_2 + cu))|$$
$$\leq LNC\|y_1 - y_2\|_{r+m}.$$

Concerning the factors V, we write them in the form

$$V(t) = \prod_{i=1}^{\mu} D^{s_i}(D^{\alpha_i}(y + cu)),$$

and we note that $|\alpha_i| \leq m$, hence $D^{\alpha_i}(y + cu) \in H_{r+m-|\alpha_i|}$, hence $D^{\alpha_i}(y + cu) \in H_r$, $i = 1, \ldots, \mu$, while $|s_1| + \cdots |s_\mu| \leq r$. By Schauder's theorem on Sobolev spaces, we know that the transformation $(D^{s_1}y, \ldots, D^{s_\mu}y) \to V$, as a map $H_r \times \cdots \times H_r$ (μ times) into $H_0 = L_2(G)$, is continuous. Since the transformations $y \to D^{\alpha_1}y, \ldots, y \to D^{\alpha_\mu}y$ are also continuous as maps from H_{r+m} into H_r, we conclude that the transformation $y \to V$ as a map from H_{r+m} into $H_0 = L_2(G)$, is Lipschitz continuous. In other words, there are constants

$$\sup_{t\in G} |U(t)| \leq \bar{M}, \quad \sup_{t\in G} |U_1(t) - U_2(t)|$$
$$\leq \bar{C}\|y_1 - y_2\|_{r+m},$$
$$\|V(t)\|_0 \leq \bar{M}, \quad \|V_1(t) - V_2(t)\|_0 \leq \bar{C}\|y_1 - y_2\|_{r+m},$$

where we have denoted by U_1, U_2, V_1, V_2 the functions U, V corresponding to elements $y_1, y_2 \in K_0$. By writing any difference $U_1V_1 - U_2V_2$ in the form $(U_1 - U_2)V_1 + U_2(V_1 - V_2)$, and majorizing, we conclude that

$$\|D_t^\alpha f(t, \mathcal{B}(y_1(t) + cu(t)) - D_t^\alpha f(t, \mathcal{B}(y_2(t) + cu(t)))\|_0$$
$$\leq C_1\|y_1 - y_2\|_{r+m}$$

for all $y_1, y_2 \in K_0$ and $0 \leq |\alpha| \leq r$. Thus

$$\|f(t,\mathcal{D}(y_1 + cu)) - f(t,\mathcal{D}(y_2 + cu)\|_r \leq C_2\|y_1 - y_2\|_{r+m},$$

and also

$$\|f(t,\mathcal{D}(y_1 + cu)\|_r \leq M_2$$

for some constants C_2, M_2 and all $y, y_1, y_2 \in K_0$. Thus, if $\wedge$ denotes the norm of the linear operator $H(I - P)$ from H_r to H_{r+m}, we have $\|Ty\|_{r+m} \leq |\varepsilon| \wedge M_2$ for $y \in K_0$. Analogously, if $y_1, y_2 \in K_0$ and $z_i = Ty_i$, $i = 1,2$, we also have $\|z_1 - z_2\|_{r+m} \leq |\varepsilon| \wedge C_2\|y_1 - y_2\|_{r+m}$. If we take $\varepsilon_1 = \min[(\wedge M_2)^{-1}R_0, (2 \wedge C_2)^{-1}]$, then for $|\varepsilon| \leq \varepsilon_1$, we have $Ty \in H_{r+m}$, $\|Ty\|_{r+m} \leq R_0$, $PTy = 0$, and $\|Ty_1 - Ty_2\|_{r+m} \leq 2^{-1}\|y_1 - y_2\|_{r+m}$. In other words $T:K_0 \to K_0$ and T is a contraction. There is, therefore, a unique fixed point $y \in K_0$, $y = Ty$, and $y = y(c,\varepsilon)$ depends on c and ε for $c \in R^p$, $|c| \leq R_1$, $|\varepsilon| \leq \varepsilon_1$.

17. <u>The Bifurcation Equation, and Examples</u>. There is no difficulty now in proving that the unique solution $y(c,\varepsilon)$ of the auxiliary equation (16.3), that is, the fixed point $y = Ty$, is a continuous function of (c,ε) for $|c| \leq R_1$, $|\varepsilon| \leq \varepsilon_1$. If we replace y by $y(c,\varepsilon)$ in the bifurcation equation (15.8), we have the equation

$$K(c,\varepsilon)v = QN(y(c,\varepsilon) + cu) = 0$$

with $K(c,\varepsilon) = (K_1,\ldots,K_q) \in R^q$. Thus, the bifurcation equation reduces now to the system of q equations in the p unknowns $c_1,\ldots,c_p$:

$$K_i(c,\varepsilon) = 0, \quad i = 1,\ldots,q.$$

It is not difficult to prove that the $K_i(c,\varepsilon)$ are continuous and continuously differentiable functions of $c_1, \dots, c_p, \varepsilon$ for $|c| < R_1$, $|\varepsilon| < \varepsilon_1$.

(17.i) Under the same previous assumptions, if $p \geq q$, if there is some $c_0 \in R^p$, $|c_0| < R_1$, such that $K_i(c_0, 0) = 0$, $i = 1, \dots, q$, and if the $q \times p$ matrix

$$[\partial K_i(c_0, 0)/\partial c_j, \quad i = 1, \dots, q, \; j = 1, \dots, p] \tag{17.1}$$

has maximum rank q, then there is $\bar{\varepsilon} > 0$, $0 < \bar{\varepsilon} \leq \varepsilon_1$, and for every $|\varepsilon| \leq \bar{\varepsilon}$, there is some $c = c(\varepsilon)$, $|c| \leq R_1$, such that $K_i(c(\varepsilon), \varepsilon) = 0$, $i = 1, \dots, q$, and thus the given problem $Ex = Nx$ has the solution

$$x(\varepsilon) = c(\varepsilon)u + y(c(\varepsilon), \varepsilon). \tag{17.2}$$

Actually, for $q = p$ this solution is unique, while for $q < p$, there is a $(p - q)$-parameter family of such solutions.

Remark. Let $p = q$ and for any $\varepsilon \in [-\varepsilon_1, \varepsilon_1]$ let us consider the map $K_\varepsilon : \gamma = K(c, \varepsilon)$, from the ball $B = [c \in R^p \,|\, |c| \leq R_1]$ into the γ-space R^p, $\gamma = (\gamma_1, \dots, \gamma_p)$. Instead of the specific hypotheses in (17.i), let us assume that, if we restrict c to the sphere $\partial B = [c \,|\, |c| = R_1]$, then $K(c, 0) \neq 0$ for $c \in \partial B$, and that the topological degree $\deg(K_0, B, 0) \neq 0$. Then by continuity, there is $\bar{\varepsilon}$ such that for all $|\varepsilon| \leq \bar{\varepsilon}$, we have $K(c, \varepsilon) \neq 0$ for $c \in \partial B$, and $\deg(K_\varepsilon, B, 0) \neq 0$. Then, for every $|\varepsilon| \leq \bar{\varepsilon}$, the equations $K_i(c, \varepsilon) = 0$, $i = 1, \dots, p$, have at least a solution $c = c(\varepsilon)$, and hence the given problem $Ex = Nx$ has the solution $x(\varepsilon) = c(\varepsilon)u + y(c(\varepsilon), \varepsilon)$.

For $p = q = 1$, the condition just stated may be reduced to

$$[\int_G u(t)f(t,\mathcal{B}(c_1u))dt][\int_G u(t)f(t,\mathcal{B}(c_2u))dt] < 0 \qquad (17.3)$$

for some scalars c_1, c_2, $|c_1|, |c_2| < R_1$.

The following examples, among others, have been studied in detail by Shaw [120] by the present approach.

1. $\Delta x + 2x = k - \arc\tan x$ on $G = [(\xi,\eta),\ 0 \leq \xi,\eta \leq \pi]$, $x = 0$ on ∂G.

Here $\lambda = 2$ is an eigenvalue of the operator Δ on G with Dirichlet boundary conditions, and eigenfunction $\sin\xi \sin\eta$, $(\xi,\eta) \in G$. It was shown in [120] that the problem is solvable if and only if $|k| < \pi/2$. Moreover, for $|k| < \pi/2$, the solution is then of the form $x(\xi,\eta) = v_d(\xi,\eta) + d \sin\xi \sin\eta$, with $\int_G v_d(\xi,\eta)\sin\xi \sin\eta\, d\xi d\eta = 0$, $\|v_d\|_{L_2} \leq \pi^2/3$, and a suitable value of the constant d.

2. $\Delta x = \epsilon[h(\xi,\eta) + x^2_{\xi\eta} + x^3]$ on any bounded connected open set G of the $\xi\eta$-plane R^2, with smooth boundary, $\partial x/\partial n = 0$ on ∂G. Here $N = h + x^2_{\xi\eta} + x^3$ depends on the derivative $x_{\xi\eta}$ of order two. We have $p = q = 1$, we take $u = 1$, $v = 1$, and the analysis of the previous no. applies in the space H^4. We consider a ball of some fixed radius $D > \|h\|^{1/3}$ (the norm of h being taken in L_1). Then, the auxiliary equation is solvable for ϵ sufficiently small, say $|\epsilon| \leq \epsilon_0$. For the bifurcation equation we apply the criterion (17.3) which here takes the form

$$\int_G (h - c_1^3)d\xi d\eta \cdot \int_G (h - c_2^3)d\xi d\eta < 0$$

and this relation is certainly satisfied for $c_1 = -D$, $c_2 = D$. Problem 2 has a solution of the form $x(\xi,\eta) = c + O(\varepsilon)$ for $|\varepsilon|$ sufficiently small.

3. $\Delta x + 5x = \varepsilon(h(\xi,\eta) + e^x)$, $(\xi,\eta) \in G = [0,\pi] \times [0,\pi]$, with Dirichlet boundary conditions $x = 0$ on ∂G. Here we have $p = q = 2$, $u_1 = v_1 = \sin\xi \sin 2\eta$, $u_2 = v_2 = \sin 2\xi \sin \eta$. As before we shall consider only the bifurcation equation to which we apply criterion (17.i) with $c_0 = (0,0)$. Here we assume that $\int v_1 h \, d\xi d\eta = \int v_2 \, h \, d\xi d\eta = 0$. The 2×2 matrix (17.1) is now the identity matrix. Thus, the problem has solutions for $|\varepsilon|$ sufficiently small.

Section 4. A Numerical Viewpoint and Error Bound Estimates

18. Error Estimates and Existence of an Exact Solution. By the alternative method it has been possible to prove in a number of situations that in some neighborhood of an approximate solution x_0 in a Banach space X, (say, a Galerkin approximation x_0) there lies an exact solution x, in other words, that an exact solution x exists satisfying a relation $\|x - x_0\| \leq \rho$ in the norm of the given space. This number ρ is then, an error estimate of the approximate solution x_0. (See Cesari [6,7,17,18].) We present here very recent work by Cesari and Bowman [29] and Ku [91], in the lines of no. 2 above. In this recent work, ρ has been obtained by electronic computer. First let us summarize the situation described in no. 2. Let X and Y be Banach spaces (over the reals), let $E:\mathcal{D}(E) \to Y$, $N:\mathcal{D}(N) \to Y$ be operators, $\mathcal{D}(E) \subset X$,

$\mathcal{D}(N) \subset X$, $\mathcal{D}(E) \cap \mathcal{D}(N) \neq \emptyset$, E linear. We are concerned with the equation $Ex = Nx$, $x \in \mathcal{D}(E) \cap \mathcal{D}(N)$. Let $P:X \to X$, $Q:Y \to Y$ be projection operators, let $X_0 = PX$, $X_1 = (I - P)X$, $Y_0 = QY$, $Y_1 = (I - Q)Y$, and let us assume that $\ker E \subset X_0$, and that there is a linear operator $H:Y_1 \to \mathcal{D}(E) \cap X_1$ such that the usual relations (k_1), (k_2), (k_3) of no. 2 hold. Then we know that the given equation $Ex = Nx$ is equivalent to the system of auxiliary and bifurcation equations

$$x = Px + H(I - Q)Nx, \tag{18.1}$$

$$Q(Ex - Nx) = 0. \tag{18.2}$$

As usual we denote by F and T the operators $F = H(I - Q)N$ and $T = P + F$, so that the auxiliary equation takes the form $x = Tx$.

Let x_0 be an element of X_0, that we consider now as an approximate solution of equation $Ex = Nx$, and thus $\Delta x_0 = Ex_0 - Nx_0$ is the corresponding error in the same equation. Then, $\delta x_0 = H(I - Q)\Delta x_0$ is an element of X_1 that we call the integrated error. We have

$$H(I - Q)Ex_0 = H(I - Q)Nx_0 + \delta x_0,$$

and by (k_1) also $x_0 = Px_0 + Fx_0 + \delta x_0$. Since $x_0 \in X_0$, then $x_0 = Px_0$ and we have $Fx_0 + \delta x_0 = 0$.

(18.i) <u>Theorem</u>. Let $x_0 \in X_0$, let W be a neighborhood of x_0 in X with $W \subset \mathcal{D}(N)$, and let N be Lipschitzian in W with constant L, i.e. $\|Nx - Ny\| \leq L\|x - y\|$ for $x,y \in W$. Let $\|H(I - Q)\| = k$ and $\|\delta x_0\| = \rho$, and let us assume that $kL < 1$. Let c,d be real numbers, $0 < c < d$, such that $(1 - kL)^{-1}(c + \rho) \leq d$, and let us assume that W contains the ball of center x_0 and radius

d in X. Let $W_0 = [x^* \in X_0 | \|x^* - x_0\| \leq c]$, and for $x^* \in W_0$ let $V^* = [x \in X | Px = x^*, \|x - x_0\| \leq d]$, so that $W_0 \subset W$, and $V^* \subset W$ for any $x^* \in W_0$. Then, T maps V^* into itself, T is a contraction on V^*, and hence T has a unique fixed point $x = Tx = \mathfrak{T}x^* \in V^*$, satisfying

$$\|x - x^*\| \leq kLd + \rho \leq d - c.$$

Moreover, the map $\mathfrak{T}: W_0 \to X$ mapping each $x^* \in W_0$ into the fixed points $x = Tx = \mathfrak{T}x^*$ of T in V, $P\mathfrak{T}x^* = x^*$, is continuous.

<u>Proof</u>. If $x \in V^*$, then $PTx = P(x^* + Fx) = x^* + PH(I - Q)Nx = x^*$, and also

$$\begin{aligned}\|Tx - x_0\| &= \|x^* + Fx - (x_0 + Fx_0 + \delta x_0)\| \\ &\leq \|x^* - x_0\| + \|Fx - Fx_0\| + \|\delta x_0\| \\ &\leq c + kLd + \rho \leq d.\end{aligned}$$

Thus, T maps V^* into itself. If $y_1 = Tx_1$, $y_2 = Tx_2$ for $x_1, x_2 \in V^*$, then

$$y_i = Tx_i = Px_i + H(I - Q)Nx_i, \quad Px_1 = Px_2 = x^*,$$

$$\|Tx_1 - Tx_2\| = \|y_1 - y_2\| = \|H(I - Q)(Nx_1 - Nx_2)\| \leq kL\|x_1 - x_2\|$$

where $kL < 1$, and thus T, restricted to V^*, is a contraction. By Banach's fixed point theorem, there is a unique fixed point $x = Tx$ for T in V^*. For this fixed point we have

$$\begin{aligned}\|x - x^*\| &= \|(x^* + Fx) - x^*\| = \|Fx - Fx_0 - \delta x_0\| \\ &\leq \|Fx - Fx_0\| + \|\delta x_0\| \leq kLd + \rho \leq d - c. \quad (18.3)\end{aligned}$$

Finally, let us prove that $\mathfrak{T}$ is a continuous map in W_0. Indeed, for $x_1, x_2 \in W_0$ and $y_1 = \mathfrak{T}x_1$, $y_2 = \mathfrak{T}x_2$, then

$$\mathfrak{T}x_i = y_i = Ty_i = x_i + H(I - P)Ny_i, \quad i = 1,2,$$

$$\|y_1 - y_2\| \le \|x_1 - x_2\| + \|H(I - P)(Ny_1 - Ny_2)\|$$
$$\le \|x_1 - x_2\| + kL\|y_1 - y_2\|,$$

and hence

$$\|\mathfrak{T}x_1 - \mathfrak{T}x_2\| = \|y_1 - y_2\| \le (1 - kL)^{-1}\|x_1 - x_2\|.$$

The proof of Theorem (18.i) is complete.

We consider now the bifurcation equation which takes the form

$$Q(E - N)\mathfrak{T}x^* = 0, \quad x^* \in W_0 \subset X_0.$$

First we note that, for $x = Tx = \mathfrak{T}x^*$ we have $x = Px + H(I - Q)Nx$, $Px = x^*$, hence $Ex = EPx + EH(I - Q)Nx$, and by (k_{23}) also $Ex = QEx + (I - Q)Nx$. Thus, for $x^* \in W_0$ we have

$$(E - N)\mathfrak{T}x^* = Q(E - N)\mathfrak{T}x^* = EP\mathfrak{T}x^* - QN\mathfrak{T}x^* = Ex^* - QN\mathfrak{T}x^*,$$
$$Q(E - N)x^* = EPx^* - QNx^* = Ex^* - QNx^*.$$

Thus

$$Q(E - N)\mathfrak{T}x^* - Q(E - N)x^* = -Q(N\mathfrak{T}x^* - Nx^*),$$
$$\|Q(E - N)\mathfrak{T}x^* - Q(E - N)x^*\| \le \|Q\|L\|\mathfrak{T}x^* - x^*\|. \tag{18.4}$$

We assume now that both X_0 and Y_0 are finite dimensional, $p = \dim X_0$, $q = \dim Y_0$. We take a base $\varphi = (\varphi_1, \ldots, \varphi_p)$ in X_0 and a base $\psi = (\psi_1, \ldots, \psi_q)$ in Y_0. Then every element $x^* \in X_0$ has a unique representation $x^* = b_1\varphi_1 + \cdots + b_p\varphi_p$, $b = (b_1, \ldots, b_p) \in R^p$, or briefly $x^* = b\varphi$. Every element $y \in Y_0$ has a unique representation $y = u_1\psi_1 + \cdots + u_q\psi_q$, $u = (u_1, \ldots, u_q) \in R^q$, or briefly $y = u\psi$.

Concerning the finite dimensional space Y_0, we denote by $\sigma > 0$ a constant such that, for any element $y = b\psi = b_1\psi_1 + \cdots + b_q\psi_q$ of Y_0, we have $|b| \le \sigma\|b\psi\| = \sigma\|y\|$, where

$|\ |$ is the Euclidean norm in R^q and $\|\ \|$ is the norm in Y. Thus

$$(\Sigma\, b_i^2)^{1/2} \leq \sigma \|b_1\psi_1 + \cdots + b_q\psi_q\|.$$

For $q = 1$, then we can take $\sigma = \|\psi_1\|^{-1}$. If it happens that Y_0 can be thought of as contained in a Hilbert space with inner product $(\ ,\ )$ so that $\|y\| = (y,y)^{1/2}$ for every $y \in Y_0$, then we can choose $\psi = (\psi_1, \ldots, \psi_q)$ to be orthonormal in Y_0 and then, of course, $\sigma = 1$.

If $x^* = b_1\varphi_1 + \cdots + b_p\varphi_p = b\varphi$, then $Q(E - N)\mathfrak{T}x^*$ and $Q(E - N)x^*$ are elements of Y_0, and hence

$$Q(E - N)\mathfrak{T}x^* = U_1(b)\psi_1 + \cdots + U_q(b)\psi_q = U(b)\psi,$$

$$Q(E - N)x^* = u_1(b)\psi_1 + \cdots + u_q(b)\psi_q = u(b)\psi,$$

where $U(b)$, $u(b) \in R^q$ are continuous functions of b. Thus by (18.4) we have

$$\begin{aligned} |U(b) - u(b)| &\leq \sigma\|(U(b) - u(b))\psi\| \\ &= \sigma\|Q(E - N)\mathfrak{T}x^* - Q(E - N)x^*\| \\ &\leq \sigma\|Q\|L\|\mathfrak{T}x^* - x^*\|. \end{aligned}$$

We note here that, whenever $p = q$ and $\varphi_1, \ldots, \varphi_p$ are eigenfunctions of the linear operator E, or $E\varphi_i + \lambda_i\varphi_i = 0$ for some real λ_i, $i = 1, \ldots, p$, then any element $x^* \in X_0$, $x^* = b_1\varphi_1 + \cdots + b_p\varphi_p = b\varphi$ such that $Q(Ex^* - Nx^*) = 0$, or $u(b) = 0$, that is, $u_1(b) = 0, \ldots, u_p(b) = 0$, is said to be a pth Galerkin approximation for some possible solution x of the equation $Ex = Nx$.

Let $p = q$, let $\Gamma(a,d)$ denote the ball in R^q of center a and radius d (or more generally a domain Γ homotopically equivalent to a ball), and let γ be the boundary of Γ. If $u(b) \neq 0$ for b on γ, and the topological degree $\deg(u,\Gamma,0) \neq 0$, then we know that $u(b)$ has at least one zero in the interior of Γ, but yet nothing can be said of $U(b)$. However, if $u(b) \neq 0$ for b on γ, if $d(u,\Gamma,0) \neq 0$, and we can show that

$$\mathrm{Max}_{b\in\gamma} |U(b) - u(b)| \leq \mathrm{Min}_{b\in\gamma} |u(b)|$$

then

$$\deg(U,\Gamma,0) = \deg(u,\Gamma,0) \neq 0,$$

and also $U(b)$ has at least one zero in the interior of Γ.

Details have been given in [6], [7], [17], where the example (c) below has been studied taking $p = q = 2$.

We restrict now to the case $p = q = 1$. We note that in this case and the problem of periodic solutions of periodic differential equations, the first Galerkin approximation is also called the harmonic balance solution.

For $p = q = 1$, let us suppose that we can find a number a and positive numbers $c_1, c_2 > 0$ such that $u(a - c_1)$ and $u(a + c_1)$ have opposite signs, and we take

$$Q_1 = |u(a - c_1)|, \quad Q_2 = |u(a + c_2)|.$$

(18.ii) Theorem. Given the assumptions above with $d_i = (c_i + \rho)(1 - kL)^{-1}$, $i = 1,2$, if $u(a - c_1)$, $u(a + c_2)$ have opposite signs, and

$$Q_i > \|\psi_1\|^{-1}\|Q\| L(d_i - c_i), \quad i = 1,2,$$

then there exists a number b such that $a - c_1 < b < a + c_2$, and b satisfies the bifurcation equation $U(b) = 0$, that is, there is an exact solution x to the equation $Ex = Nx$, with $x \in W$, $Px = x^* = b\varphi \in W_0$, and

$$|b - a| \leq \max(c_1, c_2), \quad \|x - x_0\| \leq \max(d_1, d_2).$$

Proof. We need only to show that the scalar function $U(b)$ has opposite signs at the points $a - c_1$ and $a + c_2$. Assume, for instance, that

$$-Q_1 = u(a - c_1) < 0 < u(a + c_2) = Q_2$$

Then, for $b = a + c_2$ and $x = Tx = \mathfrak{T}x_b$ we also have

$$U(b) = u(b) + (U(b) - u(b)) \geq u(b) - |U(b) - u(b)|$$
$$\geq Q_2 - \|\psi_1\|^{-1}\|Q\|L\|x - x_b\|.$$

By force of (18.3) with $\|x - x_b\| = \|x - x^*\| \leq kLd_2 + \rho = d_2 - c_2$, we have

$$U(b) \geq Q_2 - \|\psi_1\|^{-1}\|Q\|L(d_2 - c_2)$$
$$= Q_2 - \|\psi_1\|^{-1}\|Q\|L(1 - kL)^{-1}(\rho + c_2kL) > 0.$$

Similarly, we have $U(a - c_1) < 0$. Then, $U(a - c_1)$ and $U(a + c_2)$ have opposite signs, and since U is continuous, there is a root b of U with $a - c_1 < b < a + c_1$, or $|b - a| \leq \max(c_1, c_2)$. On the other hand $x = Tx = \mathfrak{T}x_b \in W$, hence $\|x - x_0\| \leq \max(d_1, d_2)$.

19. Examples, and a Method of Successive Approximations.

The following examples have been studied in the lines of this Section. For each we refer to the original paper for details of computation.

(a) The equation

$$x'' + \sin x + C \sin t = 0, \quad x(t + 2\pi) = x(t),$$

has a harmonic balance solution (or first Galerkin approximation to its 2π-periodic odd solutions) of the form $x_0(t) = a_0 \sin t$. For various values of the constant C we give below the corresponding value of a_0, and by the process of this Section with $p = q = 1$ it was then shown [29] that an exact solution $x(t)$ exists, whose Fourier series is of the form $x(t) = b_0 \sin t + b_3 \sin 3t + \cdots$, with $|b_0 - a_0| \leq \eta$, $\mathrm{Sup}_t |x(t) - x_0(t)| \leq \varepsilon$. The following values for the bounds η and ε were numerically obtained (by IBM 360)

C	a_0	η	ε
0.5	1.648	0.068	0.100
0.6	1.761	0.068	0.104
0.7	1.864	0.068	0.107
0.8	1.958	0.068	0.111
0.9	2.047	0.068	0.115
1.0	2.130	0.068	0.119
1.1	2.209	0.068	0.122
1.2	2.285	0.068	0.126
1.3	2.358	0.068	0.129
1.4	2.428	0.068	0.133
1.5	2.496	0.068	0.136

(b) The differential equation

$$x'' + Ax^3 + B \sin t = 0, \qquad x(t + 2\pi) = x(t),$$

has a harmonic balance solution again of the form $x_0(t) = a_0 \sin t$. For various values of the constants A, B we give below the corresponding value of a_0. By the process of this Section with $p = q = 1$ it was shown in [29] that an exact solution $x(t)$ exists, whose Fourier series is of the form $x(t) = b_0 \sin t + b_3 \sin 3t + \cdots$, with $|b_0 - a_0| \leq \eta$, $\mathrm{Sup}_t |x(t) - x_0(t)| \leq \varepsilon$. The following values for the bounds η and ε were numerically obtained (by IBM 360).

A	B	a_0	η	ϵ
0.2	0.2	0.2012	0.0100	0.01
0.2	0.4	0.4104	0.0095	0.01
0.2	0.6	0.6392	0.0082	0.01
0.2	0.8	0.9149	0.0143	0.02
0.4	0.2	0.2025	0.0099	0.01
0.4	0.4	0.4227	0.0089	0.01
0.4	0.6	0.7052	0.0143	0.02
0.6	0.2	0.2038	0.0098	0.01
0.6	0.4	0.4377	0.0081	0.01
0.8	0.2	0.2052	0.0097	0.01
0.8	0.4	0.4574	0.0072	0.01
1.0	0.2	0.2067	0.0095	0.01
1.0	0.4	0.4862	0.0234	0.03

(c) For the problem of the 2π-periodic solutions of the equation

$$x + x^3 = \sin t$$

(same as in (b) with $A = 1$, $B = -1$), the existence of an exact solution has been shown in [6, 17] by the process discussed above, and $p = q = 2$. The first Galerkin approximation is here $1.49 \sin t$, the second Galerkin approximation is $x_0(t) = 1.434 \sin t - 0.124 \sin 3t$. An exact 2π-periodic solution $x(t) = b_1 \sin t + b_3 \sin 3t + \cdots$, has been shown to exist with

$$\sup_t |x(t) - x_0(t)| \le 0.231,$$

$$\|x(t) - x_0(t)\|_2 \le 0.124,$$

where $\| \|_2$ is the square norm.

(d) The existence of a solution $x(t)$, $0 \leq t \leq 1$, to the boundary value problem

$$x'' + x + 0.5x^3 = 0.5t, \quad x(0) = 0, \quad x'(1) + x(1) = 0,$$

has been proved in [7] by the process discussed above and $p = q = 1$. The first Galerkin approximation is $x_0(t) = 0.11873 \sin 2.0288\, t$. An exact solution $x(t)$ has been shown to exist with

$$\operatorname{Sup}|x(t) - x_0(t)| \leq 0.083,$$

$$\|x(t) - x_0(t)\|_2 \leq 0.0038.$$

(e) In [7] an exact solution $x(t)$ was proved to exist for the boundary value problem

$$x'' + x + Ax^3 = Bt, \quad x(0) = 0, \quad x'(1) + x(1) = 0,$$

for all A, B with $|A| \leq 1$, $|B| \leq 1$.

(f) For the Dirichlet problem

$$\Delta x = g(\xi,\eta,u) \quad \text{for} \quad (\xi,\eta) \in G = [\xi^2 + \eta^2 < 1],$$

$$x = 0 \quad \text{for} \quad (\xi,\eta) \in \partial G = [\xi^2 + \eta^2 = 1],$$

the existence of an exact solution $x(\xi,\eta)$, $(\xi,\eta) \in \bar{G}$, was shown in [91] to exist by the process discussed above, with $p = q = 1$, and some functions g. Here are some of the results (by IBM 360) with first Galerkin approximation $x_0(\xi,\eta)$, and error bounds $\epsilon \geq \|x - x_0\|_{sup}$ in the Sup norm in G,

$$g = |u| + 1, \quad x_0 = 0.33493\, J_0(2.4048r), \quad \epsilon = 0.0413;$$

$$g = u^3 + 1, \quad x_0 = 0.27914\, J_0(2.4048r), \quad \epsilon = 0.0313,$$

where $r = (\xi^2 + \eta^2)^{1/2}$ and J_0 denotes the Bessel function of index zero.

(g) In [18] an exact solution $x(\xi,\eta)$, $(\xi,\eta) \in \bar{G}$, was shown to exist by the same process, for the Dirichlet problem

$$\Delta x = g(\xi,\eta,x) \quad \text{for} \quad (\xi,\eta) \in G = [\xi^2 + \eta^2 < 1],$$

$$x = 0 \quad \text{for} \quad (\xi,\eta) \in \partial G = [\xi^2 + \eta^2 = 1],$$

whenever $g(\xi,\eta,x) = f(\xi,\eta)\varphi(x) + h(\xi,\eta)$, f,h bounded and measurable on G, $|f(\xi,\eta)| \leq \alpha$, $|h(\xi,\eta)| \leq \beta$, and $\varphi(x)$, $-\infty < x < +\infty$, absolutely continuous together with $\varphi'(x), \varphi''(x)$, satisfying $\varphi(0) = 0$, $\varphi'(0) = 0$, $|\varphi'(x)| \leq 3x^2$, $|\varphi''(x)| \leq 6|x|$, and $\alpha = 0.1$, $\beta = 0.8$. Thus, in particular, for $\varphi(x) = x^3$ and all f,h with $|f| \leq 0.1$ and $|h| \leq 0.8$.

(h) A method of successive approximations. Banfi and Casadei [2,3,4] devised a process of successive approximations converging to an exact solution of selfadjoint perturbation problems, in association with a given decomposition in auxiliary and bifurcation equations.

Later Sanchez [118] considered a general selfadjoint problem $Ex = Nx$, for which the decomposition into auxiliary equation $x = Tx$ and bifurcation equation $Gx = P(Ex - Nx) = 0$ has already been made so that T is a contraction in suitable domains.

Here, as usual in selfadjoint problems (nos. 2,3,4) we have $X = Y$, $P = Q$, $X_0 = PX$, X a real Hilbert space, $X_0 = PX$, $X_1 = (I - P)X$, $X_0 \cap X_1 = \{0\}$.

If $x_k = y_k + z_k$, $y_k \in X_0$, $z_k \in X_1$, $k = 0,1,2,\ldots,$ are the successive approximations, we assume that $z_0 = 0$, that is, $x_0 = y_0$, and that $Gx_0 = 0$, with $Gx = P(Ex - Nx)$, that is, x_0 is a Galerkin approximation relative to the projection operator P. Let Dy_0 denote the Fréchet derivative of the operator G at y_0, and we assume that $(Dy_0)^{-1}$

exists as a bounded linear operator. Sanchez [118] proposed the following method of successive approximations

$$z_{k+1} = H(I - P)Nx_k, \quad k = 0,1,2,\ldots$$
$$y_{k+1} = y_k - (DGy_0)^{-1} P[Ey_k - N(y_k + z_{k+1})].$$

Sanchez stated quantitative conditions under which the process is convergent to an exact solution of $Ex = Nx$. The process reduces to the Banfi-Casadei approach for the perturbation problems taken into consideration by these authors, and in their examples, the process was rapidly convergent. For extensions of Sanchez's work to nonselfadjoint problems we refer to Ku [91].

Section 5. Solutions in the Large of Liénard Systems with Forcing Terms

20. Introductory Remarks. Let us consider for a moment the second-order differential equation

$$x'' + g(x)x' + f(x) = e(t), \qquad (20.1)$$

where $e(t)$ is periodic of given period τ, and where $f(x)$, $g(x)$ are given continuous functions of the scalar variable x. In 1943 Lefschetz [166] noted that by direct analysis of the direction field associated to (20.1) in the phase plane and Brouwer's fixed point theorem, it is possible to prove the existence of periodic solutions of (20.1) under mild qualitative assumptions on g and f. N. Levinson [168], by essentially the same argument of Lefschetz, proved that the differential equation

$$x'' + g(x,x')x' + f(x) = e(t)$$

has at least a periodic solution of period τ, under the sole assumptions: (a) e, f, g are continuous functions, and e is periodic in t of period τ and of mean value zero; (b) there are constants $a, m, M > 0$ such that $g(x,v) > m$ for both $|x| > a$, $|v| > a$, and $g(x,v) > -M$ otherwise; (c) $xf(x) > 0$ for $|x| > a$; (d) $f(x)/F(x) \to 0$ as $|x| \to \infty$, where $F(x) = \int_0^x f(\alpha)d\alpha$. This result includes the case of the van der Pol equation with forcing term, $x'' - \varepsilon(1 - x^2)x' + x = e(t)$, for any value of the parameter $\varepsilon > 0$, in particular for ε large (relaxation oscillations).

Graffi [155] in 1950 considered the equation

$$Lx'' + g'(x)x' + f(x) = e(t) \tag{20.2}$$

where $g'(x) = dg(x)/dx$, where L is a positive constant, e, f are continuous functions, and g is of class C^1. Graffi proved that (20.2) has at least a periodic solution of period τ under the assumptions: (a) $e(t)$ is periodic in t of period τ and of mean value zero; (b) $g(x)/x \to R > 0$ as $|x| \to \infty$; (c) $f(x)/x \to 1/C > 0$ as $|x| \to \infty$. Graffi proved also [155], by essentially the same argument, that the system

$$\begin{aligned} L_1x_1'' + Mx_2'' + g'(x_1)x_1' + C_1^{-1}x_1 &= e_1(t), \\ Mx_1'' + L_2x_2'' + g_2'(x_2)x_2' + C_2^{-1}x_2 &= e_2(t), \end{aligned} \tag{20.3}$$

has at least a periodic solution of period τ under just these assumptions: (a) e_1, e_2 are continuous and periodic in t of period τ, and of mean value zero; (b) L_1, L_2, C_1, C_2 are positive constants, M a constant, $L_1L_2 > M^2$; (c) g_1, g_2 are of class C^1 with $g_1(x_1)/x_1 \to R_1$, $g_2(x_2)/x_2 \to R_2$ as $|x_1|, |x_2| \to \infty$, $R_1, R_2 > 0$ constants; (d) $M^2(L_1L_2)^{-1} < 4R_1C_1R_2C_2(R_1C_1 + R_2C_2)^{-2}$. Graffi proved also that the same statement holds where C_1^{-1}, C_2^{-1} are replaced by continuous functions $f_1(x_1)$, $f_2(x_2)$ with $f_i(x_i)/x_i \to c_i^{-1}$ as $|x_i| \to \infty$, $i = 1,2$, and that an analogous statement holds for a system of n equations.

Finally, Mizohata [175], by essentially Graffi's same argument, extended previous work by Mizohata and Yamaguti [176], to general systems of the form

$$\sum_{j=1}^{m} L_{ij}x_j'' + g_i'(x)x_i' + f_i(x_i) = e_i(t), \quad i = 1,\ldots,n. \quad (20.4)$$

Mizohata [175] proved the existence of at least a periodic solution of period τ for system (20.4) under the following assumptions: (a) L_{ij} constants, $[L_{ij}]$ definite positive; (b) $e_i(t)$ continuous, periodic in t of given period τ, and of mean value zero; (c) there is a constant $a > 0$ such that $x_i f_i(x_i) > 0$ for $|x_i| \geq a$, $i = 1,\ldots,m$; (d) $g_i(x_i)$ of class C^1 with $g_i(x_i) \to +\infty$ as $|x_i| \to \infty$, $i = 1,\ldots,n$; (e) there are constants $b,c > 0$ such that $\sum_{ij}\tilde{L}_{ij}f_i(x_i)[g_j(x_j) - E_j(t)] \geq b$ for all t and for all $|x| \geq c$, where $|x| = (x_1^2 + \cdots + x_n^2)^{1/2}$, $E_i(t) = \int_0^t e_i(\alpha)d\alpha$, and $[\tilde{L}_{ij}][L_{ij}] = I$, the identity matrix.

All previous results have an element in common. Indeed, it is proved that the only possible periodic solutions must lie in a suitable ball of large radius R in the 2n-dimensional phase space.

In [155], [156] Graffi noticed that, in a number of situations, it is possible to establish <u>a priori</u> bounds for the square norms of x and $\dot{x}$ of the possible periodic solutions. This work was continued by Mawhin [102,103,106], who combined this remark with Leray-Schauder type topological degree argument to obtain existence theorems for the same periodic solutions.

21. Formulation of the Main Theorems. We consider here Liénard systems, with periodic forcing terms, of the type

$$x'' + (d/dt)\operatorname{grad} G(x(t)) + (d/dt)V(x(t),t)$$
$$+ Ax(t) + g(x(t)) = e(t),$$
$$x = (x_1,\dots,x_n),\ V(x,t) = (V_1,\dots,V_n),\ e(t) = (e_1,\dots,e_n),$$
$$g(x) = (g_1,\dots,g_n), \qquad (21.1)$$

or

$$x_i'' + (d/dt)(\partial G(x(t))/\partial x_i) + (d/dt)V_i(x(t),t)$$
$$+ \sum_{j=1}^{n} a_{ij}x_j(t) + g_i(x(t)) = e_i(t), \quad i = 1,\dots,n,$$

where $G:R^n \to R$, $V:R^{n+1} \to R^n$, $A = [a_{ij}]$, $g:R^n \to R^n$, $e:R^1 \to R^n$, A is an $n \times n$ constant matrix, V and e are periodic in t of period 2π, and $g(x)/|x| \to 0$ as $|x| \to \infty$.

We are interested in the existence of periodic solutions $x(t) = (x_1,\dots,x_n)$ of period 2π of (21.1). We state below two theorems (i and ii) recently proved by Cesari and Kannan [35,36]. The conditions are only qualitative. The proofs are based on the alternative method, and use is made of the Borsuk-Ulam theorem, and therefore of topological degree arguments and of the Leray-Schauder theory [167], of which this theorem is essentially a particular form. Finally, we establish a priori bounds for the periodic solutions under consideration, in L_2 for the derivatives and in C for the solutions themselves.

We denote by A_{-1} the transpose of any matrix A.

General Assumptions: Let $e(t) = (e_1,\dots,e_n)$ be a 2π-periodic continuous function of mean value zero; let $A = [a_{ij}]$ be a constant $n \times n$ matrix; let $G:R^n \to R$, or $G(x)$, $x = (x_1,\dots,x_n) \in R^n$, be a real-valued function of class C^2 in R^n; let $V:R^{n+1} \to R^n$, or $V(x,t) = (V_1,\dots,V_n)$ be an n-vector valued function in R^{n+1}, 2π-periodic in t for every x, such that all V, $\partial V_i/\partial x_j$, $i,j = 1,\dots,n$, are of class C^1 in R^{n+1}; let $g:R^n \to R^n$, or $g(x) = (g_1,\dots,g_n)$, $x \in R^n$, be an n-vector valued continuous function in R^n.

(21.i) Theorem [35]. Under the assumptions above, with $g \equiv 0$, system (21.1) has at least one 2π-periodic solution $x(t) = (x_1,\dots,x_n)$, $-\infty < t < +\infty$, provided one of the following assumptions holds: either (a) $G(x)$ is homogeneous in x of degree $2p$, of constant sign, satisfying $G(x) \geq c|x|^{2p}$ for some integer $p \geq 2$, constant $c > 0$, and all $x \in R^n$; $|V(x,t)| \leq C|x|^p + D$ for some constants $C,D \geq 0$ and all $(x,t) \in R^{n+1}$, A is an arbitrary constant matrix; or (b) the same as (a) with $p = 1$ and $2c > C + \|(A - A_{-1})/2\|$; or (c) the same as (a) with $p \geq 1$, with $V(x)$ depending on x only of the form $V(x) = \text{grad}\, W(x)$, W of class C^2 in R^n, $|V(x)| \leq C|x|^{2p-2} + D$ for some constants $C,D \geq 0$ and all $x \in R^n$, A an arbitrary constant matrix, and $2c > \|(A - A_{-1})/2\|$ if $p = 1$; or (d) $V \equiv 0$, $G:R^n \to R$ an arbitrary function of class C^2, A definite negative; or (e) $V \equiv 0$, $G:R^n \to R$ an arbitrary function of class C^2, A an arbitrary constant matrix with $\|A\| < 1$.

(21.ii) Theorem [36]. Under the assumptions above, with $g:R^n \to R^n$ an arbitrary continuous function with $g(x)/|x| \to 0$

as $|x| \to \infty$, system (21.1) has at least one periodic solution $x(t) = (x_1,\ldots,x_n)$, $-\infty < t < +\infty$, provided one of the following holds: either (A) G and V as in (a) with $p \geq 2$, A nonsingular; or (B) G and V as in (a) with $p = 1$, $2c > C + \|(A - A_{-1})/2\|$, and A nonsingular; or (C) G and V as in (c) with $p \geq 1$, A nonsingular, and $2c > \|(A - A_{-1})/2\|$; or (D) $V \equiv 0$, $A \equiv 0$, $n = 1$, $G:R \to R$ an arbitrary function of class C^2 in R, $xg(x) \geq 0$ for all $x \geq a$ and some constant a; or (E) $V \equiv 0$, $A \equiv 0$, $n > 1$, $G:R^n \to R$ an arbitrary function of class C^2 in R^n, $g(x) \neq 0$ for all $|x| \geq a$, and the geometrical angle between x and g(x) in R^n is $\leq \pi/2 - \sigma$, for some constants $a > 0$, $0 < \sigma < \pi/2$, and all $x \in R^n$, $|x| \geq a$.

Under the conditions of Theorem (21.i) the solutions of which we prove the existence have mean value zero. In other words, we take for the bifurcation equation only those solutions of mean value zero, and thus the bifurcation equation is trivially satisfied, and we need discuss only the auxiliary equation.

Under the conditions of Theorem (21.ii) the solutions may have mean value different from zero, both auxiliary and bifurcation equations must be discussed, and they are indeed solved as a system.

For the sake of brevity we present below (nos. 25-27) only a proof of Theorem (21.i) and we refer to [35,36] for a proof of Theorem (21.ii).

22. <u>The Canonic Decomposition</u>. We denote by xy the inner product in R^n and by $|x|$ the Euclidean norm in R^n. We shall write (21.1) in the form

$$Ex = Nx, \tag{22.1}$$

where E is the linear operator defined by $Ex = x''$ with boundary conditions $x(0) = x(2\pi)$, $x'(0) = x'(2\pi)$, and N is the nonlinear operator defined by

$$(Nx)(t) = -(d/dt)\text{grad } G(x(t)) - (d/dt)V(x(t),t)$$
$$- Ax(t) - g(x(t)) + e(t). \tag{22.2}$$

Here E has eigenvalues $0, 1^2, 2^2, \ldots,$ each with suitable multiplicity, and the elements of the usual trigonometric series as eigenfunctions. As usual, we denote by λ_i the eigenvalues and by φ_i the eigenfunctions, $E\varphi_i + \lambda_i\varphi_i = 0$, $i = 1,2,\ldots$, $0 = \lambda_1 \le \lambda_2 \le \cdots$, $\lambda_i \le \lambda_{i+1}$, $\lambda_i \to +\infty$, $\{\varphi_i, i = 1,\ldots,n\}$ a complete orthonormal system in the Hilbert space $S = (L_2[0,2\pi])^n$. We denote by xy the inner product in R^n, by $(x,y) = \int_0^{2\pi} xy\,dt$ the inner product in S, and by $\|x\| = (x,x)^{1/2}$ the norm in S.

Let us denote by m the smallest integer with $\lambda_{m+1} = 1^2$, thus $\lambda_1 = \cdots = \lambda_m = 0$, and let S_0 denote the subspace of S spanned by $\varphi_1,\ldots,\varphi_m$, so that S_0 is a space of constant functions. Let $P: S \to S_0$ denote the projector operator defined by $Px = (2\pi)^{-1}\int_0^{2\pi} x(t)dt$.

Thus, Px is the constant n-vector whose entries are the mean values of the components of x. Moreover, $S_0 = P(S)$ is the range of P, and $S = S_0 + S_1$ (direct sum) where $S_1 = (I - P)S$. Any element x of S has Fourier series

$x = \sum_1^\infty c_i\varphi_i$, $c_i = (x,\varphi_i)$, with $\|x\|^2 = \sum_1^\infty c_i^2$, and $Px = \sum_1^m c_i\varphi_i$. Any element $x \in S_1$ has Fourier series $x = \sum_{m+1}^\infty c_i\varphi_i$, and we define the operator $H: S_1 \to S_1$ by taking $Hx = -\sum_{m+1}^\infty c_i\lambda_i^{-1}\varphi_i$. Thus,

$$\|Hx\| = (\sum_{m+1}^\infty c_i^2\lambda_i^{-2})^{1/2} \le \lambda_{m+1}^{-1}\|x\| = \|x\|, \tag{22.3}$$

$$(-Hx, x) = \sum_{m+1}^\infty c_i^2\lambda_i^{-1} \ge \lambda_{m+1}^{-1}\|Hx\| = \|Hx\|. \tag{22.4}$$

If D denotes the subspace of all $x \in S$ such that $x(t)$, $x'(t)$ are absolutely continuous (AC) in $[0,2\pi]$ with $x'' \in (L_2[0,2\pi])^n$, $x(0) = x(2\pi)$, $x'(0) = x'(2\pi)$, then D is the domain of E, or $\mathfrak{D}(E) = D$, and $\mathfrak{R}(E) = S_1$, $\mathfrak{R}(H) = D \cap S_1$. Moreover, the usual relations hold (a) $H(I - P)Ex = (I - P)x$ for $x \in D$; (b) $PEx = EPx$ for all $x \in D$; and (c) $EH(I - P)y = (I - P)y$ for all $y \in S$ (cf. no. 3). Thus, problem (22.1) is equivalent to the system of auxiliary and bifurcation equations

$$x - H(I - P)Nx = x^*, \quad x^* \in S_0, \tag{22.5}$$

$$P(Ex - Nx) = 0, \tag{22.6}$$

where $x = x^* + x_1$, $x^* \in S_0$, $x_1 \in S_1$, and $Px = x^*$. It shall be noted that, for $x \in D$, $Ex = x'' \in S = L_2$, necessarily $PEx = 0$, so that (22.6) reduces here to $PNx = 0$. Thus, the bifurcation equation (22.6) for problem (21.1) is now

$$P[-(d/dt)\text{grad } G(x(t)) - (d/dt)V(x(t),t)]$$
$$- PAx(t) - Pg(x(t)) + Pe(t) = 0, \tag{22.7}$$

where $Pe = 0$ by hypothesis, and the term in brackets certainly has mean value zero. Since A is a constant matrix we have $PAx(t) = APx(t) = Ax^*$. Thus, we are left with the equation $-Ax^* - Pg(x(t)) = 0$, and system (22.5-6) reduces to

$$x - H(I - P)Nx = x^*, \tag{22.8}$$

$$Ax^* + Pgx = 0. \tag{22.9}$$

For $g \equiv 0$, as in Theorem I, equation (22.9) reduces to $Ax^* = 0$ which is certainly satisfied by taking $x^* = 0$. For A nonsingular, $Ax^* = 0$ implies $x^* = 0$, but in Theorem (21.i) we do not require A to be nonsingular. Thus, we take the trivial solution $x^* = 0$ of the bifurcation equation, and we show that there are solutions x with $Px = x^* = 0$.

We terminate this section with a remark which concerns G. Since G is homogeneous of degree $2p$, we have

$$2p\ G(x) = \sum_{i=1}^{n} x_i(\partial G/\partial x_i) = x \text{ grad } G(x). \tag{22.10}$$

Hence, by the hypotheses on G, we have

$$2pc|x|^{2p} \leq 2p|G(x)| = |x \text{ grad } G(x)| \leq |x|\ |\text{grad } G(x)|,$$

$$|\text{grad } G(x)| \geq 2pc|x|^{2p-1} \quad \text{for all} \quad x \in R^n.$$

We note that grad $G(x)$ is a homogeneous n-vector function in R^n of degree $2p - 1$, by hypothesis continuous in R^n. Thus, $|\text{grad } G(x)|$ is continuous and bounded on the unit sphere $|x| = 1$ of R^n. If K is the maximum of $|\text{grad } G(x)|$ on the unit sphere, then

$$|\text{grad } G(x)| \leq K|x|^{2p-1} \quad \text{for all} \quad x \in R^n. \tag{22.11}$$

23. <u>The Operators J, J^*.</u> We denote by H^0 the Hilbert space $H^0 = S = (L_2[0,2\pi])^n$, by H^1 the Hilbert space of all AC n-vector functions $x(t) = (x_1,\ldots,x_n)$, with $x' \in S = (L_2[0,2\pi])^n$, $x(0) = x(2\pi)$, and by H^2 the Hilbert space of all AC n-vector functions $x(t)$ with x' AC and $x'' \in S$,

$x(0) = x(2\pi)$, $x'(0) = x'(2\pi)$, and we take in H^1, H^2 the inner products and norms

$$(x,y)_1 = (x,y) + (x',y'), \quad \|x\|_1 = (\|x\|^2 + \|x'\|^2)^{1/2},$$

$$(x,y)_2 = (x,y) + (x',y') + (x'',y''),$$

$$\|x\|_2 = (\|x\|^2 + \|x'\|^2 + \|x''\|^2)^{1/2},$$

where, as stated earlier, $\|x\|$ is the L_2-norm. As it is well known the following norms

$$\|x\|_1' = \mathrm{Sup}_t |x(t)| + \|x'\|,$$

$$\|x\|_2' = \mathrm{Sup}_t |x(t)| + \mathrm{Sup}_t |x'(t)| + \|x''\|$$

are equivalent to $\|x\|_1$ in H^1, and to $\|x\|_2$ in H^2. Note that for the domain $D = \mathfrak{D}(E)$ for the operator $E:D \to S$ we have now $D = H^2$.

For functions $x \in S_1$, that is, $x \in S = H^0 = (L_2[0,2\pi])^n$ with $Px = 0$, we denote by $y = J_1 x$ the usual function y which is the unique primitive of x of mean value zero (cf. [10]). Thus, $y \in H^1$, $Py = 0$, and $J_1 : S_1 \to H^1 \cap S_1$. With both x and y extended to $(-\infty,+\infty)$ by 2π-periodicity, y has the very simple representation

$$y(t) = (2\pi)^{-1} \int_t^{t+2\pi} \alpha x(\alpha)\, d\alpha, \quad \text{where} \quad Px = 0. \tag{23.1}$$

If $z(t) = \int_0^t x(\alpha)\, d\alpha$, then z is 2π-periodic because of $Px = 0$, and then we have also

$$y = (I - P)z = \int_0^t x(\alpha)\, d\alpha - (2\pi)^{-1} \int_0^{2\pi} d\beta \int_0^{\beta} x(\alpha)\, d\alpha. \tag{23.2}$$

It is easy to see that $J_1 : S_1 \to H^1 \cap S_1$ is one-one and onto. We shall also denote by J and J^* the maps $J = J_1(I - P)$, and $J^* = -J$. Then $J, J^* : S \to H^1 \cap S_1$, and the following relation holds for the operator $K = -H(I - P)$:

$$-H(I - P) = J^* J : S \to H^2 \cap S_1.$$

This decomposition is similar to the ones discussed by Kannan and Locker in [82,83].

From (23.1) and Schwarz inequality it is also easy to see that any family Λ of elements $x \in S_1$ with equibounded L_2-norms is mapped by J_1 into a family $J_1(\Lambda)$ of equibounded and equiabsolutely continuous functions. Hence, $J_1(\Lambda)$ is relatively compact in $(C[0,2\pi])^n$, and of course in $S = H^0 = (L_2[0,2\pi])^n$ also.

We note that, if $x(t) = (x_1,\dots,x_n) \in S_1$, $y(t) = (y_1,\dots,y_n) = J_1 x$, then each x_i and corresponding y_i have Fourier series

$$x_i(t) = \sum_1^\infty (a_k \cos kt + b_k \sin kt),$$

$$y_i(t) = \sum_1^\infty ((-b_k/k) \cos kt + (a_k/k) \sin kt),$$

and therefore

$$\|y_i\|^2 = \pi \sum_1^\infty (a_k^2 + b_k^2)k^{-2} \le \pi \sum_1^\infty (a_k^2 + b_k^2) = \|x_i\|^2.$$

Hence $\|y\| \le \|x\|$, or $\|J_1 x\| \le \|x\|$ for every $x \in S_1$. Concerning the norm $\| \ \|_1$ in H^1 we also have

$$\|J_1 x\|_1 \le \|J_1 x\| + \|x\| \le \|x\| + \|x\| = 2\|x\|.$$

In conclusion we have the statement:

(23.i) J and J^* are compact maps from S to S, and bounded maps from S to H^1.

24. <u>A Topological Degree Argument and the Borsuk-Ulam Theorem</u>. Let G be any open set of a real linear topological space X, let ∂G denote the boundary of G, and let $F: \bar{G} \to X$ be any continuous map from $\bar{G} = G \cup \partial G$ into X. If X is infinite dimensional we shall further assume that F is of the form

$F = I - T$, I the identity in X, T compact, or completely continuous (i.e., $F - I$ is compact). Let p be any point of X not on the image of ∂G, or $p \notin F(\partial G) = [z \in X | z = Fx, x \in \partial G]$. Then a topological degree (or Brouwer degree) $d[F,G,p]$ has been defined, for X finite dimensional, by topological considerations (cf., e.g., Cronin [142]), as well as by analytical considerations (Nagumo [177]). For X infinite dimensional, a topological degree $d[F,G,p]$ has been defined by Leray and Schauder [167], Nagumo [178], and Rothe [184].

A degree in a different context has been introduced by Rothe in this same volume. The topological degree has very important properties of which we mention here a few.

(a) $d[F,G,p]$ is a constant finite integer ($\gtreqless 0$) on each component of $X - F(\partial G)$. Also, $d[I,G,p] = 1$ for $p \in G$; $= 0$ for $p \in X - \bar{G}$. Moreover, $d[F,G,p]$ depends only on the values of F on ∂G, that is, if $F,H:\bar{G} \to X$ and $F = H$ on ∂G, $p \notin F(\partial G) = H(\partial G)$, then $d[F,G,p] = d[H,G,p]$.

(b) $d[F,G,p] \neq 0$, ($p \notin F(\partial G)$ implies $p \in F(G)$, that is, $p = Fx$ for some $x \in G$.

(c) If G is divided into finitely many open sets $G_1,\ldots,G_k$, that is, $G \supset G_1 \cup \cdots \cup G_k$, $\bar{G} = \bar{G}_1 \cup \cdots \cup \bar{G}_k$, $G_i \cap G_j = \emptyset$ for $i \neq j$, then $d[F,G,p] = \sum_1^k d[F,G_i,p]$ for all $p \notin \bigcup_i F(\partial G_i)$.

(d) If $F(x,\lambda)$ is a continuous map from $G \times [0,1]$ into X, (and for every λ, $F - I$ is compact if X is infinite dimensional), and if $p \notin F(x,\lambda)(\partial G)$ for $0 \leq \lambda \leq 1$, and

$F_0 = F(x,0)$, $F_1 = F(x,1)$, then $d[F_0,G,p] = d[F_1,G,p]$ (invariance of degree by homotopy).

Existence of fixed points for a given map, and existence of solutions to given equations are equivalent problems. Indeed, for $\Phi:\bar{G} \to X$ and $F = I - \Phi$, we have $x = \Phi x$ for some $x \in G$ if and only if $Fx = 0$ for some $x \in G$. Thus, $0 \notin F(\partial G)$, and $d[F,G,0] \neq 0$ implies that $Fx = 0$ and $x = \Phi x$ for some $x \in G$.

If X is any real Banach space, if $F:X \to X$ is a continuous map (and $F - I$ is compact if X is infinite dimensional), and we know that $((1 - \lambda)I + \lambda F)x \neq 0$ for $\|x\| = R$ and $0 \leq \lambda \leq 1$, then $Fx = 0$ for some $\|x\| < R$. Indeed, for $G = [x \in X | \|x\| < R]$ we have $d[F,G,0] = d[I,G,0] = 1 \neq 0$. Equivalently, if $(I - \lambda\Phi)x \neq 0$ for $\|x\| = R$ and $0 \leq \lambda \leq 1$, then $x = \Phi x$ for some $\|x\| < R$. This argument was used extensively by Leray and Schauder [167] in problems of elliptic partial differential equations (cf. Cronin [142]).

The following statements are based on degree arguments.

(24.i) (A Lemma). Let S be a real Hilbert space, let $F:S \to S$ be a compact operator such that for some $R > 0$ we have $(Fx,x) \geq -\|x\|^2$ for all $x \in S$ with $\|x\| = R$. Then the equation $(I + F)x = 0$ has at least one solution $\|x\| \leq R$.

Proof. If G is the open ball $G = [x \in S | \|x\| < R]$ in S, then its boundary is $\partial G = [x \in S | \|x\| = R]$. If $(I + F)x = 0$ has a solution $x \in \partial G$, the case is trivial. Assume that $(I + F)x \neq 0$ for all $x \in \partial G$. Let us prove that $(I + \lambda F)x \neq 0$ for all $x \in \partial G$, and $0 \leq \lambda \leq 1$. There are no solutions

$x \in \partial G$ for $\lambda = 0$ and $\lambda = 1$. If we had $x + \lambda Fx = 0$ for some $x \in \partial G$ and $0 < \lambda < 1$, we would also have

$$0 = (Fx, x + \lambda Fx) \geq -\|x\|^2 + \lambda\|Fx\|^2 = (\lambda^{-1} - 1)\|x\|^2,$$

a contradiction. We have proved that $(I + \lambda F)x \neq 0$ for all $x \in \partial G$ and $0 \leq \lambda \leq 1$. By force of (a) and (e) we have

$$0 \neq \deg[I, G, 0] \neq \deg[I + \lambda F, G, 0]$$

for all $0 \leq \lambda \leq 1$. In particular, $\deg[I + F, G, 0] \neq 0$, and hence $(I + F)x = 0$ has at least one solution $x \in G$.

(24.ii) (A Lemma). Let S be a real Hilbert space, let $F: S \to S$ be an operator such that $F - I$ is compact, and such that for some $R > 0$ we have $(Fx, x) \geq 0$ for all $x \in S$ with $\|x\| = R$. Then, $Fx = 0$ for at least one $x \in S$, $\|x\| \leq R$.

Proof. Apply (24.i) to $\Phi = F - I$.

We now state here, without proofs, the following well known statements concerning odd maps (cf. [36] for references).

(24.iii) The Borsuk-Ulam Theorem. Let X be a Banach space, $F: X \to X$ a compact map, F odd outside some ball (that is, there is an $R > 0$ such that $F(-x) = -F(x)$ for all $x \in X$, $\|x\| \geq R$), and let us assume that $(I + F)^{-1}$ maps bounded sets of X into bounded sets of X. Then $(I + F)X = X$.

(24.iv) An Extended Form of Borsuk-Ulam Theorem. Let X be a Banach space, let $F_0, F_1: X \to X$ be compact maps, F_0 odd outside some ball in X, let $F = F_0 + F_1$, and let us assume that there is some function $k(\xi) \geq 0$, $0 \leq \xi \leq \infty$, such that whenever $x + (F_0 + \lambda F_1)x = w$ for some $x, w \in X$, $0 \leq \lambda \leq 1$, then we have also $\|x\| \leq k(\|w\|)$. Then $(I + F)X = X$.

25. Proof of Parts (a) and (b) of Theorem (21.i). The auxiliary equation (22.8) is here $x - H(I - P)Nx = 0$ with $x \in S = L_2$. We know already that $Px = 0$, and thus we treat the auxiliary equation as an equation in $S_1 = (I - P)S$. But now $-H(I - P) = J^*J$, where $J:S \to H^1 \subset S$, $N:H^1 \to S$, and $J^*:S \to H^1 \subset S$. Therefore, the auxiliary equation can be written as

$$x + J^*JNx = 0, \tag{25.1}$$

and thus, any solution x, if any, must belong to $\Re(J^*)$. Thus, we take $x = J^*y$, $y \in S$, and the equation becomes $J^*y + J^*JN(J^*y) = 0$. It is enough we solve the equation

$$y + JN(J^*y) = 0. \tag{25.2}$$

Obviously, any solution y of (25.2) gives rise to a solution $x = J^*y$ of (25.1). On the other hand, for any solution y of (25.2) we have $y \in \Re(J) = H^1 \cap S_1$. Hence, $Py = 0$ and certainly $y \in S_1$. Since J^* is one-one between S_1 and $\Re(J^*) = H^1 \cap S_1$, we see that for any solution x of (25.1) there is one and only one solution y of (25.2) with $x = J^*y$, $Py = 0$.

Let $N = N_0 + N_1$ with

$$\begin{aligned} N_0x &= -(d/dt)\,\mathrm{grad}\, G(x(t)), \\ N_1x &= -(d/dt)V(x(t),t) - Ax(t) + e(t), \end{aligned} \tag{25.3}$$

and equation (25.2) becomes

$$y + F_0y + F_1y = 0,$$

where $F_0y = JN_0(J^*y)$, $F_1y = JN_1(J^*y)$, $F_0, F_1:S \to S$.

Let us prove that F_0 and F_1, as maps from S to S, are compact. Indeed, any bounded sequence $[y_k]$ in S is mapped by J^* into a bounded sequence $[J^*y_k]$ in H^1 since

J^* is bounded as a map from S to H^1. Then the same sequence is mapped into bounded sequences $[N_0(J^* y_k)]$, $[N_1(J^* y_k)]$ since N_0, N_1 are bounded as maps from H^1 to S. Finally, the same sequences are mapped into sequences $[JN_0(J^* y_k)]$, $[JN_1(J^* y_k)]$ in S which must contain some convergent subsequence since J is compact from S to S.

Since G is a homogeneous function of even degree $2p \geq 2$, $G(x)$ is even, all $\partial G/\partial x_i$ are odd, all $\partial^2 G/\partial x_i \partial x_j$ are even, and

$$\begin{aligned} N_0 x(t) &= -(d/dt)\operatorname{grad} G(x(t)) \\ &= [-\sum_j (\partial^2 G(x(t))/\partial x_i \partial x_j) x_j'(t),\ i = 1,\ldots,n] \end{aligned}$$

is odd. Since J^* and J are linear and therefore odd, the operator $F_0 = JN_0J^*$ is odd.

Let us prove that N_0, N_1 map bounded subsets of H^1 into bounded subsets of S. Let Λ be any set of elements $x \in H^1$ with $\|x\|_1 \leq M$ for some constant M, and then $\|x\|$, $\|x'\| \leq M$. Because of the equivalence of the norms $\|x\|_1$ and $\|x\|_1'$, there is a constant M' such that $\|x\|_1' \leq M'$ for all $x \in \Lambda$, and then $\operatorname{Sup}_t |x(t)| \leq M'$.

Because of the hypotheses on G and V, the functions $\partial^2 G/\partial x_i \partial x_j$, $\partial V/\partial x_i$, $\partial V/\partial t$ are continuous and bounded for $x \in R^n$, $|x| \leq M'$, $0 \leq t \leq 2\pi$, say $|\partial^2 G/\partial x_i \partial x_j|$, $|\partial V_i/\partial x_j|$, $|\partial V_i/\partial t| \leq M''$ for some constant M''. Then

$$\begin{aligned} \|N_0 x(t)\| &= \|-(d/dt)\operatorname{grad} G(x(t))\| \\ &= \|-\sum_j (\partial^2 G/\partial x_i \partial x_j) x_j'(t),\ i = 1,\ldots,n\| \\ &\leq nM''\|x'\| \leq nM''M. \end{aligned}$$

Analogously, for $C = \max|a_{ij}|$, we have

$$\begin{aligned}
\|N_1x(t)\| &= \|-(d/dt)V(x(t),t) - Ax(t) + e(t)\| \\
&\leq \|\sum_j (\partial V_i/\partial x_j)x_j'(t) + \partial V_i/\partial t, \quad i = 1,\dots,n\| \\
&\quad + \|\sum_j a_{ij}x_j(t), \quad i = 1,\dots,n\| + \|e(t)\| \\
&\leq nM''\|x'\| + (2\pi n)^{1/2}M'' + nC\|x\| + \|e\| \\
&\leq nM''M + (2\pi n)^{1/2}M'' + nCM + \|e\|.
\end{aligned}$$

Let us prove now that N_0, N_1, again as maps from H^1 to S, are continuous. Indeed, we may take elements $x, y \in H^1$ in a set Λ as before with $\|x\|_1$, $\|y\|_1 \leq M$, and hence $\|x'\|$, $\|y'\| \leq M$, $\sup_t |x(t)|$, $\sup_t |y(t)| \leq M'$. Thus, $\partial^2G/\partial x_i \partial x_j$, $\partial V_i/\partial x_j$, $\partial V_i/\partial t$ are uniformly bounded and uniformly continuous with respect to x with $x \in R^n$, $|x| \leq M'$, $0 \leq t \leq 2\pi$, say in absolute value $\leq M''$ and of modulus of continuity $\omega(\eta)$, $0 \leq \eta < \infty$, $\omega(0) = 0$, ω continuous. We have now

$$\begin{aligned}
\|N_0x(t) - N_0y(t)\| &= \|-(d/dt)\text{grad } G(x(t)) \\
&\quad + (d/dt)\text{grad } G(y(t))\| \\
&= \|-\sum_j (\partial^2G(x(t))/\partial x_i \partial x_j)x_j'(t) \\
&\quad + \sum_j (\partial^2G(y(t))/\partial x_i \partial x_j)y_j'(t), \\
&\quad i = 1,\dots,n\| \\
&\leq \|\sum_j [(\partial^2G(x(t))/\partial x_i \partial x_j) \\
&\quad - (\partial^2G(y(t))/\partial x_i \partial x_j)]x_j'(t), \\
&\quad i = 1,\dots,n\| \\
&\quad + \|\sum_j (\partial^2G(y(t))/\partial x_i \partial x_j)(x_j'(t) \\
&\quad - y_j'(t)), \quad i = 1,\dots,n\|
\end{aligned}$$

$$\leq n\omega(\text{Sup}_t |x(t) - y(t))| \, \|x'\|$$
$$+ nM'' \|x' - y'\|$$
$$\leq nM\omega(\|x - y\|_1') + nM'' \|x - y\|_1'.$$

Let us prove that there is a function $k(\zeta) \geq 0$, $0 \leq \zeta < \infty$, such that, for any $0 \leq \lambda \leq 1$, and any two elements $y, w \in S$, if any, with $y + F_0 y + \lambda F_1 y = w$, we have $\|y\| \leq k(\|w\|)$. First, we assume that w is an element of H^1. Then

$$y + J(N_0 + \lambda N_1)J^* y = w, \tag{25.4}$$

and since the ranges of F_0, F_1 are in H^1, we have $y \in H^1$. It is not restrictive to assume $Pw = 0$, $Py = 0$. If $x = J^* y$, we have $x \in H^2$, $y = -x'$, and by differentiation, also

$$-x'' + (N_0 + \lambda N_1)x = w',$$

or

$$x''(t) + (d/dt)\text{grad } G(x(t)) + (d/dt)\lambda V(x(t),t)$$
$$+ \lambda A x(t) - \lambda e(t) + w'(t) = 0. \tag{25.5}$$

Let $B = 2^{-1}(A + A_{-1})$, $B^* = 2^{-1}(A - A_{-1})$, so that $A = B + B^*$, B is symmetric and B^* is antisymmetric, $A = [a_{ij}]$, $B = [b_{ij}]$, $B^* = [b^*_{ij}]$.

Let u denote the unique primitive of x of mean value zero. By multiplication of (25.5) by u, and by integration by parts, we obtain

$$-\int_0^{2\pi} \text{grad } G(x(t))x(t)dt - \lambda\int_0^{2\pi} V(x(t),t)x(t)dt$$
$$+ \lambda\int_0^{2\pi} (B^* x(t))u(t)dt$$
$$+ \lambda\int_0^{2\pi} e(t)u(t)dt$$
$$- \int_0^{2\pi} w(t)x(t)dt = 0. \tag{25.6}$$

By the hypotheses on G and V we have now, for any $p \geq 1$,

$$(2pc)(2\pi)^{-1}\int_0^{2\pi}|x(t)|^{2p}dt \le (2p)(2\pi)^{-1}\int_0^{2\pi}|G(x(t))|dt$$

$$= (2p)(2\pi)^{-1}\left|\int_0^{2\pi} G(x(t))dt\right|$$

$$= (2\pi)^{-1}\left|\int_0^{2\pi}(\text{grad } G(x(t)))x(t)dt\right|$$

$$\le (2\pi)^{-1}\left|\int_0^{2\pi} V(x(t),t)x(t)dt\right|$$

$$+ (2\pi)^{-1}\int_0^{2\pi} e(t)u(t)dt|$$

$$+ \left|(2\pi)^{-1}\int_0^{2\pi} \sum_{i\ne j} b^*_{ij}x_j(t)u_i(t)dt\right|$$

$$+ \left|(2\pi)^{-1}\int_0^{2\pi} w(t)x(t)dt\right| \le (2\pi)^{-1}\int_0^{2\pi}(C|x(t)|^p$$

$$+ D)|x(t)|dt + \left|(2\pi)^{-1}\int_0^{2\pi} e(t)u(t)dt\right|$$

$$+ (2\pi)^{-1}\|B^*\|\int_0^{2\pi}|x(t)|\ |u(t)|dt$$

$$+ \left|(2\pi)^{-1}\int_0^{2\pi} w(t)x(t)dt\right|. \tag{25.7}$$

Since u is the primitive of x of mean value zero, we have $\|u\| \le \|x\|$, and

$$\left|(2\pi)^{-1}\int_0^{2\pi} e(t)u(t)dt\right|$$

$$\le ((2\pi)^{-1}\int_0^{2\pi}|e(t)|^2dt)^{1/2}((2\pi)^{-1}\int_0^{2\pi}|u(t)|^2dt)^{1/2}$$

$$\le ((2\pi)^{-1}\int_0^{2\pi}|e(t)|^2dt)^{1/2}((2\pi)^{-1}\int_0^{2\pi}|x(t)|^2dt)^{1/2}$$

$$\le ((2\pi)^{-1}\int_0^{2\pi}|e(t)|^2dt)^{1/2}((2\pi)^{-1}\int_0^{2\pi}|x(t)|^{2p}dt)^{1/2p}.$$

Analogously,

$$\left|(2\pi)^{-1}\int_0^{2\pi} w(t)x(t)dt\right|$$

$$\le ((2\pi)^{-1}\int_0^{2\pi}|w(t)|^2dt)^{1/2}((2\pi)^{-1}\int_0^{2\pi}|x(t)|^{2p}dt)^{1/2p}.$$

Also we have

$$(2\pi)^{-1}\int_0^{2\pi}|x(t)|^{p+1}dt \le ((2\pi)^{-1}\int_0^{2\pi}|x(t)|^{2p}dt)^{(p+1)/2p},$$

$$(2\pi)^{-1}\int_0^{2\pi}|x(t)|dt \le ((2\pi)^{-1}\int_0^{2\pi}|x(t)|^{2p}dt)^{1/2p},$$

$$(2\pi)^{-1}\int_0^{2\pi}|x(t)|\,|u(t)|dt \le (2\pi)^{-1}\int_0^{2\pi}|x(t)|^2dt$$

$$\le ((2\pi)^{-1}\int_0^{2\pi}|x(t)|^{2p}dt)^{1/p}.$$

If $Z = ((2\pi)^{-1}\int_0^{2\pi}|x(t)|^{2p}dt)^{1/2p}$, then relation (25.7) yields

$$(2pc)Z^{2p} \leq CZ^{p+1} + (D + (2\pi)^{-1/2}(\|e\| + \|w\|))Z + \|B^*\|Z^2. \tag{25.8}$$

If $p > 1$, we take

$$L_1 = (2pc)^{-1}C, \quad L_2 = (2pc)^{-1}\|B^*\|,$$
$$L_3 = (2pc)^{-1}(D + (2\pi)^{-1/2}(\|e\| + \|w\|)),$$
$$M = \max[(2L_1)^{1/(p-1)}, \ (2L_1^{-1}L_2)^{1/(p-1)}, \ L_2^{-1}L_3],$$

and then (25,8) yields

$$Z^p(Z^{p-1} - L_1) \leq L_2Z + L_3.$$

We see that, either $Z \leq (2L_1)^{1/(p-1)} \leq M$, or $Z \geq (2L_1)^{1/(p-1)}$, hence $Z^{p-1} - L_1 \geq L_1$, and $Z^p \leq L_1^{-1}L_2Z + L_1^{-1}L_3$, or $Z(Z^{p-1} - L_1^{-1}L_2) \leq L_1^{-1}L_3$. Again, either $Z \leq (2L_1^{-1}L_2)^{1/(p-1)}$, or $Z \geq (2L_1^{-1}L_2)^{1/(p-1)}$, and then $L_1^{-1}L_2Z \leq L_1^{-1}L_3$ and $Z \leq L_2^{-1}L_3$. In any case we have $Z \leq M$, and then

$$((2\pi)^{-1}\int_0^{2\pi}|x(t)|^2dt)^{1/2} \leq ((2\pi)^{-1}\int_0^{2\pi}|x(t)|^{2p}dt)^{1/2p} = Z,$$
$$\|x\| \leq (2\pi)^{1/2}Z \leq (2\pi)^{1/2}M, \tag{25.9}$$
$$\int_0^{2\pi}|x(t)|^{2p}dt \leq 2\pi Z^{2p} \leq 2\pi M^{2p}.$$

If $p = 1$, then we assume $2c > C + \|B^*\|$, or $L_1 = (2c)^{-1}(C + \|B^*\|) < 1$, relation (25.8) becomes $(1 - L_1)Z \leq L_3$, and by taking $M = (1 - L_1)^{-1}L_3$ we have as before $Z \leq M$.

Analogously, if $\eta(t)$ is the unique primitive of $e(t)$ of mean value zero, then by multiplication of (25.5) by x and integration, we obtain

$$-\int_0^{2\pi}|x'(t)|^2dt - \lambda\int_0^{2\pi} V(x(t),t)x'(t)dt + \lambda\int_0^{2\pi} \eta(t)x'(t)dt - \int_0^{2\pi} w(t)x'(t)dt + \int_0^{2\pi}(Ax(t))x(t)dt = 0,$$

$$\|x'\|^2 \leq \int_0^{2\pi} (C|x(t)|^p + D)\,|x'(t)|\,dt + \|\eta\|\,\|x'\| + \|w\|\,\|x'\| + \|A\|\,\|x\|^2, \qquad (25.10)$$

where

$$\int_0^{2\pi} |x'(t)|\,dt \leq (2\pi)^{1/2} \left(\int_0^{2\pi} |x'(t)|^2 dt\right)^{1/2} = (2\pi)^{1/2} \|x'\|,$$

$$\int_0^{2\pi} |x(t)|^p |x'(t)|\,dt \leq \left(\int_0^{2\pi} |x(t)|^{2p} dt\right)^{1/2} \left(\int_0^{2\pi} |x'(t)|^2 dt\right)^{1/2} \leq (2\pi)^{1/2} M^p \|x'\|.$$

Thus,

$$\|x'\|^2 \leq ((2\pi)^{1/2}(CM^p + D) + \|\eta\| + \|w\|)\,\|x'\| + 2\pi \|A\| M^2.$$

If L denotes the expression in parenthesis, and $M^* = \max[2L, 2\pi L^{-1}\|A\|M^2]$, then we see that, either $\|x'\| \leq 2L \leq M^*$, or $\|x'\| \geq 2L$, and then $\|x'\| \leq 2\pi L^{-1}\|A\|M^2 \leq M^*$. In any case, $\|x'\| \leq M^*$. We think of M^* as a function of $\|w\|$, or $M^* = k(\|w\|)$.

We conclude that, for any $0 \leq \lambda \leq 1$, and any pair $y \in S$, $w \in H^1$ (if any) satisfying

$$y + J(N_0 + \lambda N_1)J^* y = w,$$

we certainly have

$$\|y\| = \|x'\| \leq k(\|w\|), \quad y \in S, \quad w \in H^1, \qquad (25.11)$$

where $\|w\|$ is the norm of w in the larger space $S = (L_2[0, 2\pi])^n$.

Let us prove that the relation $\|y\| \leq k(\|w\|)$ holds for any pair $y \in S$, $w \in S$ (if any) satisfying (25.4). Indeed, assume that

$$y + J(N_0 + \lambda N_1)J^* y = w, \quad y, w \in S = H^0 = (L_2[0, 2\pi])^n,$$

and take as before $x = J^* y$ so that $y = -x'$, $x \in H^1$,

$x' \in S = H^0$, and x'' is a distribution, namely, $x'' \in H^{-1}$. Analogously, $w' \in H^{-1}$. As before, by differentiation, we still obtain relation (25.5), where now x'' and w' are distributions in H^{-1}. Here $x \in H^1$, x has mean value zero, and we denote by $u \in H^2$ the unique primitive of x of mean value zero. By applying the operations in the first member of (25.5) to $u \in H^2$, and by integration by parts by the rules of the distributions in H^α (cf. [35]), we have $x' \in H^0$, $u \in H^2$, $2 + (0) = 2 > 1$, and $x''u = x'x = 0$, as well as $w \in H^0$, $u \in H^2$, and $w'u = wx$. We thus obtain relation (25.6) and the proof is as before yielding relations (25.9). Now $x \in H^1$ and we apply the operations in the first member of (25.5) to x. Then $w \in H^0$, $x \in H^1$, $0 + (1) = 1$, and by the distribution rule for integration by parts, we have $w'x = wx'$. Thus, again we obtain relation (25.10), and the argument remains the same yielding (25.11).

We are now in a position to apply Borsuk-Ulam Theorem (24.iv), to conclude that the equation

$$y + J(N_0 + N_1)(J^*y) = 0$$

has at least one solution $y \in S = (L_2[0,2\pi])^n$. The range of J is $H^1 \cap S_1$, so that $y \in H^1$, and if we take $x = J^*y$, then $x \in H^2$, and, as we have seen before, x satisfies the original equation (21.1). We have proved Theorem (21.i) under hypotheses (a) and (b).

26. <u>Proof of Part (c) of Theorem (21.i)</u>. Under hypotheses (c) equation (21.1) becomes

$$x'' + (d/dt)(\mathrm{grad}\ G(x(t)) + (d/dt)(\mathrm{grad}\ W(x(t))) + Ax(t) = e(t). \tag{26.1}$$

The argument is essentially the same. Here N_1 is still defined as in (25.3), while

$$N_2x = -(d/dt)(\mathrm{grad}\ W(x)) - Ax(t) + e(t). \tag{26.2}$$

Relation (25.7) now becomes

$$\begin{aligned}(2pc)(2\pi)^{-1}\int_0^{2\pi}|x(t)|^{2p}dt &\le (2\pi)^{-1}\left|\int_0^{2\pi}(\mathrm{grad}\ G(x(t))x(t)dt\right| \\ &\le (2\pi)^{-1}\left|\int_0^{2\pi}(\mathrm{grad}\ W(x(t))x(t)dt\right| \\ &\quad + \left|(2\pi)^{-1}\int_0^{2\pi} e(t)u(t)dt\right| \\ &\quad + \left|(2\pi)^{-1}\int_0^{2\pi} \sum_{i\neq j} b_{ij}^* x_j(t)u_i(t)dt\right| \\ &\quad + \left|(2\pi)^{-1}\int_0^{2\pi} w(t)x(t)dt\right| \\ &\le (2\pi)^{-1}\int_0^{2\pi}(C|x(t)|^{2p-2} + D)|x(t)|dt \\ &\quad + (2\pi)^{-1}\|B^*\|\int_0^{2\pi}|x(t)||u(t)|dt \\ &\quad + \left|(2\pi)^{-1}\int_0^{2\pi} e(t)u(t)dt\right| + \left|(2\pi)^{-1}\int_0^{2\pi} w(t)x(t)dt\right|,\end{aligned}$$

with

$$(2\pi)^{-1}\int_0^{2\pi}|x(t)|^{2p-1}dt \le ((2\pi)^{-1}\int_0^{2\pi}|x(t)|^{2p}dt)^{(2p-1)/2p}.$$

With the same choice of Z we have now the relation

$$(2pc)Z^{2p} \le CZ^{2p-1} + (D + (2\pi)^{-1/2}(\|e\| + \|w\|))Z + \|B^*\|Z^2. \tag{26.3}$$

If $p > 1$, we take L_1, L_2, L_3 as in no. 25, and

$$M = \max[2L_1, (2L_1^{-1}L_2)^{1/(2p-3)}, L_2^{-1}L_3],$$

relation (26.3) becomes $Z^{2p-2}(Z - L_1) \le L_2Z + L_3$, and as before we see that, in any case, we have $Z \le M$. If $p = 1$, then we assume $2c > \|B^*\|$, or $L_2 = (2c)^{-1}\|B^*\| < 1$, relation (26.3) becomes $(1 - L_2)Z \le L_1 + L_3$, and by taking $M = (1 - L_2)^{-1}(L_1 + L_3)$, again (26.3) yields $Z \le M$.

Then we have, as in parts (a) and (b),

$$\|x\| \leq (2\pi)^{1/2}M, \quad \int_0^{2\pi} |x(t)|^{2p}dt \leq 2\pi M^{2p}.$$

Relation (25.10) now simplifies since $\int_0^{2\pi}$ grad $W(x(t))x'(t)dt = 0$, and instead of (25.10) we simply have

$$-\int_0^{2\pi} |x'(t)|^2 dt - \lambda\int_0^{2\pi} \eta(t)x'(t)dt - \int_0^{2\pi} w(t)x'(t)dt$$
$$+ \int_0^{2\pi} (Ax(t)x(t)dt = 0,$$

and then

$$\|x'\|^2 \leq (\|\eta\| + \|w\|)\|x'\| + 2\pi\|A\|M^2.$$

If $L' = \|\eta\| + \|w\|$ and $M'^* = \max[2L', 2\pi L'^{-1}\|A\|M^2]$, we conclude that $\|x'\| \leq M'^*$. The argument is then exactly as for parts (a) and (b).

27. Proof of Parts (d) and (e) of Theorem (21.i). Again we write the auxiliary equation in the form

$$y + JNJ^*y = 0, \quad y \in S = (L_2[0,2\pi])^n, \tag{27.1}$$

where

$$Nx = N_1x - Ax + e(t), \quad N_1x(t) = -(d/dt)\text{grad } G(x(t)).$$

If $F = JNJ^*$ is thought of as an operator from S to S, we know already from part (a) that $F:S \to S$ is compact. Moreover, for $z = J^*y$, $z \in H^1$, $\|z\| \leq \|y\|$, we have

$$(Fy,y) = (JNJ^*y,y) = (NJ^*y,J^*y) = (Nz,z)$$
$$= (N_1z,z) - (Az,z) + (e,z),$$
$$(N_1z,z) = -\int_0^{2\pi} (d/dt)(\text{grad } G(z(t))z(t)dt$$
$$= \int_0^{2\pi} \text{grad } G(z(t))z'(t)dt = [\text{grad } G(z(t))]_{t=0}^{t=2\pi}$$
$$= 0.$$

If A is definite negative, we have $(Az,z) \leq 0$, and finally

$$(Fy,y) \geq (e,z) \geq -\|e\|\|z\| \geq -\|e\|\|y\|.$$

For $\|y\| = R = \|e\|$ we certainly have $(Fy,y) \geq -\|y\|^2$. By

(25.i) (topological degree argument) we conclude that the equation $y + Fy = 0$, that is, equation (27.1) has at least one solution $y \in S$, $\|y\| \leq R = \|e\|$, and the theorem is proved under hypotheses (d).

If $\|A\| < 1$, then $|(Az,z)| \leq \|A\|\|z\|^2$ and

$$(Fy,y) \geq -\|A\|\|z\|^2 - \|e\|\|z\| \geq -\|A\|\|y\|^2 - \|e\|\|y\|.$$

Thus, for $\|A\| < 1$, if $R > 0$ is so chosen that $\|A\| + \|e\|R^{-1} < 1$, then again, for $\|y\| = R$ we have

$$(Fy,y) \geq -R^2(\|A\| + \|e\|R^{-1}) > -\|y\|^2.$$

Again, by (25.i), the equation $y + Fy = 0$ has at least a solution $y \in S$, $\|y\| \leq R$, and the theorem is proved under hypotheses (e).

Section 6. The Alternative Method With Monotone Operators

28. Monotone and Maximal Monotone Operators. Let X be a real Hilbert space. An operator $T:D(T) \to X$, with domain $\mathcal{D}(T) \subset X$ and range $\mathcal{R}(T) \subset X$, is said to be monotone if

$$(x_1 - x_2, Tx_1 - Tx_2) \geq 0 \quad \text{for all } x_1, x_2 \in \mathcal{D}(T).$$

Here $(\ ,\ )$ denotes the inner product in H. If we allow T to be set-valued, or multi-valued, i.e. $T:\mathcal{D}(T) \to 2^X$, each $Tx, x \in \mathcal{D}(T)$, being some subset of X, then such a map $T:\mathcal{D}(T) \to 2^X$ is said to be monotone provided

$$(x_1 - x_2, u_1 - u_2) \geq 0 \quad \text{for all } x_1, x_2 \in \mathcal{D}(T),$$
$$u_1 \in Tx_1, \quad u_2 \in Tx_2.$$

It is clear that this is actually a property of the graph $\Gamma = [(x,u), x \in \mathcal{D}(T), u \in Tx] \subset X \times X$, as a subset of $X \times X$.

We shall say that any subset Γ of $X \times X$ is a monotone set provided $(x_1 - x_2, u_1 - u_2) \geq 0$ for $(x_1,u_1),(x_2,u_2) \in \Gamma$. At this level of generality we see that any (set-valued) map $T:\mathcal{D}(T) \to 2^X$ has an inverse $T^{-1}:\mathcal{R}(T) \to 2^X$ (also a set-valued map), and that T is monotone if and only if T^{-1} is monotone, and if and only if their graph Γ is a monotone set.

A monotone set Γ is said to be maximal monotone if Γ is not a proper subset of another monotone set $\Gamma' \supset \Gamma$ in $X \times X$. A (set-valued) map $T:\mathcal{D}(T) \to 2^X$ is said to be maximal monotone if its graph is a maximal monotone set.

In other words, if we partially order the subsets of $X \times X$ by set inclusion, then a set Γ is said to be maximal monotone if Γ is monotone, and in any ordered collection of monotone subsets of $X \times X'$ containing Γ, Γ is the maximal element. We see, by Zorn's lemma, that any monotone subset Γ of $X \times X$ is certainly contained in some maximal monotone set Γ' ($\Gamma' = \Gamma$ if Γ is maximal monotone).

The last remarks can be reworded by saying that for any map $T:\mathcal{D}(T) \to 2^X$ which is monotone, but not maximal monotone, there is another map $T':\mathcal{D}(T') \to 2^X$ which is maximal monotone, with $\mathcal{D}(T) \subset \mathcal{D}(T')$, and $Tx \subset T'x$ for all $x \in \mathcal{D}(T)$.

Thus, a map $T:\mathcal{D}(T) \to 2^X$ is maximal monotone if and only if its inverse T^{-1} is maximal monotone.

A map $T:\mathcal{D}(T) \to 2^X$ is said to be strictly monotone if

$$(x_1 - x_2, u_1 - u_2) > 0 \quad \text{for all } x_1, x_2 \in \mathcal{D}(T),$$
$$u_1 \in Tx_1, \quad u_2 \in Tx_2, \quad x_1 \neq x_2.$$

A map $T:\mathcal{D}(T) \to 2^X$ is said to be strongly monotone if there is a constant $c > 0$ such that

$(x_1 - x_2, u_1 - u_2) \geq c\|x_1 - x_2\|^2$ for all $x_1, x_2 \in \mathcal{D}(T)$, $u_1 \in Tx_1$, $u_2 \in Tx_2$.

A map $T: \mathcal{D}(T) \to 2^X$ is said to be coercive if $\|x\|^{-1}(x,u) \to +\infty$ as $\|x\| \to \infty$, $x \in \mathcal{D}(T)$, $u \in Tx$; briefly, if $\|x\|^{-1}(Tx,x) \to \infty$ as $\|x\| \to \infty$. The definition of course is of interest only if $\mathcal{D}(T)$ is unbounded. It is easy to prove that a strongly monotone map is coercive. We also mention here that any (nonlinear) operator T, or N, is said to be bounded if it maps bounded sets into bounded sets.

For the theory of maximal monotone operators in a Hilbert space X we refer to Brézis [133].

We only mention here a few statements which we shall need below.

(28.i) (Minty [172]). A monotone map $T: \mathcal{D}(T) \to 2^X$, $\mathcal{D}(T) \subset X$, is maximal monotone if and only if $\mathcal{R}(I + T) = X$.

(28.ii) (Minty [172,173]). If $T: \mathcal{D}(T) \to X$ is maximal monotone and coercive, then $\mathcal{R}(T) = X$.

The sum $A + B$ of monotone maps with $\mathcal{D}(A) \cap \mathcal{D}(B) \neq \emptyset$ is monotone, as a map defined on $\mathcal{D}(A) \cap \mathcal{D}(B)$. If both A and B are maximal monotone, then $A + B$ is also maximal monotone under a number of alternate mild hypotheses. The following statement will be used below.

(28.iii) (cf. Brézis, Crandall, and Pazy [134], and Brézis [133]). If both $A: \mathcal{D}(A) \to 2^X$ and $B: \mathcal{D}(B) \to 2^X$, $\mathcal{D}(A), \mathcal{D}(B) \subset X$, are maximal monotone, and if $\operatorname{Int} \mathcal{D}(A) \cap \mathcal{D}(B) \neq \emptyset$, then $A + B$ is maximal monotone.

A map $T:\mathcal{D}(T) \to X$, $\mathcal{D}(T) \subset X$, is said to be hemicontinuous if T is continuous on every line segment s of $\mathcal{D}(T)$ (with respect to the strong topology on s and the weak topology on the range). The following simple theorem is of interest.

(28.iv) (cf. Brézis [133]). A monotone hemicontinuous map $T:X \to X$ is maximal monotone.

Also, we shall say with Kato [161] that an operator $T:\mathcal{D}(T) \to X$, $\mathcal{D}(T) \subset X$, is demicontinuous if $x, x_k \in \mathcal{D}(T)$, $x_k \to x$ strongly in X, implies $Tx_k \to Tx$ weakly in X.

(28.v) (Kato [161]). If $T:\mathcal{D}(T) \to X$, $\mathcal{D}(T) \subset X$, is monotone, and $\mathcal{D}(T)$ is convex, then hemicontinuity and demicontinuity are equivalent.

29. <u>The Auxiliary Equation and Monotone Maps</u>. Throughout this chapter we shall again consider an equation $Ex = Nx$ in a real Hilbert space X, and we shall make the same basic assumptions as in no. 3. Let X be a real separable Hilbert space, with inner product (x,y) and norm $\|x\| = (x,x)^{1/2}$. Let E be a linear operator in X with real eigenvalues λ_i and eigenelements φ_i, $i = 1,2,\ldots$, $\lambda_i \le \lambda_{i+1}$, $E\varphi_i + \lambda_i\varphi_i = 0$, with $\lambda_i \to +\infty$ as $i \to \infty$, and $\{\varphi_i, i = 1,2,\ldots\}$ a complete orthonormal system in X.

In applications, E may be a selfadjoint linear differential operator on some domain $G \subset E^\nu$ with given homogeneous linear boundary conditions on the boundary ∂G of G, and then $X = L_2(G)$.

Thus, every element $x \in X$ has a Fourier series $x = \sum_1^\infty c_i\varphi_i$, $c_i = (x,\varphi_i)$, which is convergent in X, and if $y = \sum_1^\infty d_i\varphi_i$ is any other element, then

$$(x,y) = \sum_1^\infty c_i d_i, \quad \|x\| = (x,x)^{1/2} = (\sum_1^\infty c_i^2)^{1/2}.$$

Let m be chosen so that $\lambda_{m+1} > 0$, and let P and H be defined as in no. 3 so that $X_0 = PX = \mathrm{span}(\varphi_1,\ldots,\varphi_n)$, $X_1 = (I - P)X$. Let $N:X \to X$ be a given operator, not necessarily linear. As in no. 3, we split the equation $Ex = Nx$ into the system of auxiliary and bifurcation equations

$$x = x^* + H(I - P)Nx, \quad x^* = Px, \tag{29.1}$$

$$P(E - N)x = 0. \tag{29.2}$$

We shall write the auxiliary equation as either

$$x - H(I - P)Nx = x^*, \quad \text{or} \tag{29.1$'$}$$

$$[I - H(I - P)N]x = x^*, \tag{29.1$''$}$$

that is, in the form $x + KNx = x^*$ of a Hammerstein equation. As mentioned, a number of results from Hammerstein equation theory are pertinent. We present here results obtained by the alternative method and by maximal monotone operator theory.

(29.i) Theorem (Cesari and Kannan [34]). Under the above hypotheses, if N is monotone and hemicontinuous, then the auxiliary equation (29.1) has always one and only one solution $x = x(x^*)$, $x \in x^* + X_1$, for every $x^* \in X_0$.

Proof. We consider here the operators $N:X \to X$ and $-H(I - P):X \to X_1$. Actually, $-H$ is a one-to-one map from X_1 onto X_1. If, for any $z \in X_1$ we write $z = -Hy$, $y \in X_1$, then $-H(I - P)(y + X_0) = z$, and $[-H(I - P)]^{-1}z = y + X_0$.

Thus, $[-H(I - P)]^{-1}$ is a set-valued map defined for each $z \in X_1$ with values $y + X_0$. Also, we see that (29.1)′ certainly implies

$$[-H(I - P)]^{-1}(x - x^*) + Nx \ni 0. \tag{29.3}$$

Conversely, if (29.3) holds, then $x - x^* - H(I - P)Nx$ is in X_0 and since $x - x^* \in X_1$ and $X_0 \cap X_1 = \{0\}$, we derive $x - x^* - H(I - P)Nx = 0$. Thus equations (29.1) and (29.3) are equivalent.

Equation (29.3) is of the form $(A + B)x \ni 0$, where $A:x^* + X_1 \to X$ and $B:X \to X$ are the operators defined by $Ax = [-H(I - P)]^{-1}(x - x^*)$ for $x \in x^* + X_1$, and $Bx = Nx$ for $x \in X$. Thus A is a set-valued map. Since $-H(I - P):$ $X \to X_1 \subset X$ is monotone and continuous, $-H(I - P)$ is maximal monotone, and so is its inverse $A = [-H(I - P)]^{-1}:X_1 \to X$. By hypothesis, N is monotone and hemicontinuous. Thus, $B = N$ is also maximal monotone. Finally, $\mathfrak{D}(A) = X_1 \subset X = \mathfrak{D}(B)$, where X is the whole space. By (28.iii), A + B is maximal monotone. Let us prove that A is coercive. Indeed, for $z \in Ay = [-H(I - P)]^{-1}y$ we have $y = -H(I - P)z$, $\|y\| = \|-H(I - P)z\|$ and, by (3.3), also $\|y\|^{-1}(z,y) = \|y\|^{-1}(z, -H(I - P)z) \geq \|y\|^{-1}\lambda_{m+1}\|-H(I - P)z\|^2 = \lambda_{m+1}\|y\| \to \infty$ as $\|y\| \to \infty$. Since A is coercive, so is $A + B$. We conclude by (28.i) that $\mathcal{R}(A + B) = X$. Hence, equation $(A + B)x \ni 0$ is solvable, that is, (29.3) is solvable and (29.1) is solvable.

Let us prove that, for every $x^* \in X_0$, (29.1) has a unique solution. Indeed, if x_1, x_2 were two solutions, then

$$x_1 - x^* = H(I - P)(Nx_1), \quad x_2 - x^* = H(I - P)(Nx_2), \quad \text{and}$$

$$0 \leq (Nx_1 - Nx_2, x_1 - x_2) = -(Nx_1 - Nx_2, -H(I - P)Nx_1 + H(I - P)Nx_2)$$
$$\leq -\lambda_{m+1}\|-H(I - P)Nx_1 + H(I - P)Nx_2\|^2 = -\lambda_{m+1}\|x_1 - x_2\|^2,$$

where $\lambda_{m+1} > 0$, and we have used property (3.3) of the operator H. Thus, $x_1 = x_2$. Statement (29.i) is thereby proved. Thus the auxiliary equation (29.1) has a unique solution $x \in x^* + X_1$ for every $x^* \in X_0$. We shall indicate it by

$$x = [I - H(I - P)N]^{-1}x^*. \tag{29.4}$$

30. The Bifurcation Equation and Monotone Maps. The analysis of the bifurcation equation will now allow us to guarantee the existence of solutions to the given problem $Ex = Nx$.

(30.i) Theorem (Cesari and Kannan [34]). Under the hypotheses of (29.i) the following holds:

(a) If $\lambda_i > 0$ for all i and $N:X \to X$ is monotone and hemicontinuous, then the given equation $Ex = Nx$ has a unique solution.

(b) If $\lambda_i \geq 0$ for all i, and $N:X \to X$ is monotone, hemicontinuous and coercive, then the given equation $Ex = Nx$ has at least one solution. The coerciveness hypothesis can be replaced here by the following alternate assumption: there is $R > 0$ such that for any $x \in X$, with $Px = x^*$, $\|x^*\| \geq R$ we have $(Nx, x^*) \geq 0$.

(c) If some λ_i are negative, say $\lambda_1 \leq \cdots \leq \lambda_m \leq 0 < \lambda_{m+1} \leq \cdots$, if $N:X \to X$ is hemicontinuous and bounded, and if $(Nx - Ny, x - y) \geq c\|x - y\|^2$ for all $x, y \in X$ and some constants $c > 0$, $c > -\lambda_1$, then the given equation $Ex = Nx$ again has at least one solution.

Case (b) covers resonance situations. In cases (a) and (c) there is no resonance.

Proof. (a) If $\lambda_i > 0$ for all i, that is, $0 < \lambda_1 \leq \lambda_2 \leq \cdots$, $\lambda_m \to +\infty$ as $m \to \infty$, then we can take, in no. 3, $m = 0$, $X_0 = \{0\}$, $X_1 = X$, and the given equation $Ex = Nx$ is reduced to only the auxiliary equation. Then our statement is a corollary of (29.i).

(b) If some λ_i are zero, $0 = \lambda_1 \leq \lambda_2 \leq \cdots$, let m be the smallest integer such that $\lambda_{m+1} > 0$. We reduced $Ex = Nx$ to the system of the auxiliary and bifurcation equations (29.1) (29.2). For any $x^* \in X_0$ the auxiliary equation has a unique solution given by (29.4). Then the bifurcation equation $P(E - N)x = 0$ becomes

$$PN[I - H(I - P)N]^{-1}x^* - PEx = 0.$$

Since $PEx = EPx = Ex^* = PEx^*$, we have the equation

$$PN[I - H(I - P)N]^{-1}x^* - PEx^* = 0,$$

or $Tx^* = PEx^*$, where $T:X_0 \to X_0$ is the operator defined by $T = PN[I - H(I - P)N]^{-1}$. Note that in this part (b), X_0 is spanned by $\varphi_1,\ldots,\varphi_m$ with $E\varphi_i = -\lambda_i\varphi_i = 0$, $\lambda_i = 0$, $i = 1,\ldots,m$; and thus $Ex^* = 0$. Since we shall need the same argument in part (c), where Ex^* may not be zero, we do not make use here of this simplification.

Let us prove that T is monotone. Indeed, if $x^*,y^* \in X_0$ and we take $u = [I - H(I - P)N]^{-1}x^*$, $v = [I - H(I - P)N]^{-1}y^*$, then $u - H(I - P)Nu = x^*$,

$v - H(I - P)Nv = y^*$, and

$$(Tx^* - Ty^*, x^* - y^*) = (PNu - PNv, x^* - y^*)$$
$$= (Nu - Nv, x^* - y^*) = (Nu - Nv, u - v)$$
$$+ (Nu - Nv, -H(I - P)Nu + H(I - P)Nv)$$
$$\geq 0, \qquad (30.1)$$

since both N and $-H(I - P)$ are monotone operators. Thus, T is monotone.

Let us prove that T is maximal monotone. It is enough to prove that $\mathcal{R}(I + T) = X_0$. In other words, we have to prove that, for every $a^* \in X_0$ the equation

$$x + PN[I - H(I - P)N]^{-1}x = a^* \qquad (30.2)$$

is solvable in X_0. Actually, we shall first show that the equation

$$x - H(I - P)Nx = a^* - PNx \qquad (30.3)$$

is solvable. Indeed, this equation is equivalent to

$$[-H(I - P) + P]^{-1}(x - a^*) + Nx \ni 0, \qquad (30.4)$$

and the same argument as for equation (29.3) shows that (30.4) and hence (30.3) is solvable. From (30.4) now, for $b^* = a^* - PNx$, we obtain $x - H(I - P)Nx = b^*$, and also

$$x = [I - H(I - P)N]^{-1}b^*,$$
$$PN[I - H(I - P)N]^{-1}b^* = PHx = a^* - b^*,$$
$$b^* + PN[I - H(I - P)N]^{-1}b^* = a^*.$$

Thus, equation (30.2) is solvable for every $a^* \in X_0$; hence $\mathcal{R}(I + T) = X_0$, and T is maximal monotone.

We have written the bifurcation equation in the form $(T - E)x^* = 0$, $x^* \in X_0$, where $T:X_0 \to X_0$ is maximal monotone and so is $-E:X_0 \to X_0$. Thus, $T - E:X_0 \to X_0$ is maximal monotone, and it remains to prove that $T - E$ is coercive. For every $x^* \in X_0$ we have now

$$(Tx^* - Ex^*, x^*) = (Tx^*, x^*) + (-Ex^*, x^*) \geq (Tx^*, x^*)$$
$$= (PN[I - H(I - P)N]^{-1}x^*, x^*) = (PNx, x^*),$$

where $x = [I - H(I - P)N]^{-1}x^*$, or $x^* = x - H(I - P)Nx$, so that, since $x^* \in X_0$, we have $x^* = Px^* = Px$. Thus

$$(Tx^* - Ex^*, x^*) \geq (PNx, x^*) = (Nx, x^*)$$
$$\geq (Nx, x) + (Nx, -H(I - P)Nx) \geq (Nx, x).$$

Also, $\|x^*\| = \|Px\| \leq \|x\|$, so that

$$\|x^*\|^{-1}(Tx^* - Ex^*, x^*) \geq \|x\|^{-1}(Nx, x).$$

Since $\|x^*\| \to \infty$ implies that $\|x\| \to \infty$, and since N is coercive, we conclude that $\|x^*\|^{-1}(Tx^* - Ex^*, x^*) \to \infty$ as $\|x^*\| \to \infty$. We have proved that $T - E$ is maximal monotone and coercive, thus $\mathcal{R}(T - E) = X_0$, and the equation $(T - E)x^* = 0$ is solvable.

If N is not coercive, but the alternate hypothesis mentioned under (b) holds, then we note that $Ex^* = 0$, and thus the bifurcation equation reduces to $Tx^* = 0$, an equation in X_0, a finite dimensional space. Thus, T and $T - I$ are certainly compact. Now the argument above shows that

$$(Tx^*, x^*) \geq (Nx, x^*) \geq (Nx, x),$$

where $x \in X$, $x^* \in X_0$, $Px = x^*$. Then, $(Tx^*, x^*) \geq 0$ for all $x^* \in S_0$ with $\|x^*\| = R$. By Lemma (24.ii), equation $Tx^* = 0$ has at least one solution $x^* \in X_0$, $\|x^*\| \leq R$.

(c) Let us assume now that some λ_i are negative, and as usual let m be the smallest integer such that $\lambda_{m+1} > 0$. The proof in (b) above remains essentially the same for what concerns T. Still we have to prove that $T - E$ is monotone, maximal monotone, and coercive. From (30.1) we have now

$$(Tx^* - Ty^*, x^* - y^*) \geq (Nu - Nv, u - v) \geq c\|u - v\|^2,$$

with $u - H(I - P)Nu = x^*$, $v - H(I - P)Nv = y^*$, hence $(u - v) - H(I - P)(Nu - Nv) = x^* - y^*$, where $x^* - y^* \in X_0$ and $-H(I - P)(Nu - Nv) \in X_1$. Thus, $\|u - v\| \geq \|x^* - y^*\|$, and

$$(Tx^* - Ty^*, x^* - y^*) \geq c\|x^* - y^*\|^2.$$

On the other hand, for $x^* = \sum_1^m c_i\varphi_i$ we have $Ex^* = -\sum_1^m c_i\lambda_i\varphi_i$, thus $(-Ex^*, x^*) = \sum_1^m c_i^2\lambda_i \geq \lambda_1\|x^*\|^2$, and finally

$$(-Ex^* + Ey^*, x^* - y^*) \geq \lambda_1\|x^* - y^*\|^2,$$

$$((T - E)x^* - (T - E)y^*, x^* - y^*) \geq (c + \lambda_1)\|x^* - y^*\|^2,$$

where $c + \lambda_1 > 0$. We have proved that $T - E$ is monotone and coercive.

Note that, if we prove that T is continuous, then since E is certainly a continuous operator restricted to X_0 we conclude that $T - E: X_0 \to X_0$ is a continuous map defined on the whole of X_0. Thus, $T - E$ is maximal monotone and coercive as proved above. Hence, $\mathcal{R}(T - E) = X_0$ and equation $(T - E)x^* = 0$ is solvable in X_0.

It remains to prove that T is continuous.

If K denotes the map $K = -H(I - P)$, then $T = PN(I + KN)^{-1}$. Let us prove that $(I + KN)^{-1}$ is bounded. To this effect, let u, u^* be such that $u + KNu = u^*$. Then, $u - u^* = -KNu$, and from (3.3) also

$$\begin{aligned}\lambda_{m+1}\|u - u^*\|^2 &= \lambda_{m+1}\|KNu\|^2 \leq (KNu, Nu) = (u^* - u, Nu)\\ &= (u^* - u, Nu - Nu^* + Nu^*)\\ &= -(u - u^*, Nu - Nu^*) + (u^* - u, Nu^*)\\ &\leq 0 + \|u^* - u\|\|Nu^*\|,\end{aligned}$$

and hence $\|u - u^*\| \le \lambda_{m+1}^{-1}\|Nu^*\|$. If $\|u^*\| \le R$, then $\|Nu^*\| \le R'$ for some R', depending solely on R since N is bounded, and $\|u\| \le \|u^*\| + \lambda_{m+1}^{-1}\|Nu^*\| \le R + \lambda_{m+1}^{-1}R'$. Thus, $(I + KN)^{-1}$ is bounded.

Note here that for any two $y, z \in X$ we certainly have $\|y + z\|^2 + \|z\|^2 \ge (1/4)\|y\|^2$. Indeed, either $\|z\| \ge (1/2)\|y\|$ and then $\|y + z\|^2 + \|z\|^2 \ge \|z\|^2 \ge (1/4)\|y\|^2$; or $\|z\| \le (1/2)\|y\|$, and then $\|y + z\| \ge \|y\| - \|z\| \ge (1/2)\|y\|$, and again $\|y + z\|^2 + \|z\|^2 \ge \|y + z\|^2 \ge (1/4)\|y\|^2$.

Let us prove now that $(I + KN)^{-1}$ is a continuous map. Indeed, if

$$u_1 + KNu_1 = u_1^*, \quad u_2 + KNu_2 = u_2^*,$$

then

$$u_1 - u_2 + KNu_1 - KNu_2 = u_1^* - u_2^*.$$

Hence, from this relation, the monotonicity of N, and (3.3) we derive

$$\begin{aligned} 0 + \lambda_{m+1}\|KNu_1 - KNu_2\|^2 &\le (Nu_1 - Nu_2, u_1 - u_2) \\ &\quad + (Nu_1 - Nu_2, KNu_1 - KNu_2) \\ &= (Nu_1 - Nu_2, u_1 - u_2 + KNu_1 - KNu_2) \\ &= (Nu_1 - Nu_2, u_1^* - u_2^*) \le \|Nu_1 - Nu_2\|\|u_1^* - u_2^*\|. \end{aligned}$$

On the other hand, from the simple relation $\|y + z\|^2 + \|z\|^2 \ge (1/4)\|y\|^2$, with $-y = u_1 - u_2$, $z = u_1^* - u_2^*$, we have

$$\begin{aligned} (1/4)\|u_1 - u_2\|^2 - \|u_1^* - u_2^*\| &\le \|u_1^* - u_2^* - u_1 + u_2\|^2 \\ &= \|KNu_1 - KNu_2\|^2. \end{aligned}$$

Thus, by multiplication by λ_{m+1}, we also have

$$(\lambda_{m+1}/4)\|u_1 - u_2\|^2 \le \lambda_{m+1}\|u_1^* - u_2^*\| + \|Nu_1 - Nu_2\|\|u_1^* - u_2^*\|.$$

If $\|u_1^*\|, \|u_2^*\| \le R$, then since $(I + KN)^{-1}$ is bounded, we have $\|u_1\|, \|u_2\| \le R'$ for some constant R', and

$\|Nu_1\|, \|Nu_2\| \leq R''$ for some constant R'' since N is bounded. Finally,

$$\|u_1 - u_2\| \leq (4 + 4\lambda_{m+1}^{-1}R'')\|u_1^* - u_2^*\|,$$

and this proves that $(I + KN)^{-1}$ is a continuous map.

Since $N:X \to X$ is defined in the whole of X and is hemicontinuous, by Kato's theorem (28.v), N is demicontinuous.

Now let us prove that the operator $T = PN(I + KN)^{-1}$: $X_0 \to X_0$ is continuous. Let $[x_k]$ be any strongly convergent sequence in X_0, say, $x_k \to x$ strongly in X_0. Then $y_k \to y$ strongly in X_0 for $y_k = (I + KN)^{-1}x_k$, $y = (I + KN)^{-1}x$, since $(I + KN)^{-1}$ is continuous. Then $Ny_k \to Ny$ weakly in X since N is demicontinuous. Finally, $PNy_k \to PNy$ weakly in X since P is continuous. But PNy_k, PNy are elements of the finite dimensional space X_0, and weak and strong convergence coincide in X_0. Thus, $x_k \to x$ strongly in X_0 implies $Tx_k \to Tx$ strongly in X_0. We have proved that T is a continuous map $T:X_0 \to X_0$. This concludes our proof of (30.i).

31. Two Statements for the Nonresonance Case. For $\lambda \neq \lambda_i$, $\mu = \min|\lambda - \lambda_i|$, the operator $E + \lambda I$ is invertible. Indeed, if $x \in X$, $x = \sum d_i\varphi_i$, $\sum d_i^2 < \infty$, then $Hx = (E + \lambda I)^{-1}x = \sum d_i(\lambda - \lambda_i)^{-1}\varphi_i$, and $\|H\| \leq \mu^{-1}$.

(31.i) If $c = 2^{-1}(\lambda_n + \lambda_{n+1})$ and $N:X \to X$ is an operator with the property that, for some γ, $0 < \gamma < 2^{-1}(\lambda_{n+1} - \lambda_n)$ and all $x_1, x_2 \in X$ we have

$$\|Nx_1 - Nx_2 + c(x_1 - x_2)\| \leq \gamma\|x_1 - x_2\|,$$

then the equation $Ex = Nx$ has a unique solution in X.

Indeed, $Ex + cx = Nx + cx$ has a unique solution $x = Ax$, $A = (E + cI)^{-1}(N + cI)$ a uniform contraction map in X.

Here is a theorem in the nonresonance case for nonlinear ordinary differential equations in $a \leq t \leq b$,

$$Ex - h(t,x,x',\ldots,x^{(n-1)})x = f(t,x,x',\ldots,x^{(n-1)}), \quad (31.1)$$

with n linearly independent homogeneous boundary conditions

$$B_i x = \sum_0^{n-1} b_{ij} x^{(j)}(a) + \sum_0^{n-1} c_{ij} x^{(j)}(b) = 0,$$

$$i = 1,\ldots,n, \quad (31.2)$$

and E the linear differential operator $Ex = \sum_1^n a_j(t)x^{(i)}$. We assume E selfadjoint with eigenelements λ_i, φ_i, $a_j \in C^\infty[a,b]$, $a_n(t) \neq 0$ and $h, f \in C([a,b] \times R^n)$.

(31.ii) If there are constants p, q, M such that $p \leq h(t,z) \leq q$, $|f(t,z)| \leq M$ and $-\lambda_i \notin [p,q]$, then problem (31.1-2) has at least a solution $x(t)$, $a \leq t \leq b$, $x \in H^n$ [84].

Analogous theorems for elliptic boundary value problems have been proved by Williams, De Figueiredo, et. al. (see [84]).

32. Schauder's Principle of Invariance of Domain.

Schauder's principle is generally stated as follows:

(32.i) If $T:X \to X$ is a weakly compact map, if $I + T$ is one-to-one, and if $(I + T)^{-1}$ is bounded, then $\mathcal{R}(I + T) = X$ and $(I + T)^{-1}$ is single valued.

Here X denotes a real Hilbert space, and by T compact we mean, as usual, that every bounded sequence $[u_k]$ is mapped into a relatively compact sequence $[Tu_k]$. By $(I + T)^{-1}$ bounded we mean that $(I + T)^{-1}$ maps bounded sets into bounded sets. The conclusion that $\mathcal{R}(I + T) = X$ and that $(I + T)^{-1}$ is single valued can be reworded by saying that for every $v \in X$ there is one and only one element $u \in X$ such that $(I + T)u = v$. The important statement (32.i) is essentially proved in the form above by Nagumo [177,178], but can be traced back to Schauder. The proof of (32.i) is based on the concept and properties of topological degree (see Nagumo [177,178]).

33. The Auxiliary Equation and Schauder's Principle.

(33.i) Theorem (Kannan [79]). Under the hypotheses of no. 3, if $N:X \to X$ is continuous and bounded, and if there are numbers p,q,m, $p > 0$, m integer, such that $\lambda_m < q \leq p < \lambda_{m+1}$, and

$$(Nx - Ny, x - y) \geq -p\|x - y\|^2 \quad \text{for all} \quad x,y \in X, \tag{33.1}$$

then the auxiliary equation is uniquely solvable. If in addition

$$(Nx - Ny, x^* - y^*) \leq -q\|x^* - y^*\|^2 \tag{33.2}$$

for all $x^*, y^* \in X_0$ and corresponding solutions x, y of the auxiliary equation, then the given equation $Ex = Nx$ also has a unique solution.

Proof. (a) Let $T = -H(I - P)N:X \to X_1$. Any bounded sequence is mapped by N into a bounded sequence and then by $-H(I - P)$ into one possessing a convergent subsequence since H is compact. Thus, T is compact.

Let us prove that $I + T$ is one-to-one. Indeed, let us assume, if possible, that we have $(I + T)x = (I + T)y$, or

$$x - H(I - P)Nx = y - H(I - P)Ny, \quad x,y \in X.$$

Then, by (33.1) and (3.3), we have

$$-p\|x - y\|^2 \le (Nx - Ny, x - y)$$
$$= -(Nx - Ny, -H(I - P)Nx + H(I - P)Ny)$$
$$\le -\lambda_{m+1}\|-H(I - P)Nx + H(I - P)Ny\|^2$$
$$= -\lambda_{m+1}\|x - y\|^2,$$

a contradiction, since $-p > -\lambda_{m+1}$. Thus, $I + T$ is one-to-one.

Let us prove that $(I + T)^{-1}$ is bounded. For $(I + T)^{-1}y = x$, we have $y = (I + T)x$, or $y = x - H(I - P)Nx$. Then, by (3.3) we have

$$\lambda_{m+1}\|y - x\|^2 = \lambda_{m+1}\|-H(I - P)Nx\|^2 \le (Nx, -H(I - P)Nx)$$
$$= (Nx, y - x) = -(Ny - Nx, y - x) + (Ny, y - x).$$

By hypothesis we have now

$$\lambda_{m+1}\|y - x\|^2 \le p\|y - x\|^2 + (Ny, y - x),$$
$$(\lambda_{m+1} - p)\|y - x\|^2 \le \|Ny\|\|y - x\|,$$
$$\|y - x\| \le (\lambda_{m+1} - p)^{-1}\|Ny\|.$$

Thus $\|x\| \le \|y\| + (\lambda_{m+1} - p)^{-1}\|Ny\|$. Since N is bounded, $\|y\| \le R$ for some R implies $\|x\| \le R'$ for some R', that is, $(I + T)^{-1}$, or $[I - H(I - P)N]^{-1}$, is a bounded map.

By force of Schauder's principle (32.i), the equation $x + Tx = x^*$, or $x - H(I - P)x = x^*$, has a unique solution x in X for every $x^* \in X_0$, and necessarily $x \in x^* + X_1$.

(b) From the previous part (a) we know that the auxiliary equation has a unique solution $x = x(x^*)$, $x \in x^* + X_1$, for every $x^* \in X_0$. It remains to show that the bifurcation equation has a unique solution.

Let $T': X_0 \to X_0$ denote the operator defined by $T'x^* = Ex^* - PN[I - H(I - P)N]^{-1}x^*$, so that the bifurcation equation reduces to $T'x^* = 0$. Let x^*, y^* be any two elements of

X_0 and let x, y denote the corresponding solutions of the auxiliary equation. Thus

$$[I - H(I - P)N]^{-1}x^* = x, \quad [I - H(I - P)N]^{-1}y^* = y,$$

$$\begin{aligned}(T'x^* - T'y^*, x^* - y^*) &= (Ex^* - Ey^*, x^* - y^*) \\ &\quad - (PNx - PNy, x^* - y^*) \\ &= (Ex^* - Ey^*, x^* - y^*) \\ &\quad - (Nx - Ny, x^* - y^*). \end{aligned} \qquad (33.3)$$

Note that, for $x^* \in X_0$, or $x^* = \sum_1^m c_i\varphi_i$, $Ex^* = -\sum_1^m c_i\lambda_i\varphi_i$, and $(Ex^*, x^*) = -\sum_1^m c_i^2\lambda_i \geq -\lambda_m\|x^*\|^2$. Hence, from (33.2) and (33.3) we derive

$$\begin{aligned}(T'x^* - T'y^*, x^* - y^*) &\geq -\lambda_m\|x^* - y^*\|^2 + q\|x^* - y^*\|^2 \\ &= (q - \lambda_m)\|x^* - y^*\|^2,\end{aligned}$$

where $q - \lambda_m > 0$. We have proved that $T': X_0 \to X_0$ is strongly monotone and therefore coercive.

Let us prove that T' is continuous on X_0. We have only to show that $[I - H(I - P)N]^{-1}$ is continuous. With the same notations as in (33.3) we have

$$[I - H(I - P)N]^{-1}x^* = x, \quad [I - H(I - P)N]^{-1}y^* = y,$$

$$x - H(I - P)Nx = x^*, \quad y - H(I - P)Ny = y^*,$$

$$x - y = (x^* - y^*) + H(I - P)(Nx - Ny),$$

where the two terms in the last member are in X_0 and X_1, respectively. Hence,

$$\|x - y\|^2 = \|x^* - y^*\|^2 + \|-H(I - P)(Nx - Ny)\|^2. \qquad (33.4)$$

By (33.1), (33.2), (3.3), and (33.4) we have now

$$\begin{aligned}-q\|x^* - y^*\|^2 \geq (Nx - Ny, x^* - y^*) &= (Nx - Ny, x - y) \\ &\quad + (Nx - Ny, -H(I - P)(Nx - Ny))\end{aligned}$$

$$\geq -p\|x - y\|^2 + \lambda_{m+1}\|-H(I - P)(Nx - Ny)\|^2$$
$$= -p\|x^* - y^*\|^2$$
$$+ (\lambda_{m+1} - p)\|-H(I - P)(Nx - NY)\|^2.$$

Thus,

$$\|-H(I - P)(Nx - Ny)\|^2 \leq (\lambda_{m+1} - p)^{-1}(p - q)\|x^* - y^*\|^2,$$

and again by (33.4) also

$$\|x - y\|^2 \leq [1 + (\lambda_{m+1} - p)^{-1}(p - q)]\|x^* - y^*\|^2.$$

We have proved that T' is continuous. Since $T': X_0 \to X_0$ is monotone and X_0 is the whole space for T', we conclude that T' is maximal monotone, and, as we have proved above, T' is also coercive. Thus $\mathcal{R}(T') = X_0$; hence, the equation $T'x^* = 0$ is solvable in X_0.

Let us prove that the bifurcation equation has only one solution. Indeed, if $x^*, y^* \in X_0$ were two solutions, that is, $T'x^* = 0$, $T'y^* = 0$, and we denote by x, y the corresponding solutions of the auxiliary equation, then

$$[I - H(I - P)N]^{-1}x^* = x, \quad [I - H(I - P)N]^{-1}y^* = y,$$
$$Ex^* - PN[I - H(I - P)N]^{-1}x^* = 0,$$
$$Ey^* - PN[I - H(I - P)N]^{-1}y^* = 0,$$
$$Ex^* - Ey^* = PNx - PNy.$$

Hence,

$$-\lambda_m\|x^* - y^*\|^2 \leq (Ex^* - Ey^*, x^* - y^*)$$
$$= (PNx - PNy, x^* - y^*)$$
$$= (Nx - Ny, x^* - y^*) \leq -q\|x^* - y^*\|^2,$$

a contradiction, since $-q < -\lambda_m$. This proves the uniqueness, and (33.i) is proved.

Section 7. Elliptic Problems at Resonance in the Large

34. Two Abstract Alternative Statements. By use of the alternative method and Schauder's fixed point theorem, Landesman and Lazer [92] proved, in 1969, an existence theorem for selfadjoint elliptic Dirichlet nonlinear problems at resonance. A version of this theorem, for periodic solutions of certain nonlinear second order ordinary differential equations, had been proved before by Lazer and Leach [94] by essentially the same method. The theorem of Landesman and Lazer for nonlinear Dirichlet problems has since been repeatedly extended, again with the help of the alternative method by Williams [125] and Necas [111], and by other methods by Hess [159], De Figueiredo [145,146], Fucik [153], Fucik, Kucera, and Necas [154].

Here we present first two very recent abstract existence theorems at resonance in the frame of the alternative method, namely (34.i) and (34.ii). The latter was proved by Cesari and Kannan [37] and by Kannan and McKenna [85]. Both of these theorems yield the theorems of Lazer and Leach, Landesman and Lazer, and Williams when applied to the corresponding context (see no. 42).

Suitable variants of these abstract theorems yield most of the above results as well as new results, all qualitative in character (existence across a point of resonance, extensions to nonselfadjoint problems), which we present in this Section. We refer to Cesari [26,27,28] and McKenna [109] for a more extensive analysis. Also, direct extensions of the

Landesman and Lazer theorem have been given by Shaw [120], and his work is presented in a paper in this volume.

Let X be a real separable Hilbert space with inner product (x,y) and norm $\|x\| = (x,x)^{1/2}$. Let $E:\mathcal{D}(E) \to X$ be a linear operator with domain $\mathcal{D}(E) \subset X$, and a non-trivial, finite dimensional null space $X_0 = \ker E$. Thus, $1 \leq \dim \ker E < +\infty$ and we are considering below a resonance problem. Let $P:X \to X$ denote the projection operator with range X_0 and null space $X_1 = (I - P)X = \ker P$, and let us assume that X_1 is also the range of E. Then the restriction $E:\mathcal{D}(E) \cap X_1 \to X_1$ is one-one and onto, and thus the partial inverse $H:X_1 \to \mathcal{D}(E) \cap X_1$ exists. We assume here that H is a linear bounded compact operator, and that E, P, H satisfy the usual axioms (h_1) $H(I - P)E = I - P$; (h_2) $EP = PE$; (h_3) $EH(I - P) = I - P$. (In no. 3 we have seen that the compactness of H and axioms are usually verified.) Let $N:X \to X$ be a continuous operator on X, not necessarily linear, and let us consider the equation

$$Ex = Nx, \quad x \in \mathcal{D}(E) \subset X. \tag{34.1}$$

From no. 3 we know that this equation is equivalent to the system of auxiliary and bifurcation equations

$$x = Px + H(I - P)Nx, \tag{34.2}$$

$$P(Ex - Nx) = 0. \tag{34.3}$$

Let $L = \|H\|$. Also note that, by $EP = PE$ and $X_0 =$ range $P = \ker E$, we have $PEx = EPx = 0$, so that (34.2), (34.3) can be written in the form

$$x = x^* + H(I - P)Nx, \quad Px = x^* \in X_0, \tag{34.4}$$

$$PNx = 0. \tag{34.5}$$

(34.i) <u>Theorem</u>. (Existence at resonance). For E, N, P, H as stated, and under the assumptions: (B) there is a constant J_0 such that $\|Nx\| \leq J_0$ for all $x \in X$; and (N_ϵ) there are constants $K \geq LJ_0$, $R_0 \geq 0$, $\epsilon > 0$ such that $(Nx,x^*) \leq -\epsilon\|x^*\|$ (or $(Nx,x^*) \geq \epsilon\|x^*\|$) for all $x \in X$, $x^* \in X_0$ with $Px = x^*$, $\|x^*\| \geq R_0$, $\|x - x^*\| \leq K$; then equation $Ex = Nx$ has at least one solution.

As we shall see in nos. 39 and 42, the specific requirements of either the Lazer and Leach [94], the Landesman and Lazer [92], or the Williams [125] theorems all imply (B) and (N_ϵ).

<u>Proof</u>. Let us assume that we always have $(Nx,x^*) \leq -\epsilon\|x^*\|$ for all x,x^* as stated. Since $X_0 = \ker E$ is finite dimensional, we can take a base $w = (w_1,\ldots,w_m)$ for X_0, and we can even assume that $w_1,\ldots,w_m$ are orthonormal. Then, for every element $x^* \in X_0$ we have $x^* = c_1w_1 +\cdots+ c_mw_m$, or briefly $x^* = cw$ with $c = (c_1,\ldots,c_m) \in R^m$ uniquely defined by x^* in X_0, and moreover $\|x^*\| = (\sum_1^m c_i^2)^{1/2}$. Then system (34.4-5) can be written in the form $x = cw + H(I - P)Nx$, $(PNx,w) = 0$, or $(PNx,w_i) = 0$, $i = 1,\ldots,m$. Let us consider now the continuous transformation $T:(x,c) \to (\bar{x},\bar{c})$, defined by

$$T:\bar{x} = cw + H(I - P)Nx, \quad \bar{c} = c + (PN\bar{x},w). \tag{34.6}$$

For $x^* = cw$, $\bar{x}^* = \bar{c}w$, relations (34.6) take the form $\bar{x} = x^* + H(I - P)Nx$, $\bar{x}^* = x^* + PN\bar{x}$. The fixed points $(x,c) = T(x,c)$ of T exactly correspond to the solutions $x,x^* = cw$, of system (34.4-5), that is, of equation (34.1). Let $R = R_0 + J_0 + J_0^2/2\epsilon$, $J = LJ_0$, and let us restrict T to the set

$$\mathfrak{C} = [(x,c) \mid x \in X,\ c \in R^m,\ \|x\| \le R + J,\ |c| \le R].$$

Let us prove that T maps $\mathfrak{C}$ into itself. For $(x,c) \in \mathfrak{C}$, $x^* = cw = \sum_1^m c_i w_i$, $\bar{x}^* = \bar{c}w = \sum_1^m \bar{c}_i w_i$, $\|x^*\| = |c| \le R$, $c = (c_1,\ldots,c_m)$, $\bar{c} = (\bar{c}_1,\ldots,\bar{c}_m)$, $PN\bar{x} \in X_0$, $PN\bar{x} = dw = \sum_1^m d_i w_i$, $d = (d_1,\ldots,d_m)$, and we certainly have

$$\|\bar{x}\| \le \|cw\| + \|H(I - P)Nx\| \le |c| + LJ_0 \le R + J.$$

Also, we have $P\bar{x} = cw = x^*$, and $\|\bar{x} - x^*\| = \|\bar{x} - P\bar{x}\| \le LJ_0 = J$. Now, for $|c| \le R_0 + J_0^2/2\varepsilon$ we have

$$|\bar{c}| \le |c| + |\bar{c} - c| \le |c| + \|N\bar{x}\| \le (R_0 + J_0^2/2\varepsilon) + J_0 \le R.$$

Now, let $R_0 + J_0^2/2\varepsilon \le |c| \le R$. As mentioned we have $P\bar{x} = x^*$, $\|\bar{x} - P\bar{x}\| = \|\bar{x} - x^*\| \le J$, $\|x^*\| > R_0$, and by (N_ε) also $(N\bar{x},x^*) \le -\varepsilon\|x^*\|$. We have now

$$\bar{c} - c = (PN\bar{x},w) = [(\sum_1^m d_j w_j, w_i),\ i = 1,\ldots,m]$$
$$= (d_i,\ i = 1,\ldots,m), \tag{34.7}$$
$$-\varepsilon|c| = -\varepsilon\|x^*\| \ge (N\bar{x},x^*) = (PN\bar{x},x^*)$$
$$= (\sum_1^m d_i w_i, \sum_1^m c_i w_i) = \sum_1^m c_i d_i = (\bar{c} - c)c. \tag{34.8}$$

On the other hand, we also have

$$J_0^2 \ge |\bar{c} - c|^2 = (\bar{c} - c, \bar{c} - c) = |\bar{c}|^2 - 2(\bar{c} - c)c - |c|^2$$
$$\ge |\bar{c}|^2 + 2\varepsilon|c| - |c|^2.$$

Since $|c| \ge R_0 + J_0^2/2\varepsilon$ we have $J_0^2 - 2\varepsilon|c| \le -2\varepsilon R_0 \le 0$; hence $|\bar{c}|^2 - |c|^2 \le 0$, or $|\bar{c}|^2 \le |c|^2$, $|\bar{c}| \le |c| \le R$. We have proved $T:\mathfrak{C} \to \mathfrak{C}$ maps $\mathfrak{C}$ into itself.

Let us prove now that T is compact. For this we consider any (bounded) sequence (x_k,c_k), $k = 1,2,\ldots,$ of points of $\mathfrak{C}$. Then the sequence Nx_k is bounded, actually $\|Nx_k\| \le J_0$, $\|y_k\| = \|H(I - P)Nx_k\| \le LJ_0 = J$, and since H is compact by hypothesis, there is a subsequence, say still $[k]$, so that y_k is convergent in X. Also, $x_k^* = c_k w$, $z_k = PN\bar{x}_k$ are bounded sequences, that is, c_k, d_k are bounded sequences in R^m, a finite dimensional space, and we can take the subsequence, say still [k], such that c_k, d_k converge in R^m. Thus, $\bar{x}_k = x_k^* + y_k$, $\bar{c}_k = c_k + d_k$ are also convergent in X and R^m respectively. Thus, T is compact.

By Schauder's fixed point theorem, there is at least one fixed point (x,c) for T in $\mathfrak{C}$, or $(x,c) = T(x,c)$. This proves that the equation $Ex = Nx$ has at least one solution $x \in S$. (Cf. [92], [94], [125] for a trace of this proof).

In case the alternate inequality holds, or $(Nx,x^*) \geq \varepsilon\|x^*\|$, then in the definition of T we take $\bar{c} = c - (PN\bar{x},w)$. Then (34.7) and (34.8) are replaced by

$$\bar{c} - c = -(PN\bar{x},w) = (-d_i,\quad i = 1,\dots,m),$$
$$\varepsilon|c| = \varepsilon\|x^*\| \leq (N\bar{x},x^*) = \sum_1^m c_i d_i = -(\bar{c} - c)c.$$

Hence again $(\bar{c} - c)c \leq -\varepsilon|c|$, and the same argument holds. Theorem (34.i) is thereby proved.

Remark. Note that Theorem (34.i) proves that equation $Ex = Nx$ has at least one solution $x \in X$ with $\|x\| \leq R + J$, or

$$\|x\| \leq R_0 + J_0 + (2\varepsilon)^{-1} J_0^2 + LJ_0,$$

where $L = \|H\|$ and $\varepsilon > 0$ is a constant for which (N_ε) holds. We shall see below analogous estimates in a larger context. An extension of (34.i) to nonselfadjoint problems is stated in no. 37.

(34.ii) Theorem. (Existence at resonance) (Cesari and Kannan [37]). For E, N, P, H as stated, and under the assumptions: (B) there is a constant J_0 such that $\|Nx\| \leq J_0$ for all $x \in X$; and (N_0) there are constants $K \geq LJ_0$, $R_0 \geq 0$ such that $(Nx,x^*) \leq 0$ (or $(Nx,x^*) \geq 0$) for all $x \in X$, $x^* \in X_0$, with $Px = x^*$, $\|x^*\| \geq R_0$, $\|x - x^*\| \leq K$; then equation $Ex = Nx$ has at least one solution.

Proof. We proceed as in the proof of (34.i), but now we define $T:(x,c) \to (\bar{x},\bar{c})$ as follows. Let R_1, R_2, R, ε be real numbers so that

$$R_0 \leq R_1 < R_2 < R, \quad R_2 + J_0 \leq R, \quad 0 < \varepsilon < R_1/2.$$

Let $J = LJ_0$ and

$$T:\bar{x} = cw + H(I - P)Nx, \quad \bar{c} = c + F(\bar{x},c),$$

$$(x,c) \in \mathfrak{C} = [(x,c) | x \in X, \ c \in R^m, \ \|x\| \leq R + J, \ |c| \leq R],$$

where $x^* = cw = \sum_1^m c_i w_i$, $\bar{x}^* = \bar{c}w = \bar{c}_1 w_1 + \cdots + \bar{c}_m w_m$, and $F(\bar{x},c) = (F_1,\ldots,F_m)$ is explicitely given below. Thus, the second relation defining T can also be written in the form

$$\bar{x}^* = x^* + F(\bar{x},c)w = x^* + \sum_1^m F_i w_i.$$

Here for $0 \leq |c| \leq R_1$ we take $F(\bar{x},c) = (PN\bar{x},w)$, or

$$\bar{x}^* = x^* + PN\bar{x}.$$

For $R_2 \leq |c| \leq R$, hence $R_2 \leq \|x^*\| = c \leq R$, we take

$$F(\bar{x},c) = [(PN\bar{x},x^*) - \varepsilon\|PN\bar{x}\|](2J_0\|x^*\|)^{-1}(x^*,w), \quad \text{or}$$

$$\bar{x}^* = x^* + [(PN\bar{x},x^*) - \varepsilon\|PN\bar{x}\|](2J_0\|x^*\|)^{-1}x^*.$$

For $R_1 \leq |c| \leq R_2$ we take

$$F(\bar{x},c) = \lambda(PN\bar{x},w) + (1 - \lambda)[(PN\bar{x},x^*) - \varepsilon\|PN\bar{x}\|](2J_0\|x^*\|)^{-1}(x^*,w), \quad \text{or}$$

$$\bar{x}^* = x^* + \lambda PN\bar{x} + (1 - \lambda)[(PN\bar{x},x^*) - \varepsilon\|PN\bar{x}\|](2J_0\|x^*\|)^{-1}x^*,$$

$$\lambda = (R_2 - R_1)^{-1}(R_2 - |c|), \quad 0 \leq \lambda \leq 1.$$

We have $\bar{c} = c$, or $\bar{x}^* = x^*$ if and only if $F(\bar{x},c) = 0$. For $|c| \leq R_1$ we have $F = 0$ if and only if $PN\bar{x} = 0$. For $R_2 \leq c \leq R$, we have

$$Fw = [(PN\bar{x},x^*) - \varepsilon\|PN\bar{x}\|](2J_0\|x^*\|)^{-1}x^*$$

with $x^* \neq 0$, and $(PN\bar{x},x^*) - \varepsilon\|PN\bar{x}\| \leq -\varepsilon\|PN\bar{x}\|$ and $F = 0$ if and only if $PN\bar{x} = 0$. For $R_1 < |c| < R_2$, hence $0 < \lambda < 1$, we have

$$(Fw,x^*) = \lambda(PN\bar{x},x^*) + (1 - \lambda)[(PN\bar{x},x^*) - \varepsilon\|PN\bar{x}\|](2J_0\|x^*\|)^{-1}\|x^*\|^2,$$

where $(PN\bar{x},x^*) \leq 0$, $(PN\bar{x},x^*) - \varepsilon\|PN\bar{x}\| \leq -\varepsilon\|PN\bar{x}\|$ and again $F = 0$ if and only if $PN\bar{x} = 0$. Thus, concerning the transformation T we have $(x,c) = T(x,c)$ if and only if $x = x^* + H(I - P)Nx$ and $PNx = 0$, that is, the fixed points (x,c) of T correspond exactly to the solutions of system (34.4-5), and thus to the solutions of the equation $Ex = Nx$.

Let us prove that T maps $\mathfrak{C}$ into itself. First for $(x,c) \in \mathfrak{C}$, we have $x^* = cw$, $\|x^*\| = \|cw\| = |c| \leq R$, and

$$\|\bar{x}\| \leq \|cw\| + \|H(I - P)Nx\| \leq |c| + LJ_0 \leq R + J.$$

Now we have $P\bar{x} = cw = x^*$, $\|\bar{x} - P\bar{x}\| = \|\bar{x} - x^*\| \leq \|H(I - P)Nx\| \leq LJ_0 = J$. Thus, $(PN\bar{x},x^*) = (N\bar{x},x^*) \leq 0$ for $\|x^*\| \geq R_0$, in particular for $R_1 \leq |c| \leq \|x^*\| \leq R$.

For $|c| \leq R_1$ we have $\bar{c} = c + F(\bar{x},c) = c + (PN\bar{x},w)$, hence

$$|\bar{c}| \leq |c| + \|PN\bar{x}\| \leq R_1 + J_0 \leq R_2 + J_0 \leq R.$$

For $R_2 \leq |c| \leq R$ we have

$$\bar{x}^* = \{1 + [(PN\bar{x},x^*) - \varepsilon\|PN\bar{x}\|](2J_0\|x^*\|)^{-1}\}x^* = \Lambda x^*,$$

where Λ is real, $(PN\bar{x},x^*) \leq 0$, $|(PN\bar{x},x^*)| \leq J_0\|x^*\|$,

$\|PN\bar{x}\| \leq J_0$, $\|x^*\| \geq R_2 > R_1$, $\epsilon/R_1 \leq 1/2$, $\frac{1}{4} = 1 - \frac{1}{2} - \frac{1}{4} \leq \Lambda \leq 1$, and $\bar{x}^*$ is on the segment between $\frac{1}{4} x^*$ and x^*. Thus $|\bar{c}| \leq |c| \leq R$. For $R_1 < |c| < R_2$ we have $0 < \lambda < 1$,

$$\bar{x}^* = \lambda(PN\bar{x}) + \{1 + (1 - \lambda)[(PN\bar{x}, x^*) - \epsilon\|PN\bar{x}\|](2J\|x^*\|)^{-1}\}x^*$$

and $\|x^*\| \leq \|PN\bar{x}\| + \|\{\ldots\}x^*\| \leq J_0 + R_2 = R$. We have proved that $T: \mathfrak{C} \to \mathfrak{C}$. The proof that T is compact is the same as for (34.i). Thus, T has at least one fixed point $(x,c) \in \mathfrak{C}$ by Schauder's fixed point theorem. Theorem (34.ii) is thereby proved.

Remark. Kannan and McKenna [85] have given an analogous proof of this theorem by using Leray-Schauder type topological degree argument instead of Schauder's fixed point theorem. For an extension of Theorem (34.ii) to the nonselfadjoint case we refer to Cesari [27].

35. One More Existence Theorem at Resonance.

Below, we shall deal with constants $L > 0$, $R_0 > 0$, $\epsilon > 0$, $0 < \lambda < 1$, a function $\varphi(r) = J_0 + J_1 r^k$, $0 \leq r < \infty$, with given k, $0 < k < 1$, $J_0 \geq 0$, $J_1 > 0$. Let us show that, as a consequence, there are constants R_1, R, R_2 such that

$$R_0 \leq R_1 < R < R_2, \quad R_1 + \varphi(R_2) = R, \quad R + L\varphi(R_2) = R_2,$$
$$(\varphi(R_2))^2 \leq 2\epsilon R_1^{1+k}, \quad R_2 \leq (1 + \lambda)R_1.$$

Indeed, for $R > 0$, $R_2 = (1 + \rho)R$, $0 \leq \rho < +\infty$, the second equation becomes $\Delta(\rho) = L\varphi((1 + \rho)R) - \rho R = 0$ with $\Delta(0) > 0$, and $\Delta(\rho) < 0$ for $\rho > 0$ large, since $L\varphi((1 + \rho)R)$ grows as ρ^k, that is, slower then ρR. Thus, $R_2 = R_2(R)$ can be determined, and then $R_1 = R - \varphi(R_2) = R - L^{-1}(R_2 - R)$. Note that, given σ, $0 < \sigma < 1$, $\sigma < (L + 1)^{-1}$, then for

R sufficiently large, $R < R_2 < R + L\sigma R_2$, or $R_2 < (1 - L\sigma)^{-1}R$, and then $R_1 > (1 - L\sigma)^{-1}[1 - (1 + L)\sigma]R$. In other words, given any η, $0 < \eta < 1$, then for $R > 0$ sufficiently large we certainly have

$$R(1 - \eta) < R_1 < R < R_2 < R(1 + \eta).$$

Thus, for η sufficiently small we also have

$$R_2 < R_1(1 + \eta)(1 - \eta)^{-1} < (1 + \lambda)R_1.$$

We can now take R sufficiently large so that

$$R_0 \leq R_1, \quad (\varphi(R_2))^2 \leq 2\varepsilon R_1^{1+k}.$$

This is possible since $R_1 \to +\infty$ as $R \to \infty$, and because, for R large we have, since $0 < k < 1$, hence $2k < 1 + k$,

$$(\varphi(R_2))^2 < [J_0 + J_1(1 + \eta)^k R^k]^2 \leq 2\varepsilon(1 - \eta)^{1+k}R^{1+k} < 2\varepsilon R_1^{1+k}.$$

(35.i) (An existence theorem at resonance) (Cesari [26]). For E, N, P, H as stated in no. 34, $L = \|H\|$, and under the assumptions: (B_k) there are constants $J_0 \geq 0$, $J_1 > 0$, $0 < k < 1$, such that $\|Nx\| \leq J_0 + J_1\|x\|^k$ for all $x \in X$; and ($N_{\varepsilon k}$) there are constants $R_0 \geq 0$, $\varepsilon > 0$, $K_0 \geq LJ_0$, $K_1 > LJ_1$ such that $(Nx, x^*) \leq -\varepsilon\|x^*\|^{1+k}$ (or always $(Nx, x^*) \geq \varepsilon\|x^*\|^{1+k}$) for all $x \in X$, $x^* \in X_0$ with $Px = x^*$, $\|x^*\| \geq R_0$, $\|x - x^*\| \leq K_0 + K_1\|x\|^k$; then equation $Ex = Nx$ has at least one solution.

<u>Proof</u>. Let $0 < \lambda < 1$ denote any number such that $LJ_1(1 + \lambda)^k \leq K_1$. Let us choose R_1, R, R_2 as indicated above, and let $T: (x, c) \to (\bar{x}, \bar{c})$ denote the transformation defined by

$$T: \bar{x} = cw + H(I - P)Nx, \quad \bar{c} = c + (PN\bar{x}, w),$$

$$(x, c) \in \mathfrak{C} = [(x, c) \mid x \in X,\ c \in R^m,\ \|x\| \leq R_2,\ |c| \leq R].$$

We need to prove that T maps $\mathfrak{C}$ into itself. First we have, for $(x,c) \in \mathfrak{C}$,

$$\bar{x} = cw + H(I - P)Nx, \quad \|x^*\| = \|cw\| = |c| \le R, \quad \|x\| \le R_2,$$

$$|\bar{x}| \le |c| + \|H(I - P)Nx\| \le |c| + L(J_0 + J_1R_2^k)$$

$$\le R + L\varphi(R_2) = R_2.$$

For $|c| \le R_1$ we have now $\bar{c} = c + PN\bar{x}$, $\|\bar{x}\| \le R_2$, and

$$|\bar{c}| \le |c| + \|PN\bar{x}\| \le |c| + (J_0 + J_1R_2^k) \le R_1 + \varphi(R_2) = R.$$

For $R_1 \le |c| \le R$ we have now

$$P\bar{x} = cw = x^*, \quad \|\bar{x} - P\bar{x}\| = \|\bar{x} - x^*\| \le L(J_0 + J_1R_2^k)$$

$$\le LJ_0 + LJ_1(1 + \lambda)^k R_1^k$$

$$\le LJ_0 + LJ_1(1 + \lambda)^k |c|^k$$

$$\le K_0 + K_1\|\bar{x}\|^k.$$

By $(N_{\varepsilon k})$ we have

$$(N\bar{x}, x^*) \le -\varepsilon\|x^*\|^{1+k}.$$

We have now as in the proof of (34.i),

$$(\bar{c} - c)c = (N\bar{x}, x^*) \le -\varepsilon|c|^{1+k},$$

and

$$(J_0 + J_1\|\bar{x}\|^k)^2 \ge \|PN\bar{x}\|^2 = \|\bar{c} - c\|^2 = (\bar{c} - c, \bar{c} - c)$$

$$= |\bar{c}|^2 - |c|^2 - 2(\bar{c} - c)c$$

$$\ge |\bar{c}|^2 - |c|^2 + 2\varepsilon|c|^{1+k}, \quad \text{or}$$

$$0 \ge (\varphi(R_2))^2 - 2\varepsilon R_1^{1+k} \ge |\bar{c}|^2 - |c|^2.$$

Thus, $|\bar{c}| \le |c| \le R$. The proof proceeds now as for (34.i).

36. Existence of Solutions Across a Point of Resonance.

(a) We shall now show that essentially the same assumptions, (N_ε) and (B) of Theorem (34.i), have indeed a much stronger implication.

(36.i) Theorem. (Existence of solutions across a point of resonance) (Cesari [26]). For E, N, P, H as stated in no. 34, $L = \|H\|$, and under the assumptions: (B) there is a constant J_0 such that $\|Nx\| \leq J_0$ for all $x \in X$; and (N_ε) there are constants $K > LJ_0$, $R_0 \geq 0$, $\varepsilon > 0$ such that $(Nx, x^*) \leq -\varepsilon\|x^*\|$ (or $(Nx, x^*) \geq \varepsilon\|x^*\|$) for all $x \in X$, $x^* \in X_0$, with $Px = x^*$, $\|x^*\| \geq R_0$ and $\|x - x^*\| \leq K$; then there are constants $\alpha_0 \geq 0$ and $C > 0$ such that, for any $|\alpha| \leq \alpha_0$, equation $Ex + \alpha x = Nx$ has at least a solution x with $\|x\| \leq C$.

Remark. Theorem (36.i) holds even for $Ex + \alpha Ax = Nx$, where $A: X \to X$ is any continuous operator mapping bounded sets into bounded sets (Cesari [26]). The same holds for (36.ii) below.

Proof. We need constants R_1, R, R_2, α_0 such that

$$R_0 \leq R_1 < R < R_2, \quad \alpha_0 > 0, \tag{36.1}$$

$$R + LJ_0 + L\alpha_0 R_2 \leq R_2, \tag{36.2}$$

$$LJ_0 + L\alpha_0 R_2 \leq K, \tag{36.3}$$

$$(1 + \alpha_0)R_1 + J_0 \leq R, \tag{36.4}$$

$$J_0^2 + 3\alpha_0 R^2 \leq 2\varepsilon(1 - \alpha_0)R_1. \tag{36.5}$$

It suffices to choose in order R_1, R, R_2, α_0 as follows

$$R_1 \geq \max[R_0, 2\varepsilon^{-1}J_0^2], \quad R \geq (3/2)R_1 + J_0, \quad R_2 = 2(R + LJ_0),$$

$$\alpha_0 = \min[2^{-1}, (2L)^{-1}, (K - LJ_0)(LR_2)^{-1}, 2^{-1}J_0^2R^{-2}].$$

Relations (36.1) and (36.4) are trivially satisfied. Now, since $0 \leq \alpha_0 < (2L)^{-1}$ we have $2^{-1} \leq 1 - L\alpha_0 < 1$,

$$R_2(1 - L\alpha_0) - R - LJ_0 \geq 2^{-1}R_2 - (R + LJ_0) = 0,$$

and (36.2) is satisfied. Since $L\alpha_0 R_2 \leq K - LJ_0$, we have

$$LJ_0 + L\alpha_0 R_2 \leq LJ_0 + (K - LJ_0) = K,$$

and (36.3) is satisfied. Finally, since $1 - \alpha_0 \geq 2^{-1}$, $\alpha_0 R^2 \leq 2^{-1} J_0^2$, we also have

$$2\epsilon(1 - \alpha_0)R_1 - J_0^2 - 2\alpha_0 R^2 \geq \epsilon R_1 - J_0^2 - 2\alpha_0 R^2$$
$$\geq 2J_0^2 - J_0^2 - J_0^2 = 0,$$

and (36.5) is satisfied. Now we define the map $T:(x,c) \to (\bar{x},\bar{c})$ by taking

$$T: \bar{x} = cw + H(I - P)[-\alpha x + Nx],$$
$$\bar{c} = c + (P[-\alpha\bar{x} + N\bar{x}],w), \quad (x,c) \text{ in the set}$$
$$\mathfrak{C} = [(x,c) \mid x \in X,\ c \in R^m,\ \|x\| \leq R_2,\ |c| \leq R]. \tag{36.6}$$

For $(x,c) \in \mathfrak{C}$ we have now, by using (36.2),

$$\|\bar{x}\| \leq \|cw\| + \|H(I - P)[-\alpha x + Nx]\| \leq |c| + L(\alpha_0\|x\| + J_0)$$
$$\leq R + LJ_0 + L\alpha_0 R_2 \leq R_2,$$

and by using (36.3) also, since $P\bar{x} = cw = x^*$,

$$\|\bar{x} - P\bar{x}\| = \|\bar{x} - x^*\| \leq \|H(I - P)[-\alpha x + Nx]\|$$
$$\leq LJ_0 + L\alpha_0 R_2 \leq K.$$

Since $P\bar{x} = cw = x^*$, the second relation (36.6) becomes

$$\bar{c} = c - \alpha(P\bar{x},w) + (PN\bar{x},w) = (1 - \alpha)c + (PN\bar{x},w).$$

From here we derive first, for $|\alpha| \leq \alpha_0$,

$$|\bar{c} - (1 - \alpha)c| \leq \|PN\bar{x}\| \leq J_0.$$

Thus, for $|c| \leq R_1$, by using (36.4) we have

$$|\bar{c}| \leq (1 - \alpha)|c| + J_0 \leq (1 + \alpha_0)R_1 + J_0 \leq R.$$

For $R_0 \leq R_1 \leq |c| \leq R$, and by denoting as usual $PN\bar{x} = \sum_1^m d_j w_j$, $x^* = \sum_1^m c_j w_j$, we have

$$\bar{c} - (1 - \alpha)c = (d_j,\ j = 1,\dots,m), \quad c = (c_j,\ j = 1,\dots,m),$$
$$(\bar{c} - (1 - \alpha)c)c = \sum_1^m c_j d_j,$$
$$(N\bar{x},x^*) = (PN\bar{x},x^*) = \sum_1^m c_j d_j = (\bar{c} - (1 - \alpha)c)c.$$

Since $P\bar{x} = cw = x^*$, $\|\bar{x} - P\bar{x}\| \leq K$, $\|x^*\| \geq R_1 \geq R_0$, we have

$$(N\bar{x}, x^*) \leq -\varepsilon\|x^*\| = -\varepsilon|c| \leq -\varepsilon R_1.$$

Now, by the identity

$$\begin{aligned}(\bar{c} - (1 - \alpha)c, \bar{c} - (1 - \alpha)c) \\ = |\bar{c}|^2 - |c|^2 - 2(1 - \alpha)(\bar{c} - (1 - \alpha)c)c \\ + (2\alpha - \alpha^2)|c|^2,\end{aligned} \tag{36.7}$$

and the relations

$$(\bar{c} - (1 - \alpha)c)c = (N\bar{x}, x^*) \leq -\varepsilon R_1 < 0, \quad |(2 - \alpha)\alpha| \leq 3\alpha_0,$$

we derive, by using (36.5),

$$\begin{aligned}|\bar{c}|^2 - |c|^2 &= |\bar{c} - (1 - \alpha)c|^2 + 2(1 - \alpha)(\bar{c} - (1 - \alpha)c)c \\ &\quad - (2\alpha - \alpha^2)|c|^2 \\ &\leq J_0^2 - 2\varepsilon(1 - \alpha_0)R_1 + 3\alpha_0 R^2 \leq 0.\end{aligned}$$

Thus, $|\bar{c}|^2 - |c|^2 \leq 0$, or $|\bar{c}| \leq |c| \leq R$. We have proved that $T:\mathfrak{C} \to \mathfrak{C}$ maps $\mathfrak{C}$ into itself. The argument is now the same as for (34.i). Theorem (36.i) is thereby proved.

Remark. The following particular case of the inequality $(Nx, x^*) \geq 0$, (or ≤ 0) will be used below. Assume that X is a space of real valued functions $x(t)$, $t \in G$, on a bounded domain G in R^ν, $\nu \geq 1$, and that $X_0 = \ker E$ is a space of constants; hence, $m = 1$ and $Px = (\text{meas } G)^{-1}\int_G x(t)dt$. Let $F:R^1 \to R^1$ be a real-valued function on R^1 with $sF(s) \geq 0$ for all s real, and take $Nx = F[x(t)]$, $t \in G$. Then, for $x(t) \geq 0$ we have $Nx = F[x(t)] \geq 0$, $x^* = PNx \geq 0$. Analogously, for $x(t) \leq 0$, we have $Nx = F[x(t)] \leq 0$, $x^* = PNx \leq 0$. In any case we have $(Nx, x^*) \geq 0$. We see that the condition $(Nx, x^*) \geq 0$ is satisfied by all functions x of constant sign on G. Analogous conclusion holds if $sF(s) \leq 0$ for all s reals.

(b) Below, we shall deal, as in no. 35, with certain constants $L > 0$, $R_0 \geq 0$, $\epsilon > 0$, $0 < \lambda < 1$, $K_0 > LJ_0$, and a function $\varphi(r) = J_0 + J_1 r^k$, $0 \leq r < +\infty$, $0 < k < 1$, $J_0 \geq 0$, $J_1 > 0$. Let us show that, as a consequence, there are constants R_1, R, R_2, α_0 such that

$$R_0 \leq R_1 < R < R_2, \quad \alpha_0 > 0, \tag{36.8}$$

$$R + L\alpha_0 R_2 + L\varphi(R_2) \leq R_2, \tag{36.9}$$

$$L\alpha_0 R_2 + LJ_0 \leq K_0, \quad R_2 \leq (1 + \lambda)R_1, \tag{36.10}$$

$$R_1 + \alpha_0 R_2 + \varphi(R_2) \leq R, \tag{36.11}$$

$$(\varphi(R_2) + \alpha_0 R_2)^2 + 2\alpha_0 R R_2 \leq 2\epsilon R_1^{1+k}. \tag{36.12}$$

To see that R_1, R, R_2, α_0 can be chosen so as to satisfy these relations, we first restrict α_0 to satisfy $L\alpha_0 R_2 \leq K_0 - LJ_0$. For any constant $\sigma > 0$ and r sufficiently large, $\varphi(r) \leq \sigma r$. We shall take $\sigma < (2L)^{-1}$, $\sigma < 2^{-1}(L + 1)^{-1}$. We shall further restrict α_0 to satisfy $\alpha_0 \leq \sigma$. Then, for $R_2 = (1 - 2L\sigma)^{-1}R$ and R sufficiently large, we have $R + 2L\sigma R_2 = R_2$, and

$$R + \alpha_0 L R_2 + L\sigma R_2 \leq R_2,$$

and hence (36.9) is satisfied. On the other hand, for $R_1 = R(1 - 2L\sigma)^{-1}(1 - 2L\sigma - 2\sigma)$, we also have

$$\begin{aligned} R_1 + \alpha_0 R_2 + \varphi(R_2) &\leq R(1 - 2L\sigma)^{-1}(1 - 2L\sigma - 2\sigma) \\ &\quad + \sigma R_2 + \sigma R_2 = R_2(1 - 2L\sigma - 2\sigma) + 2\sigma R_2 \\ &= R_2(1 - 2L\sigma) = R, \end{aligned}$$

and (36.11) is also satisfied. Note that

$$(1 - 2L\sigma)^{-1}(1 - 2L\sigma - 2\sigma)R = R_1 < R < R_2 = (1 - 2L\sigma)^{-1}R.$$

This shows that, given η, $0 < \eta < 1$, we can take $\sigma > 1$ sufficiently small so that

$$R(1 - \eta) < R_1 < R < R_2 < R(1 + \eta).$$

Thus, for η sufficiently small we also have

$$R_2 < R_1(1 + \eta)(1 - \eta)^{-1} \leq (1 + \lambda)R_1.$$

We see that R_2 grows as fast as R and R_1, and thus $(\varphi(R_2))^2$ grows as R_1^{2k} with $2k = k + k < 1 + k$, that is, slower than $2\epsilon R_1^{1+k}$. For R sufficiently large we have then $(\varphi(R_2))^2 + 2 \leq 2\epsilon R_1^{1+k}$, as well as $R_0 \leq R_1$. For $\alpha_0 R_2 \leq 1$, $2\alpha_0 RR_2 \leq 1$, and $\alpha_0 R_2(2\varphi(R_2) + 1) \leq 1$, we have then

$$(\alpha_0 R_2 + \varphi(R_2))^2 + 2\alpha_0 RR_2 \leq (\varphi(R_2))^2 + (2\varphi(R_2) + 1)\alpha_0 R_2$$
$$+ 2\alpha_0 RR_2 \leq (\varphi(R_2))^2 + 2 \leq 2\epsilon R_1^{1+k},$$

and also (36.12) is satisfied.

(36.ii) Theorem. (Existence of solutions across a point of resonance) (Cesari [26]). For E, N, P, H as stated in no. 34, $L = \|H\|$, and under the assumptions: (B_k) there are constants $J_0 \geq 0$, $J_1 > 0$, $0 < k < 1$, such that $\|Nx\| \leq J_0 + J_1\|x\|^k$ for all $x \in X$; and $(N'_{\epsilon k})$ there are constants $R_0 \geq 0$, $\epsilon > 0$, $K_0 > LJ_0$, $K_1 > LJ_1$ such that $(Nx, x^*) \leq -\epsilon\|x^*\|^{1+k}$ (or always $(Nx, x^*) \geq \epsilon\|x^*\|^{1+k}$) for all $x \in X$, $x^* \in X_0$ with $Px = x^*$, $\|x - x^*\| \leq K_0 + K_1\|x\|^k$; then there are also two constants $\alpha_0 > 0$, $C > 0$ such that, for every α real with $|\alpha| \leq \alpha_0$, equation $Ex + \alpha x = Nx$ has at least a solution $x \in X$ with $\|x\| \leq C$.

Proof. Let $0 < \lambda < 1$ be a number such that $LJ_1(1 + \lambda)^k \leq K_1$. Let us choose constants R_1, R, R_2, α_0 as stated above, and let us consider the map $T: (x, c) \rightarrow (\bar{x}, \bar{c})$ defined by

$$T: x = cw + H(I - P)[-\alpha x + Nx], \quad \bar{c} = c + (P[-\alpha\bar{x} + N\bar{x}], w),$$
$$(x, c) \in \mathfrak{C} = [(x, c) \,|\, x \in X,\ c \in R^m,\ \|x\|^m \leq R_2,$$
$$|c| \leq R]. \qquad (36.13)$$

For $(x,c) \in \mathfrak{C}$ we have now, using (36.9),

$$\|\bar{x}\| \leq \|cw\| + \|H(I - P)[-\alpha x + Nx]\|$$
$$\leq |c| + L(\alpha_0\|x\| + \varphi(\|x\|))$$
$$\leq R + L\alpha_0 R_2 + L\varphi(R_2) \leq R_2,$$

and by using (36.10) also

$$\|\bar{x} - P\bar{x}\| = \|\bar{x} - x^*\| \leq \|H(I - P)[-\alpha x + Nx]\|$$
$$\leq L\alpha_0 R_2 + L\varphi(R_2).$$

Since $P\bar{x} = cw = x^*$, the second relation (36.13) becomes

$$\bar{c} = c - \alpha(P\bar{x},w) + (PN\bar{x},w).$$

From here we derive first, for $|\alpha| \leq \alpha_0$,

$$|\bar{c} - c| \leq \alpha_0 R_2 + \varphi(R_2).$$

Thus, for $|c| \leq R_1$, by using (36.11), we have

$$|\bar{c}| \leq |c| + \alpha_0 R_2 + \varphi(R_2) \leq R_1 + \alpha_0 R_2 + \varphi(R_2) \leq R.$$

For $R_1 \leq |c| \leq R$ we have now

$$P\bar{x} = cw = x^*, \quad \|\bar{x} - P\bar{x}\| = \|\bar{x} - x^*\| \leq L\alpha_0 R_2 + L(J_0 + J_1 R_2^k)$$
$$\leq (L\alpha_0 R_2 + LJ_0) + LJ_1(1 + \lambda)^k R_1^k$$
$$\leq K_0 + K_1|c|^k \leq K_0 + K_1\|\bar{x}\|^k.$$

By $(N'_{\epsilon k})$ we have

$$(N\bar{x},x^*) \leq -\epsilon\|x^*\|^{1+k}.$$

We have now, as in the proof of (35.i),

$$\bar{c} = c - \alpha(P\bar{x},w) + (PN\bar{x},w),$$
$$(\bar{c} - c) = -\alpha(P\bar{x},x^*) + (PN\bar{x},x^*) \leq -\epsilon\|x^*\|^{1+k} + \alpha_0 RR_2,$$

and by the identity

$$(\bar{c} - c,\bar{c} - c) = |\bar{c}|^2 - |c|^2 - 2(\bar{c} - c)c,$$

we derive by using (36.12)

$$|\bar{c}|^2 - |c|^2 = |\bar{c} - c|^2 + 2(\bar{c} - c)c$$
$$\leq (\alpha_0 R_2 + \varphi(R_2))^2 - 2\epsilon R_1^{1+k} + 2\alpha_0 RR_2 \leq 0.$$

Thus, $|\bar{c}|^2 - |c|^2 \leq 0$, or $|\bar{c}| \leq |c| \leq R$. We have proved that $T:\mathfrak{C} \to \mathfrak{C}$ maps $\mathfrak{C}$ into itself. The argument is now the same as for (34.i). Theorem (36.ii) is thereby proved.

37. Extensions to Nonselfadjoint Problems. Let X and Y be Banach spaces over the reals with norms $\|\ \|_X$ and $\|\ \|_Y$, respectively. We shall write $\|\ \|$ when no explanation is needed. Let $E:\mathcal{D}(E) \to Y$, $\mathcal{D}(E) \subset X$, and $N:X \to Y$ be operators, E linear, and $P:X \to X$, $Q:Y \to Y$ be projection operators, as in no. 2. Thus $PX = X_0$, $(I - P)X = X_1$, $QY = Y_0$, $(I - Q)Y = Y_1$, and we assume here that

$$\operatorname{Ker} E = X_0, \quad \mathcal{R}(E) = Y_1, \quad 1 \leq \dim X_0 < \infty.$$

Thus, $E:\mathcal{D}(E) \cap X_1 \to Y_1$ is one-one and onto, and the partial inverse $H:Y_1 \to \mathcal{D}(E) \cap X_1$ exists. We assume that relations (k_{123}) of no. 2 hold, and we assume explicitly that H is (linear), bounded, and compact. We consider the equation

$$Ex = Nx, \quad x \in X_0.$$

We know from no. 2 that this equation is equivalent to the system of auxiliary and bifurcation equations

$$x = Px + H(I - Q)Nx, \quad Q(Ex - Nx) = 0.$$

In the present situation we can write these equations in the form

$$x = x^* + H(I - Q)Nx, \quad Px = x^*,$$

$$QNx = 0.$$

Let $L = \|H\|$ be the norm of the linear operator $H:Y_1 \to X_1$ with respect to the norms $\|\ \|_X$ and $\|\ \|_Y$. We shall now further assume that the space Y is a space of linear operators on X so that the operation $\langle y,x\rangle$, or $Y \times X \to$ reals, is defined, is linear both in x and y, and $|\langle y,x\rangle| \leq K\|y\|_Y\|x\|_X$ for some constant K and all $x \in X$, $y \in Y$. We can always choose norms in X and Y so that $K = 1$. Whenever $X \subset Y$, $X_0 \subset Y_0$ and X_0 is finite dimensional, then for the elements $x \in X_0$ the norms in X and in Y are equivalent (in X_0).

For instance, if G is any bounded domain in a space R^ν, $\nu \geq 1$, and we take $X = Y = L_2(G)$, then of course $|\langle y,x\rangle| = |\int_G y(t)x(t)dt| \leq \|y\|_2\|x\|_2$. If we take now instead, $X = L_2(G)$, and $Y = L_\infty(G)$ with norm $\|y\|_Y = (\text{meas } G)^{1/2}\|y\|_\infty$, then we have again $|\langle y,x\rangle| = |\int_G y(t)x(t)dt| \leq \|y\|_Y\|x\|_2$. If we take $X = Y = L_\infty(G)$, with norm $\|y\|_Y = (\text{meas } G)\|y\|_\infty$, then again $\langle y,x\rangle = |\int_G y(t)x(t)dt| \leq \|y\|_Y\|x\|_\infty$.

Actually, we need the operation $\langle y,x\rangle$ only for $y \in Y_0$, $x \in X_0$, and we assume that, for $y \in Y_0$ we have $y = 0$ if and only if $\langle y,x^*\rangle = 0$ for all $x^* \in X_0$.

Having assumed $1 \leq m = \dim \ker E < \infty$, let $w = (w_1,\ldots,w_m)$ be a basis for the finite dimensional space $X_0 = \ker E = PX$. Then, for $x^* \in X$, we have $x^* = \sum_i^m c_i w_i$, or briefly $x^* = cw$, $c = (c_1,\ldots,c_m) \in R^m$, and there are constants $0 < \gamma' < \gamma < +\infty$ such that $\gamma'|c| \leq \|cw\| \leq \gamma|c|$, where $|\ |$ denotes the Euclidean norm in R^m. We shall choose the basis w in such a way that $\gamma = 1$, hence $0 < \gamma' \leq \gamma = 1$. Here P, Q are linear, bounded, idempotent operators. Let $k_0 = \|P\|$, $k' = \|I - P\|$, $\chi = \|Q\|$, $\chi' = \|I - Q\|$, k,k', $\chi,\chi' \geq 1$. Whenever X, Y are Hilbert spaces and P, Q are orthogonal projections, then $k_0 = k' = \chi = \chi' = 1$, and if w is orthonormal in X_0, also $\gamma = \gamma' = 1$.

(37.i) Theorem. (An existence theorem at resonance) (Cesari [27]). Under the assumptions above on X, Y, E, N, P, Q, H, if (B) there is a constant $J_0 \geq 0$ such that $\|Nx\| \leq J_0$ for all $x \in X$; and if (N_0) there are constants $R_0 \geq 0$, $K_0 \geq L\chi' J_0$ such that $\langle QNx, x^* \rangle \leq 0$ (or $\langle QNx, x^* \rangle \geq 0$) for all $x \in X$, $x^* \in X_0$ with $Px = x^*$, $\|x^*\| \geq R_0$, $\|x - x^*\| \leq K_0$, then equation $Ex = Nx$ has at least a solution $x \in \mathcal{D}(E) \subset X$.

The proof is similar to the one for (34.ii) and is omitted.

Let $A: X \to Y$ be any continuous not necessarily linear operator mapping bounded subsets of X into bounded subsets of Y, or equivalently such that $\|Ax\| \leq \omega(\|x\|)$ for all $x \in X$ and some monotone function $\omega(\xi) \geq 0$, $0 \leq \xi < +\infty$.

(37.ii) Theorem. (Existence of solutions across a point of resonance) (Cesari [27]). Under the assumptions above on X, Y, E, N, P, Q, H, if (B) there is a constant $J_0 > 0$ such that $\|Nx\| \leq J_0$ for all $x \in X$; and if (N_ε) there are constants $R_0 \geq 0$, $\varepsilon > 0$, $K_0 > L\chi' J_0$ such that $\langle QNx, x^* \rangle \leq -\varepsilon \|x^*\|$ (or $\langle QNx, x^* \rangle \geq \varepsilon \|x^*\|$) for all $x \in X$, $x^* \in X_0$ with $\|x^*\| \geq R_0$, $Px = x^*$, $\|x - x^*\| \leq K_0$, then there are constants $\alpha_0 > 0$, $C > 0$ such that, for every real α with $|\alpha| \leq \alpha_0$, the equation $Ex + \alpha Ax = Nx$ has at least a solution $x \in \mathcal{D}(E) \subset X$ with $\|x\| \leq C$.

The proof is similar to the one for (36.i) and is omitted.

(37.iii) Theorem. (Existence of solutions across a point of resonance) (Cesari [27]). Under the assumptions above on X, Y, E, N, A, P, Q, H, if (B_k) there are constants $0 < k < 1$, $J_0 \geq 0$, $J_1 > 0$ such that $\|Nx\| \leq J_0 + J_1\|x\|^k$ for all $x \in X$; and if $(N_{\epsilon k})$ there are constants $R_0 \geq 0$, $\epsilon > 0$, $K_0 > L\chi' J_0$, $K_1 > L\chi' J_1((\gamma')^{-1}k_0)^k$ such that $\langle QNx, x^*\rangle \leq -\epsilon\|x^*\|^{1+k}$ [or always $\langle QNx, x^*\rangle \geq \epsilon\|x^*\|^{1+k}$] for all $x \in X$, $x^* \in X_0$ with $Px = x^*$, $\|x^*\| \geq R_0$, $\|x - x^*\| \leq K_0 + K_1\|x\|^k$; then there are also constants $\alpha_0 > 0$, $C > 0$ such that, for every real α with $|\alpha| \leq \alpha_0$, the equation $Ex + \alpha Ax = Nx$ has at least a solution $x \in \mathcal{D}(E) \subset X$ with $\|x\| \leq C$.

The proof is similar to the one for (36.ii) and is omitted.

38. Some Theorems in the Uniform Topology. Let us assume now $X = L_\infty(G)$, the space of all measurable essentially bounded real-valued functions $x(t)$, $t \in G$, with $\|x\|_0 = \text{Ess Sup}[|x(t)|, t \in G]$, over a bounded domain G of R^ν, $\nu \geq 1$. We take for Y the space $Y = L_\infty(G)$ and norm $\|y\|_0 = (\text{meas } G)\|y\|_\infty$ with $\langle y, x\rangle = \int_G y(t)x(t)dt$ so that $|\langle y,x\rangle| \leq \|y\|_0\|x\|_\infty$. We assume that $X_0 = PX = \ker E$ is the space of constants, and so is $Y_0 = QY$, with

$$Px = (\text{meas } G)^{-1}\int_G x(t)dt, \quad Qy = (\text{meas } G)^{-1}\int_G y(t)dt.$$

Let $X_1 = (I - P)X$, $Y_1 = (I - Q)Y$. Let $F(t,s), (t,s) \in G \times R^1$, be a continuous real-valued function, and let N be the Nemitsky operator defined by $Nx = F(t,x(t))$, $t \in G$. For F bounded on $G \times R^1$, then certainly $N:X \to Y$. We understand here that, for any $h(t)$, $t \in G$, $h \in Y = L_\infty(G)$, the operator $H(I - Q)$ yields a unique element $x(t) = H(I - Q)h$ which belongs to $\mathcal{D}(E) \cap X_1 \subset L_\infty(G)$ with $Ex = (I - Q)h$, and that H is a linear bounded compact operator $H:Y_1 \to \mathcal{D}(E) \cap X_1$, and that conditions (k_{123}) are satisfied.

The classical condition $sF(t,s)$ of constant sign for $|s|$ large (see Section 5), has here the following simple implication (38.i). We shall give below a stronger form (38.ii) of this statement. (See [28] for other forms).

(38.i) (An existence theorem at resonance) (Cesari [28]). Under the above hypotheses, let us assume that (B) there is a constant $J_0 > 0$ such that $|F(t,s)| \leq J_0$ for all $(t,s) \in G \times R^1$; and that (F) there is a constant $R_0 \geq 0$ such that $sF(t,s) \geq 0$ for all $t \in G$ and $|s| \geq R_0$ (or $sF(t,s) \leq 0$ for $t \in G$, $|s| \geq R_0$). Let $h(t)$, $t \in G$, be any element of Y such that $\int_G h(t)dt = 0$. Then equation $Ex = h(t) + F(t,x(t))$, $t \in G$, has at least a solution $x \in \mathcal{D}(E) \subset L_\infty(G)$.

The proof is similar to the one for Theorem (34.ii). The same conclusion holds even if (B) is replaced by the slow growth condition: $|F(t,s)| \leq a|s|^k + b$ for given constants $a > 0$, $b \geq 0$, $0 < k < 1$ (cf. [28]).

Let $g(t,s), (t,s) \in \bar{G} \times R^1$, be any real valued continuous function on $\bar{G} \times R^1$.

If we assume that, not only $F(t,s)$ has the sign of s (or the opposite one) for $|s|$ large, but actually F is larger than some constant η_2 for s large, and F is below some constant η_1 for $-s$ large, $\eta_1 < 0 < \eta_2$, then the existence "across a point of resonance" can be proved.

(38.ii) (An existence theorem across a point of resonance) (Cesari [28]). Under the same hypotheses as above, let us assume that (B) there is a constant $J_0 > 0$ such that $|F(t,s)| \leq J_0$ for all $(t,s) \in G \times R^1$; and that (F_η) for suitable constants $\sigma_1 < \sigma_2$, $\eta_1 < \eta_2$, we have, either

$F(t,s) \geq \eta_2$ for $t \in G$, $s \geq \sigma_2$;

$F(t,s) \leq \eta_1$ for $t \in G$, $s \leq \sigma_1$; or

$F(t,s) \leq -\eta_2$ for $t \in G$, $s \geq \sigma_2$;

$F(t,s) \leq -\eta_1$ for $t \in G$, $s \geq \sigma_1$.

Let $h(t)$, $t \in G$, be any element of Y such that, either

$$-\eta_2 < (\text{meas } G)^{-1}\int_G h(t)dt < -\eta_1 , \tag{38.1}$$

or

$$\eta_1 < (\text{meas } G)^{-1}\int_G h(t)dt < \eta_2 , \tag{38.2}$$

correspondingly to the two cases above. Then, there are constants $\alpha_0 > 0$ and $C > 0$ such that for every real α with $|\alpha| \leq \alpha_0$, equation $Ex + \alpha g(t,x(t)) = h(t) + F(t,x(t))$, $t \in G$, has at least a solution $x \in \mathcal{D}(E) \subset L_\infty(G)$ with $\|x\|_0 \leq C$.

The proof is similar to that of Theorem (36.i).

39. Applications to Problems of Oscillation at Resonance.

(a) First, we consider the nonlinear differential equation

$$x'' + h(x) = p(t), \tag{39.1}$$

where h and p are real-valued continuous functions, and p is T-periodic.

(39.i) If p is T-periodic and continuous with $\int_0^T p(t)dt = 0$, if h is continuous with $sh(s) \geq 0$ for all s [or $sh(s) \leq 0$ for all s], and if $|h(s)| \leq J$ for some constant J, then (39.1) has at least a T-periodic solution. The same occurs if $|h(s)| \leq a|s|^k + b$ for given constants $a,b \geq 0$, $0 < k < 1$, and all s.

If, finally, $|h(s)| \leq a|s|^k + b$ for constants $a,b \geq 0$, $k > 1$, and all s, then the same occurs provided a is sufficiently small.

This statement is a consequence of (38.i), (38.ii) and remarks (cf. [28]).

Estimates for $\|x\|_0$ and, in the third case, for a, can be easily obtained from those proved in [28].

(b) Let us consider the nonlinear differential equation

$$x'' + m^2x + h(x) = p(t), \tag{39.2}$$

where m is an integer, h is a real-valued function of the real variable x, and $p(t)$ is 2π-periodic, $p \in L_2[0,2\pi]$.

(39.ii) <u>Theorem</u>. (Lazer and Leach [94]). If $p(t)$ is 2π-periodic and continuous; if h is continuous with $|h(x)| \leq M$ for all real x; and if there are constants $c < d$, $C < D$ such that $h(x) \leq C$ for $x \leq c$, $h(x) \geq D$ for $x \geq d$, and $(A^2 + B^2)^{1/2} < 2(D - C)$ where $A = \int_0^{2\pi} p(s)\cos ms\, ds$, $B = \int_0^{2\pi} p(s)\sin ms\, ds$, then (39.2) has a periodic solution of period 2π.

<u>Proof</u>. A proof could be given by using statement (34.ii). However, it is convenient to prove both theorem (39.ii) and the following one (39.iii) by using statement (37.i). Let X denote the space of all 2π-periodic functions $x(t)$, which are continuous in $(-\infty,+\infty)$, absolutely continuous (AC) in $[0,2\pi]$, with derivative $x' \in L_2[0,2\pi]$. Thus X is a Sobolev type space H^1, a Hilbert space, with usual inner product and norm $\|x\|_1$, or $\|x\|_X$. Let $\mathcal{D}(E) \subset X$ denote the set of all functions $x \in X$, which are continuous in $(-\infty,+\infty)$ and AC in $[0,2\pi]$ together with x', and with $x'' \in L_2[0,2\pi]$. Let $Y = L_2[0,2\pi]$ (the functions $y \in Y$ extended to $(-\infty,+\infty)$ by 2π-periodicity). Let E and N be defined by $Ex = x'' + m^2x$, $Nx = -h(x(t)) + p(t)$, so that the equation takes the form $Ex = Nx$, and we associate to E the boundary conditions $x(0) = x(2\pi)$, $x'(0) = x'(2\pi)$. Thus, $E:\mathcal{D}(E) \to Y$, $N:X \to Y$. For $x \in X$, $y \in Y$ we take $(y,x) = \int_0^{2\pi} y(t)x(t)dt$, so that $|(y,x)| \leq \|x\|\|y\| \leq \|x\|_1\|y\|$, where $\|\ \|$ denote L_2 norms. Let $X_0 = Y_0$ be the space spanned by $\cos mt$, $\sin mt$, and P, Q the usual projections of X and Y on X_0 and Y_0 respectively. Let $X_1 = (I - P)X$, $Y_1 = (I - Q)Y$. Every element $x \in Y_1$ has the Fourier series representation

$$x = (1/2)a_0 + \sum_{k\geq 1, k\neq m} (a_k \cos kt + b_k \sin kt)$$

and we define $H:Y_1 \to \mathcal{D}(E) \cap X_1$ by taking

$$Hx = (1/2m^2)a_0 + \sum_{k\geq 1, k\neq m} (m^2 - k^2)^{-1}(a_k \cos kt + b_k \sin kt).$$

Thus H is a bounded map from Y_1 into H^2; hence a compact map from Y_1 into X.

Since $|h(x)| \leq M$, $|p(t)| \leq M_1$, M, M_1 constants, we see that $-h(x(t)) + p(t)$ is a periodic function of period 2π with a bound $|-h(x(t)) + p(t)| \leq M + M_1$ independent of $x(t)$. Thus, we have $\|Nx\| \leq (2\pi)^{1/2}(M + M_1) = J_0$ for all $x \in X$. This proves (B). Let K_0 be any constant as in (37.i), and let

$$2\mu = \pi^{-1/2}[2(D - C) - (A^2 + B^2)^{1/2}].$$

On the finite dimensional space $X_0 = Y_0$ the norms of X and Y are equivalent. For $x \in X$, $Px = x^*$, $R = \|x^*\|$ (square norm), $x^* = Ry^*$, $\|y^*\| = 1$, we have $y^* = a \cos mt + b \sin mt$ with $\pi(a^2 + b^2) = 1$. For ξ such that $(a^2 + b^2)^{1/2} \sin \xi = -a$, $(a^2 + b^2)^{1/2} \cos \xi = b$, we have also $y^* = \pi^{-1/2} \sin(mt - \xi)$.

For $\|x - x^*\|_1 \leq K_0$, then the function $\gamma(t) = x(t) - x^*(t)$ has a bound $|\gamma(t)| \leq L$ which is independent of x. We have now

$$\begin{aligned}(h(x(t)),y^*) &= \int_0^{2\pi} h(x(t))y^*(t)dt \\ &= \int_0^{2\pi} h[R\pi^{-1/2} \sin(mt - \xi) \\ &\quad + \gamma(t)]\pi^{-1/2} \sin(mt - \xi)dt.\end{aligned}$$

By the change of variable $mt - \xi = mr$, and because of the periodicity of period 2π, we also have

$$\begin{aligned}(h(x(t)),y^*) &= \int_0^{2\pi} h[R\pi^{-1/2} \sin mr \\ &\quad + \gamma(r + \xi/m)]\pi^{-1/2} \sin mr dr.\end{aligned}$$

Let δ be any number $0 < \delta < \pi/2m$. For $(\pi/m)2k + \delta \leq r \leq (\pi/m)(2k + 1) - \delta$ we have $2\pi k + m\delta \leq mr \leq (2k + 1)\pi - m\delta$, and for $R \geq \pi^{1/2}(L + d)(\sin m\delta)^{-1}$, we also have

$$R\pi^{-1/2} \sin mr + \gamma(r + \xi/m) \geq R\pi^{-1/2} \sin m\delta - L \geq d.$$

Analogously, for $(\pi/m)(2k + 1) + \delta \leq r \leq (2\pi/m)(k + 1) - \delta$ and $R \geq \pi^{1/2}(L - c)(\sin m\delta)^{-1}$, we have

$$R\pi^{-1/2} \sin mr + \gamma(r + \xi/m) \leq -R\pi^{-1/2} \sin m\delta + L \leq c.$$

Thus

$$\int_{(\pi/m)2k+\delta}^{(\pi/m)(2k+1)-\delta} h[R\pi^{-1/2} \sin mr + \gamma(r + \xi/m)]\pi^{-1/2} \sin mr dr$$

$$\geq \int_{(\pi/m)2k+\delta}^{(\pi/m)(2k+1)-\delta} D\pi^{-1/2} \sin mr dr,$$

$$\int_{(\pi/m)(2k+1)-\delta}^{(2\pi/m)(k+1)-\delta} h[R\pi^{-1/2} \sin mr + \gamma(r + \xi/m)]\pi^{-1/2} \sin mr dr$$

$$\geq \int_{(\pi/m)(2k+1)-\delta}^{(2\pi/m)(k+1)-\delta} C\pi^{-1/2} \sin mr dr.$$

We have now

$$\begin{aligned}(h(x(t),y^*) &= \sum_{k=0}^{m-1} \Big(\int_{(\pi/m)2k}^{(\pi/m)(2k+1)} + \int_{(\pi/m)(2k+1)}^{(2\pi/m)(k+1)}\Big)[h[R\pi^{1/2} \sin mr \\ &\quad + \gamma(r + \xi/m)]\pi^{-1/2} \sin mr dr \\ &\geq \pi^{-1/2}[m(2/m)D + m(-2/m)C] + Q(\delta) \\ &= 2\pi^{-1/2}(D - C) + Q(\delta),\end{aligned}$$

where $Q(\delta) \to 0$ as $\delta \to 0+$.

If we denote by $p(t) = \alpha_0 + \sum_{k=1}^{\infty} (\alpha_k \cos kt + \beta_k \sin kt)$ the Fourier series of $p(t)$, then $\alpha_m = A/\pi$, $\beta_m = B/\pi$, and

$$\begin{aligned}(p(t),y^*(t)) &= (p(t),a \cos mt + b \sin mt) = \pi(\alpha_m a + \beta_m b) \\ &= Aa + Bb \leq (A^2 + B^2)^{1/2}(a^2 + b^2)^{1/2} \\ &= \pi^{-1/2}(A^2 + B^2)^{1/2}.\end{aligned}$$

Finally,

$$\begin{aligned}(Nx,y^*) &= (-h(x(t)),y^*) + (p(t),y^*) \\ &\leq \pi^{-1/2}[-2(D - C) + (A^2 + B^2)^{1/2}] - Q(\delta).\end{aligned}$$

Since $2\mu = \pi^{-1/2}[2(D - C) - (A^2 + B^2)^{1/2}]$, if we take $\delta > 0$ sufficiently small, so that $0 < \delta < \pi/2m$ and $|Q(\delta)| \leq \mu/2$, then $\langle Nx,y^*\rangle \leq -\mu$, and $\langle Nx,x^*\rangle \leq -\mu R = -\mu\|x^*\|$. Condition (N_ε) is thereby proved, and so is statement (39.ii).

(c) (39.iii) (Existence of periodic solutions across a point of resonance). Under the conditions of (39.ii), there are constants $\alpha_0 > 0$ and $\Lambda > 0$, such that for every α, β real, $|\alpha| \le \alpha_0$, $|\beta| \le \alpha_0$, the equation

$$x'' + \beta x' + (m^2 + \alpha)x + h(x) = p(t)$$

has at least a periodic solution $x(t)$ of period 2π with $\|x\|_1 \le \Lambda$.

<u>Proof</u>. In the proof of (39.ii) we have seen that the operators E and N satisfy the properties required in the statements (37.i) and (37.ii). Let $A:X \to Y$ denote the map defined by $Ax = \alpha x + \beta x'$. Then the equation has the form $Ex + Ax = Nx$, with $\|Ax\| \le |\alpha|\,\|x\| + |\beta|\,\|x'\| \le (|\alpha| + |\beta|)\,\|x\|_1$. Thus A is a continuous operator satisfying the requirements of no. 37. Theorem (39.ii) now follows from (37.ii). Theorem (39.ii) holds even if the linear perturbation $\alpha x + \beta x'$ is replaced by any continuous nonlinear operator from X into Y satisfying the requirements of no. 37.

40. <u>Linear Elliptic Operators in R^ν</u>. For a multi-index $\alpha = (\alpha_1,\dots,\alpha_\nu)$ of integers $\alpha_i \ge 0$, we denote by $|\alpha|$ the integer $|\alpha| = \alpha_1 + \dots + \alpha_\nu$. We consider below a bounded domain G in the t-space, R^ν, $t = (t_1,\dots,t_\nu)$, $\nu \ge 1$, defined formally by

$$Ex = \sum_{\substack{|\alpha|\le m \\ |\beta|\le m}} (-1)^{|\beta|} D^\beta (a_{\alpha\beta}(t) D^\alpha x).$$

We shall need the following assumptions:

(E_1) The coefficients $a_{\alpha\beta}(t)$ are real-valued bounded measurable functions defined on G for all $0 \le |\alpha|, |\beta| \le m$.

(E_2) E is uniformly strongly elliptic, that is, there is a constant $c > 0$ such that

$$\sum_{|\alpha|=|\beta|=m} a_{\alpha\beta}(t)\xi^{\alpha}\xi^{\beta} \ge c|\xi|^{2m} \tag{40.2}$$

for all $\xi = (\xi_1,\ldots,\xi_\nu) \in R^\nu$ and all $t = (t_1,\ldots,t_\nu) \in G$.

The adjoint E^* of E is the operator defined by $E^*x = \sum_{\alpha,\beta} (-1)^{|\alpha|} D^{\alpha}(a_{\alpha\beta}(t)D^{\beta}x)$, and $E = E^*$ if and only if $a_{\alpha\beta} = a_{\beta\alpha}$ for all α and β.

Then the Dirichlet form associated to E is

$$A[\varphi,\psi] = \sum_{|\alpha|,|\beta|\le m} \int_G a_{\alpha\beta}(x)D^{\alpha}\psi(x)D^{\beta}\varphi(x)\,dx. \tag{40.3}$$

As usual we denote by $(\ ,\ )$ and $\|\ \|$ the inner products and norms in $L_2(G)$:

$$(\varphi,\psi) = \int_G \varphi(t)\psi(t)\,dt, \quad \|\varphi\| = (\varphi,\varphi)^{1/2}, \quad \varphi,\psi \in L_2(G).$$

For any integer $N \ge 0$ we denote by $(\varphi,\psi)_N$ and $\|\varphi\|_N$ the inner product and norm defined by

$$(\varphi,\psi)_N = \sum_{|\alpha|\le N} (D^{\alpha}\varphi, D^{\alpha}\psi), \quad \|\varphi\|_N = (\varphi,\varphi)_N^{1/2}, \quad \varphi,\psi \in C_0^\infty(G).$$

As usual, we denote by $H^N(G)$, $[H_0^N(G)]$ the (Sobolev) space obtained by completion of $C^\infty(E)$ $[C_0^\infty(G)]$ with respect to the norm $\|\ \|_N$. Thus, $H^N(G)$ and $H_0^N(G)$ are Hilbert spaces with inner product $(\ ,\ )_N$ and norm $\|\ \|_N$. We may denote them simply by H^N and H_0^N. In particular, $H_0^0 \subset H^0 = L_2$.

In view of (E_1), $A[\varphi,\psi]$ is certainly a bounded bilinear form in H^m as well as H_0^m. We shall assume below that G satisfies usual regularity conditions (see Agmon [126]).

Under assumptions (E_1) and (E_2), Gårding's inequality holds, that is, there are constants $c_0 > 0$ and k real such that

$$A[\varphi,\varphi] \geq c_0\|\varphi\|_m^2 - k\|\varphi\|_0^2 \quad \text{for all} \quad \varphi \in H_0^m.$$

Whenever ∂G is sufficiently smooth, so that an (inner) normal n is associated to the points of ∂G, then by the Dirichlet boundary conditions associated to an operator E of order $2m$ we mean the conditions

$$x = 0, \quad \partial x/\partial n = 0, \ldots, \partial x^{m-1}/\partial n^{m-1} = 0 \quad \text{on} \quad \partial G,$$

or zero data on ∂G.

Let $f \in L_2(G)$ be any given function on G. We say that x is a (weak) solution for the Dirichlet problem on G with zero data on ∂G provided $x \in H_0^m$ and

$$A[x,\varphi] = (f,\varphi) \quad \text{for all} \quad \varphi \in H_0^m.$$

We say that two functions x,y are equal if $x = y$ almost everywhere in G. Thus $x = 0$ on G will mean $x = 0$ a.e. in G.

(40.i) (Fredholm alternative theorem). Under hypotheses (E_1) and (E_2) and for E selfadjoint, or $a_{\alpha\beta} = a_{\beta\alpha}$, either of the following two cases (α) or (β) occurs:

(α) There are no nonzero elements $x \in H_0^m$ such that $A[x,\varphi] = 0$ for all $\varphi \in H_0^m$. Then, for any $f \in L_2(G)$, there is one and only one element $x \in H_0^m(G)$ such that $A[x,\varphi] = (f,\varphi)$ for all $\varphi \in H_0^m$. The single-valued operator $K: f \to x$, or $K: L_2 \to H_0^m$, so defined, is linear and bounded. Since the inclusion map $j: H_0^m \to H_0^{m-1}$ is compact, the same operator K, thought of as a map from L_2 to H_0^{m-1}, is compact. In particular, so is the map K thought of from L_2 to L_2.

(β) There are nonzero elements $x \in H_0^m$ such that $A[x,\varphi] = 0$ for all $\varphi \in H_0^m$. Then, the subspace W of all elements $x \in H_0^m$ with $A[x,\varphi] = 0$ for all $\varphi \in H_0^m$, is finite dimensional. Then W is the kernel or null space of E, and we denote by D the orthogonal complement of W in L_2, by Q and $I - Q$ the orthogonal projections of L_2 onto W and D respectively, and then $W^{\perp} = D \cap H_0^m$ is the orthogonal complement of W in H_0^m.

Statement (β) can now be completed as follows:

(β^*) For every $f \in D \subset L_2$, or $f \in L_2(G)$ with $(f,w) = 0$ for all $w \in W$, there is one and only one element $x \in W^{\perp} \subset H_0^m$ such that $A[x,\varphi] = (f,\varphi)$ for all $\varphi \in H_0^m$ (and hence also $A[x + w,\varphi] = (f,\varphi)$ for all $\varphi \in H_0^m$ and $w \in W$). The single-valued operator $K: f \to x$, or $K: D \to H_0^m$, is linear and bounded with domain $\mathcal{D}(K) = D \subset L_2$ and range $\mathcal{R}(K) \subset W^{\perp} \subset H_0^m$.

As before, K, thought of as a map from D into H_0^{m-1}, is compact. In particular, K as a map from $D \subset L_2$ into L_2 is compact. Furthermore the following regularity statement holds.

(γ) If the selfadjoint elliptic operator E satisfies (E_1) and (E_2), if each coefficient $a_{\alpha\beta} \in C^{|\beta|}(\bar{G})$, and ∂G is smooth, then for every $f \in L_2$ and $x \in H_0^m$ satisfying $A[x,\varphi] = (f,\varphi)$ for all $\varphi \in H_0^m$, the element x actually belongs to the Sobolev class H^{2m}, thus $x \in H_0^m \cap H^{2m}$. In particular $Kx \in H_0^m \cap H^{2m}$, and $\|Kx\|_{2m} \leq C\|f\|_0$ for some constant C independent of f, ($f \in L_2$ in case (α); $f \in D$ in case (β)) [146].

Thus, in case (β), for every $f \in L_2(G)$, the operator $K(I - Q)f$ is uniquely defined, and yields the unique solution $x \in H_0^m \cap H^{2m}$, $x \in W^\perp \subset H_0^m$ or $Px = 0$, of the Dirichlet problem $Ex = (I - Q)f$ with zero data on ∂G. Here $(I - Q)f$ is an element of D, the orthogonal complement of W in $L_2(G)$. Moreover $\|K(I - Q)f\| \leq C\|f\|$ for all $f \in L_2(G)$. For E selfadjoint, satisfying (E_1), (E_2), it is well known that we may have $A[x,\varphi] + \lambda(x,\varphi) = 0$ for all $\varphi \in H_0^m$, $x \in H_0^m$, x nonidentically zero, only for countably many λ, λ real, say $\lambda = \lambda_1, \lambda_2, \ldots$, with $\lambda_i \leq \lambda_{i+1}$, $\lambda_i \to +\infty$ as $i \to +\infty$.

41. <u>A Nonlinear Dirichlet Problem at Resonance</u>. Let G be a bounded domain in R^ν, $\nu \geq 1$, let E be a selfadjoint elliptic operator of order $2m$ as in no. 40, let $f(t)$, $t \in G$, a given real-valued function, $f \in L_2(G)$, and $g(x)$, $-\infty < x < +\infty$, a continuous bounded real-valued function defined on R^1. Let us consider the Dirichlet problem on G:

$$Ex = f(t) + g(x(t)), \quad t \in G, \quad \text{(a.e.)},$$
$$x = 0, \quad \partial x/\partial n = 0, \ldots, \partial x^{m-1}/\partial n^{m-1} = 0 \quad \text{on } \partial G, \qquad (41.1)$$

where n is the inner normal. We shall say that x is a weak solution of (41.1) provided $x \in H_0^m$ and

$$A[x, \psi] = (Nx, \psi) \quad \text{for all} \quad \psi \in H_0^m. \tag{41.2}$$

In (41.1) we shall assume that g possesses finite limits $R = g(+\infty)$ and $r = g(-\infty)$. Also, we shall assume that the homogeneous problem $Ex = 0$, $x \in H_0^m$, has nonzero solutions, and that the set W of these solutions has dimension $k \geq 1$. For any nonzero element $w \in W$ let G^+, G^- denote the sets of all $t \in G$ where $w(t) > 0$ and $w(t) < 0$ respectively, and let

$$w^+ = \int_{G^+} |w|dt, \quad w^- = \int_{G^-} |w|dt. \tag{41.3}$$

(41.i) <u>Theorem</u>. (Landesman, Lazer, Williams). If the elliptic selfadjoint operator E satisfies (E_1) and (E_2), and W has dimension $k \geq 1$, if $f \in L_2(G)$, if $g(x)$, $-\infty < x < +\infty$, is a bounded continuous real-valued function possessing finite limits $R = g(+\infty)$, $r = g(-\infty)$, and $r \leq g(x) \leq R$ for all $x \in R^1$, then a necessary condition for (41.1) to have a weak solution $x \in H_0^m$ is that for any nonzero element $w \in W$ we have

$$rw^- - Rw^+ \leq \int_G fwdt \leq Rw^- - rw^+. \tag{41.4}$$

A sufficient condition for (41.1) to have a weak solution $w \in H_0^m$ is that

$$rw^- - Rw^+ < \int_G fwdt < Rw^- - rw^+. \tag{41.5}$$

Note that if $g = 0$, and then $r = R = 0$, then the necessary condition (41.4) reduces to $\int_G fwdt = 0$, which is the well known necessary and sufficient condition in order that the linear Dirichlet problem $Ex = f(t)$ in G with zero boundary data has a weak solution $x \in H_0^m$. Statement (41.i) holds also for (strong, or usual) solutions $x \in H_0^m \cap H^{2m}$

provided the coefficients $a_{\alpha\beta}$ of E and the boundary ∂G of G are smooth. The sufficiency part of Theorem (41.i) was proved by the alternative method by Landesman and Lazer [92] for $m = 2$, and by the same method and a simpler argument by Williams [125]. Below we shall derive from (36.ii) stronger forms of the Landesman and Lazer, and Williams theorems, and of the De Figueiredo extension (for strong solutions). (Note that (41.i) holds also with $g(+\infty), g(-\infty)$ interchanged).

42. <u>More on Nonlinear Dirichlet Problems at Resonance</u>. Let G be a bounded domain in R^ν, $\nu \geq 1$, with smooth boundary, and let E be a linear selfadjoint elliptic operator of order $2m$, as in no. 40, satisfying (E_1) and (E_2) and with smooth coefficients. Thus $a_{\alpha\beta} \in C^{|\beta|}(\bar{G})$, $a_{\alpha\beta} = a_{\beta\alpha}$. For any function $x(t)$, $t \in G$, $x \in H^{2m-1}(G)$, let $Dx(t)$ denote the set of x and all partial derivatives of x up to and including $2m - 1$. Thus $Dx(t) = [D^\alpha x(t),\ t \in G,\ 0 \leq |\alpha| \leq 2m - 1]$. Let M be the total number of these derivatives, $M = M(\nu,m) = 1 + \nu + \nu(\nu + 1)/2 + \cdots$. Let $F(x,v)$ denote any given real-valued function defined on $G \times E^M$. We consider now the Dirichlet boundary value problem

$$Ex = F(t,Dx(t)), \quad t \in G, \tag{42.1}$$

$$x = 0, \quad \partial x/\partial n = 0, \ldots, \partial x^{m-1}/\partial n^{m-1} = 0. \tag{42.2}$$

Let Nx denote the Nemitsky operator defined by $(Nx)(t) = F(t,Dx(t))$, $t \in G$. We shall look for (strong) solutions x of problem (42.1-2), or $x \in H_0^m \cap H^{2m}$. Thus, the problem can be written in the form

$$Ex = Nx, \quad x \in H_0^m \cap H^{2m}. \tag{42.3}$$

We shall explicitly assume below that the null space W of E is nontrivial, and thus of dimension k, $1 \leq k < +\infty$, and we denote by $w = (w_1, \ldots, w_k)$ an arbitrary basis for W. It is not restrictive to assume that the w_i are orthonormal in $L_2(G)$. For $c = (c_1, \ldots, c_k) \in R^k$, then $v(t) = cw = c_1 w_1 + \cdots + c_k w_k$ spans W. Let Σ denote the unit sphere in R^m, or $\Sigma = [\eta \in R^m \mid |\eta| = 1]$.

(42.i) <u>Theorem</u>. Let E be selfadjoint, with smooth coefficients, and satisfy (E_1), (E_2), with null space W of dimension $k \geq 1$. Let us assume:

(F_1) For every $t \in G$, $\eta \in \Sigma$, and sequences (r_s, η_s) with $r_s > 0$, $\eta_s \in \Sigma$, $r_s \to +\infty$, $\eta_s \to \eta$ we have

$$\lim_{s\to\infty} F(t, r_s \eta_s) = h(t, \eta). \tag{42.4}$$

(F_2) The function $F(t,v)$, $(t,v) \in G \times R^M$ is continuous, and there are functions $k_0(t), k_1(t)$, $t \in G$, $k_0, k_1 \in L_2(G)$, such that

$$\begin{aligned} &|h(t,\eta)| \leq k_1(t) \quad \text{for all} \quad t \in G, \quad \eta \in \Sigma, \\ &|F(t,v)| \leq k_0(t) \quad \text{for all} \quad t \in G, \quad v \in R^M. \end{aligned} \tag{42.5}$$

(F_3) For every $v(t) = bw = b_1 w_1 + \cdots + b_k w_k$, $|b| = 1$, $t \in G$, we have

$$\int_G h\left(t, \frac{Dv(t)}{|Dv(t)|}\right) v(t)\,dt > 0, \tag{42.6}$$

[or always < 0].

Then, there are constants $\alpha_0 > 0$, $C > 0$ such that for every real $|\gamma| \leq \alpha_0$, the problem

$$Ex + \gamma x = F(t, D(x(t))), \quad t \in G, \quad \text{(a.e.)}, \tag{42.7}$$

with boundary conditions (42.2), has at least one solution $x \in H_0^m \cap H^{2m}$, $\|x\|_{2m} \leq C$.

<u>Proof</u>. Let $X = H^{2m-1}$, $\mathcal{D}(E) = H^{2m} \cap H_0^m \subset X$, $Y = L_2$. Then, $E:\mathcal{D}(E) \to Y$, and if $X_0 = \ker E$, $Y_0 = X_0$, and $P:X \to X$, $Q:Y \to Y$ are the usual orthogonal projections with $PX = X_0$, $QY = Y_0$, and we take $X_1 = (I - P)X$, $Y_1 = (I - Q)Y = \mathcal{R}(E)$, then for the map K of no. 40 we have, with the present notations, $K:Y_1 \to \mathcal{D}(E) \cap X_1$, and K is a linear bounded compact operator. For any $x \in H^{2m-1}$, in particular for any $x \in H^{2m}$, or $x \in H_0^m \cap H^{2m}$, every derivative $D^\alpha x$, $0 \leq |\alpha| \leq 2m - 1$, certainly is an element of L_2. Hence $(Nx)(t) = F(t,Dx(t))$ is defined a.e. in G, and $|(Nx)(t)| \leq k_0(t)$, $t \in G$, $k_0 \in L_2$. Thus, $\|Nx\|_0 \leq \int_G k_0(t)dt = J$, where J is a fixed constant. Condition (B) of (37.ii) is satisfied. We need prove (N_ε). Let us assume that (42.6) is satisfied. Then, there is a constant $\varepsilon > 0$ such that

$$\int_G h(t, \frac{Dv(t)}{|Dv(t)|}) v(t)dt \geq 2\varepsilon \tag{42.8}$$

for all $v = bw = b_1w_1 + \cdots + b_kw_k$, $|b| = 1$.

For any element $w \in W$ we have $w = c_1w_1 + \cdots + c_kw_k$, $c = (c_1,\ldots,c_k) \in R^k$, and thus $b = c/|c|$, $b = (b_1,\ldots,b_k)$, $|b| = 1$. For $r = |c|$, $c = rb$, $v(t) = bw = b_1w_1 + \cdots + b_kw_k$, we have

$$\begin{aligned}(F(t,Dx(t)),x^*) &= (F(t,Dx(t)),cw)\\ &= |c|\int_G F(t,Dx(t))v(t)dt.\end{aligned}$$

By comparison with (42.8) it is enough we prove that there is $R_0 > 0$ sufficiently large so that, for $r > R_0$, we have

$$\Delta = \int_G |h(t, \frac{Dv(t)}{|Dv(t)|}) - F(t,Dx(t))||v(t)|dt < \varepsilon. \tag{42.9}$$

Indeed, then (42.8), (42.9) yield $(QNx,x^*) \geq |c|\varepsilon = \varepsilon\|x^*\|$, that is, (N_ε) of statements (34.ii), (37.ii), provided (42.9) holds for all $x \in X$, $x^* \in X_0$, $x^* = cw = rv$, $Px = x^*$,

$\|x^*\| \geq R_0$, $\|x - x^*\| \leq K_0$ for some $R_0 \geq 0$ and some constant K_0 as in (37.ii). Thus, we write $x - x^* = x(t) - rv(t) = \sigma(t)$, and then

$$\Delta = \int_G \left| h\left(t, \frac{Dv(t)}{|Dv(t)|}\right) - F(t, rDv(t) + D\sigma(t)) \right| |v(t)| dx.$$

Let G_1 denote the set of all points $t \in G$ where the scalar function $v(t)$ is not defined or zero. Obviously, the set G_1 has no bearing on the value of Δ.

Let G_2 denote the set of points $t \in G$ where either the M-vector $Dv(t) = [D^\alpha v(t),\ 0 \leq |\alpha| \leq 2m - 1]$, or $D\sigma(t)$ are not defined. This set G_2 has measure zero in G and again has no bearing on the value of Δ.

Then, for $t \in G' = G - (G_1 \cup G_2)$, the M-vector $Dv(t)$ is defined and certainly not identically zero since its first component $D^0 v(t) = v(t)$ is $\neq 0$. For $t \in G'$ we take

$$\eta(t) = \frac{Dv(t)}{|Dv(t)|}, \quad \rho_r(t) = |rDv(t) + D\sigma(t)|,$$

$$\eta_r(t) = \frac{Dv(t) + r^{-1}D\sigma(t)}{|Dv(t) + r^{-1}D\sigma(t)|}.$$

Since Σ denotes the unit sphere in R^M, or $\Sigma = [\eta \in R^M \,|\, |\eta| = 1]$, $\eta(t), \eta_r(t) \in \Sigma$, $\eta_r(t) \to \eta(t)$, $\rho_r(t) \to +\infty$ as $r \to \infty$, and this holds for all $t \in G'$. Moreover,

$$\Delta = \int_{G'} |h(t, \eta(t)) - F(t, \rho_r(t)\eta_r(t))| |v(t)| dt. \tag{42.10}$$

Here $|h(t, \eta)| \leq k_1(t)$ for all $(t, \eta) \in G \times \Sigma$ and some $k_1 \in L_2(G)$. Also, $|F(t,v)| \leq k_0(t)$ for all $(t,v) \in G \times R^M$ and some $k_0 \in L_2(G)$. Thus, there is $\delta > 0$ such that for every $b = (b_1, \ldots, b_k)$, $|b| = 1$, $v(t) = bw(t)$, and any subset V of G of measure $\leq \delta$ we have

$$\int_V (k_0(t) + k_1(t)) |v(t)| dt \leq \epsilon/2.$$

Let $L_\varepsilon > 0$ be so chosen that for all $b \in R^k$ with $|b| = 1$ and $v(t) = bw(t)$ we have

$$\int_G |v(t)|dt < L_\varepsilon \varepsilon/2.$$

Now we note that, for every $t \in G'$ we have $\rho_r(t) \to +\infty$, $\eta_r(t) \to \eta(t)$, and hence $h(t,\eta(t)) - F(t,\rho_r(t)\eta_r(t)) \to 0$ as $r \to \infty$ for every $t \in G'$. This convergence is pointwise in G', hence in measure in G'. That is, given $\bar{\varepsilon} = 1/L_\varepsilon$ and δ as chosen, there is R so large that for $r \geq R$ we have

$$|h(t,\eta(t)) - F(t,\rho_r(t)\eta_r(t))| \leq 1/L_\varepsilon$$

for all $t \in G'$ except on a set Λ of measure $\leq \delta$. Thus, if $\chi_r(t)$ is the integrand in (42.10) we have

$$\Delta = \int_G \chi_r(t)dt = \int_{G'} \chi_r(t)dt = (\int_{G'-\Lambda} + \int_\Lambda)\chi_r(t)dt$$

$$\leq (1/L_\varepsilon)\int_{G'} |v(t)|dt + \int_\Lambda (k_0(t) + k_1(t))|v(t)|dt$$

$$\leq \varepsilon/2 + \varepsilon/2 = \varepsilon.$$

Here R can be chosen independently of σ because of $\|\sigma\| \leq K_0$.

Theorem (42.i) now follows from statement (37.ii). The term γx in equation (42.7) can be replaced by any sum $\sum_{|\alpha|\leq 2m-1} b_\alpha c_\alpha(t)D^\alpha x$, where c_α are fixed bounded measurable functions in G, and the real coefficients b_α, $|\alpha| \leq 2m - 1$, can be made as small as we want.

<u>Remark</u>. Let us show that (42.i) contains (41.i) (for strong solutions). Let F be of the form $F = f(t) + g(x)$ with $f \in L_2$, and g continuous in $(-\infty,+\infty)$ with finite limits $R = g(+\infty)$, $r = g(-\infty)$.

Let us assume that (42.6) holds, or $\int_G hv(t) > 0$. Let w be any nonzero element $w \in W = \ker E$, let G^+, G^- be the sets of all $t \in G$ where $w(t) > 0$ and $w(t) < 0$ respectively, so that $w/|w| = +1$ in G^+, $= -1$ in G^-, and (42.6) becomes

$$\int_{G^+} [f(t) + R]w(t)dt + \int_{G^-} [f(t) + r]w(t)dt > 0.$$

By replacing w by $-w$, we also have

$$\int_{G^+} (f(t) + r)(-w(t))dt + \int_{G^-} [f(t) + R](-w(t))dt > 0,$$

and then

$$rw^- - Rw^+ < \int f(t)w(t)dt < Rw^- - rw^+.$$

(42.ii) (Theorem). Let E be selfadjoint and satisfy (E_1) and (E_2). Let us assume that the space W of all solutions w of $Ew = 0$ has dimension $k \geq 1$, and let $w = (w_1,\ldots,w_k)$ be any basis for W. Let us assume that, for some k, $0 \leq k < 1$, we have:

(F_{1k}) For every $t \in G$, $\eta \in \Sigma$ and sequences (r_n, η_n) with $r_n > 0$, $\eta_n \in \Sigma$, $r_n \to +\infty$, $\eta_n \to \eta$ we have

$$\lim_{n\to\infty} \frac{F(t, r_n\eta_n)}{r_n^k} = h(t,\eta) \tag{42.11}$$

(F_{2k}) The function $F(t,v), (t,v) \in G \times R^M$, is continuous, and there are functions $g(t)$, $g_0(t)$, $g_1(t)$, $t \in G$, $g_0 \in L_2$, $g, g_1 \in L_{2(1-k)^{-1}}$, and a constant M such that

$$|h(t,\eta)| \leq g(t) \quad \text{for all } t \in G, \ \eta \in \Sigma,$$

$$|F(t,v)| \leq g_0(t) + |v|^k g_1(t) \quad \text{for all } t \in G, \ v \in R^M. \tag{42.12}$$

(F_{3k}) For every $v(t) = bw = b_1w_1 + \cdots + b_kw_k$, $|b| = 1$,

$$\int_G h\left(t, \frac{Dv(t)}{|Dv(t)|}\right) |Dv(t)|^k v(t)dt > 0, \tag{42.13}$$

[or always < 0].

Then, there are constants $\alpha_0 > 0$, $C > 0$ such that, for any real $|\gamma| \le \alpha_0$, the problem $Ex + \gamma x = F(t,Dx(t))$, $t \in G$, (a.e.) has at least a solution $x \in H_0^m \cap H^{2m}$, $\|x\| \le C$.

<u>Proof</u>. For the sake of brevity, we only sketch the reduction of the theorem to (37.iii). Since the integral which appears in (42.13) is > 0 for every $v(t) = bw = b_1w_1 + \cdots + b_kw_k$, $|b| = 1$, there is $\varepsilon > 0$ such that

$$\int_G h\left(t, \frac{Dv(t)}{|Dv(t)|}\right) |Dv(t)|^k \; v(t)dt > 2\varepsilon.$$

Let us choose X, Y, P, Q, K as in the proof of (42.i), and let $\|\ \|$, $\|\ \|_p$ denote the norms in X and in L_p respectively. For $s' = 2(2 - k)^{-1}$, by (42.12) and Hölder inequality, we have

$$|Nx(t)| = |F(t,Dx(t))| \le g_0(t) + g_1(t)|Dx(t)|^k,$$
$$\|Nx\|_2 \le \|g_0\|_2 + \|g_1\|_{s'}\|Dx(t)\|_2^k \le \|g_0\|_2 + \|g_1\|_{s'}\|x\|^k,$$

or $\|Nx\|_2 \le J_0 + J_1\|x\|^k$ for all $x \in X$. Now we have to show that $(QNx,x^*) \ge \varepsilon\|x^*\|^{1+k}$ for all $x \in X$, $x^* \in X_0$, $Px = x^*$, $\|x^*\| \ge R_0$, $\|x - x^*\| \le K_0 + K_1\|x\|^k$ for some $R_0 \ge 0$ and any two constants K_0, K_1 as in (37.iii).

For any $\eta > 0$, $R \ge ((1 + \eta)\eta^{-1}K_1)^{1/(1-k)}$, and $\|x^*\| \ge R$, we have $\|x\| \ge \|x^*\| \ge R$, $K_1\|x\|^k \le K_1\|x\|R_0^{k-1} \le \eta(1 + \eta)^{-1}\|x\|$, and $\|x^*\| \le \|x\| \le (1 + \eta)(\|x^*\| + K_0)$. Thus, $\|x\| \le K_0 + K_1\|x\|^k$, $\|x^*\| \ge R$ implies $\|x - x^*\| \le \eta\|x^*\| + (1 + \eta)K_0$ and also $\|x\| \le K_0' + K_1'\|x^*\|^k$ for $K_0' = K_0 + K_1K_0^k(1 + \eta)^k$, $K_1' = K_1(1 + \eta)^k$.

For $x^* = cw = rv$ as in the proof of (42.i) we have again

$$(QNx,x^*) = r\int_G F(t,rDv(t) + D\sigma(t))v(t)dt,$$

where $x - x^* = \sigma(t)$. We define the sets $G_1, G_2, G' = G - (G_1 \cup G_2)$, as in the proof of (42.i), and for every $r \ge R$

we denote by G_{1r} the set of all $t \in G'$ where either $rDv(t) + D\sigma(t) = 0$, or $|F(t, rDv(t) + D\sigma(t))| \leq g_0(t)$, and we take $G_{2r} = G' - G_{1r}$. Then meas $G_{1r} \to 0$ as $r \to \infty$, and both integrals

$$\int_{G_{1r}} h(t, \eta(t))\, |Dv(t)|^k\, v(t)\,dt,$$

$$r^{-k}\int_{G_{1r}} F(t, rDv(t) + D\sigma(t))v(t)\,dt$$

approach zero as $r \to \infty$. For $v \in R^M$ we write $F(t,v) = \tilde{F}(t,v) + F_1(t,v)$, where $\tilde{F}(t,v) = F(t,v)$ and $F_1(t,v) = 0$ whenever $|F(t,v)| \leq g_0(t)$, and $\tilde{F}(t,v) = 0$, $F_1(t,v) = F(t,v)$ otherwise. Then $|\tilde{F}(t,v)| \leq g_0(t)$, $|F_1(t,v)| \leq |v|^k g_1(t)$, and as before $r_n^{-k} F_1(t, r_n\eta_n) \to h(t,\eta)$. Now the integral

$$r^{-k}\int_{G_{2r}} \tilde{F}(t, rDv(t) + D\sigma(t))v(t)\,dt$$

approaches zero as $r \to \infty$, since its absolute value is $\leq r^{-k}\|g_0\|_2$. Thus, all we have to prove is that, for

$$\Delta = \Big|\int_{G_{2r}} [h(t,\eta(t))\,|Dv(t)|^k v(t)$$
$$- \rho_r^{-k}(t) F_1(t, \rho_r(t)\eta_r(t))\,|Dv(t) + r^{-1}D\sigma(t)|^k v(t)]\,dt\Big|,$$

we have $\Delta \leq \epsilon$ for all r sufficiently large. We shall decompose Δ into the sum $\Delta_1 + \Delta_2$ of the two expressions below. First

$$\Delta_1 = \Big|\int_{G_{2r}} h(t,\eta(t))[\,|Dv(t) + r^{-1}D\sigma(t)|^k - |Dv(t)|^k]v(t)\,dt\Big|$$
$$\leq \int_{G_{2r}} g(t) r^{-k} |D\sigma(t)|^k\, |v(t)|\,dt,$$

where $g \in L_{2(1-k)^{-1}}$, $|D\sigma(t)|^k \in L_{2/k}$, $v \in L_2$ with $(1 - k)/2 + k/2 + 1/2 = 1$. By the Hölder inequality for three integrals and $s = 2(1 - k)^{-1}$, we have

$$\Delta_1 \leq \|g\|_s r^{-k} \|D\sigma(t)\|_2^k \|v\|_2 = \|g\|_s r^{-k}\|\sigma\|^k$$
$$\leq \|g\|_s r^{-k}(\eta r + (1 + \eta)K_0)^k \leq \|g\|_s(\eta^k + (1 + \eta)^k K_0^k r^{-k}).$$

By taking η so that $\|g\|_s \eta^k < \varepsilon/4$, we have $\Delta_1 < \varepsilon/2$ for r sufficiently large. Now

$$\Delta_2 = |\int_{G_{2r}} [h(t,\eta(t)) - \rho_r^{-k} F_1(t,\rho_r(t)\eta_r(t))] |Dv(t)$$
$$+ r^{-1} D\sigma(t)|^k v(t)dt| \le \int_{G_{2r}} (g(t) + g_1(t))[|Dv(t)|^k$$
$$+ r^{-k}|D\sigma(t)|^k] |v(t)|dt,$$

and the last expression is therefore the sum of two parts, one finite and independent of r, and one $\le \varepsilon/4$ as we have seen before. There is $\delta > 0$ such that for any measurable subset V of G' of measure $< \delta$ we have

$$\int_V (g(t) + g_1(t))|Dv(t)|^k |v(t)|dt \le \varepsilon/4.$$

Now, for every $t \in G'$ we have $\rho_r(t) \to +\infty$, $\eta_r(t) \to \eta(t)$, and hence

$$h(t,\eta(t)) - \rho_r^{-k}(t) F_1(t,\rho_r(t)\eta_r(t)) \to 0$$

as $r \to \infty$ for $t \in G'$. This convergence is pointwise in G', hence in measure in G'. If L_0 denote the integral of $|Dv(t)|^k|v(t)|$ on G, then, given $\bar{\varepsilon} = \varepsilon/4L_0$ and $\delta > 0$ as above, we have

$$|h(t,\eta(t)) - \rho_r^{-k} F(t,\rho_r(t)\eta_r(t))| \le \varepsilon/4L_0$$

for all r sufficiently large and all $t \in G'$ except on a set Λ of measure $< \delta$. As in the proof of (42.i) we conclude that $\Delta_2 < \varepsilon/2$ and $\Delta \le \varepsilon$ for r large. Theorem (42.ii) follows now from (37.iii).

The same remark concerning the term γx holds as for theorem (42.i).

Section 8. Hyperbolic Partial Differential Equations

43. Work of Cesari on Doubly Periodic Solutions of the Canonic Hyperbolic Equation and Related Plane Wave Equation.

Let us consider first the nonlinear problem of the doubly T-periodic solutions $x(t,s)$ of the canonic equation in R^2:

$$x_{ts}(t,s) = f(t,s,x,x_t,x_s), \quad (t,s) \in R^2, \tag{43.1}$$

$$x(t+T,s) = x(t,s) = x(t,s+T) \tag{43.2}$$

where $x = (x_1,\ldots,x_n)$, $f = (f_1,\ldots,f_n)$, and f is assumed to be T-periodic in t and s. The underlying homogeneous linear problem is here $x_{ts} = 0$ with x satisfying (43.2). The class W of solutions to this linear homogeneous problem is ∞-dimensional (all constants, and all functions of t only and of s only). The corresponding nonhomogeneous linear problem is, therefore, $x_{ts} = h(t,s)$, $(t,s) \in R^2$, and x satisfying (43.2). If we assume that h is T-periodic in t and s and continuous, then we know that $x_{ts} = h$ has solutions $x(t,s)$ which are T-periodic in t and s if and only if

$$\int_0^T\int_0^T h(\alpha,\beta)\,d\alpha d\beta = 0, \quad \int_0^T h(t,\beta)\,d\beta = 0, \quad 0 \le t \le T,$$

$$\int_0^T h(\alpha,s)\,d\alpha = 0, \quad 0 \le s \le T.$$

Let C denote the linear space of all continuous functions $y(t,s)$, $(t,s) \in R^2$, which are T-periodic in t and s. Let $Q{:}C \to C$ denote the projection operator defined by $Qy = y_0 = \mu + m(t) + n(s)$, where

$$\mu = T^{-2}\int_0^T\int_0^T y(\alpha,\beta)\,d\alpha d\beta,$$

$$m(t) = T^{-1}\int_0^T (y(t,\beta) - \mu)\,d\beta, \tag{43.3}$$

$$n(s) = T^{-1}\int_0^T (y(\alpha,s) - \mu)\,d\alpha,$$

and thus m, n are continuous T-periodic functions of t and s respectively, and of mean value zero.

Let $C_0 = QC$ and $C_1 = (I - Q)C$, so that for every $y \in C_1$ we have $Qy = 0$, or $\mu = 0$, $m(t) \equiv 0$, $n(s) \equiv 0$.

For $y \in C$ of class C^2, $y_0 = Qy \in C^2$, $y_0 = m(t) + n(s) + \mu$, m,n T-periodic, and $y_{0st} = 0$.

Let S denote the linear space of all continuous functions $x(t,s)$, $(t,s) \in R^2$, which are T-periodic in t and s, and which possess continuous partial derivatives x_t, x_s, x_{ts}. Let $P:S \to S$ denote the projection operator defined by $Px = x_0 = u(t) + v(s)$, where

$$u(t) = x(t,0), \quad v(s) = x(0,s) - x(0,0), \quad v(0) = 0.$$

Let $S_0 = PS$ and $S_1 = (I - P)S$, so that for every $x \in S_1$ we have $Px = 0$, or $x(t,0) \equiv 0$, $x(0,s) \equiv 0$, and hence also $x(t,T) \equiv 0$, $x(T,s) \equiv 0$.

For every $y \in C_1$ the operator H defined by $H\varphi = y(t,s) = \int_0^t\int_0^s \varphi(\alpha,\beta)\,d\alpha d\beta$ yields an element $y \in S_1$ with $y_{ts} = \varphi$ in R^2. Here H is a linear operator $H:C_1 \to S_1$, and the main axioms of no. 2 are satisfied, or (k_1) $H(I - Q)Ex = (I - P)x$, (k_2) $QEx = EPx$ for every $x \in S$, (k_3) $EH(I - Q)y = (I - Q)y$ for every $y \in C$, and actually we have $QEx = EPx = 0$ for every $x \in S$.

If we now write problem (43.1-2) in the form $Ex = Nx$, where the boundary conditions (43.2) are included in the operator E, then the auxiliary equation of (43.1-2) can be written in the form

$$x(t,s) = u(t) + v(s) + \int_0^t\int_0^s [f(\alpha,\beta),x(\alpha,\beta),x_t(\alpha,\beta),x_s(\alpha,\beta))$$
$$- m(\alpha) - n(\beta) - \mu]d\alpha d\beta,$$
$$\mu = T^{-2}\int_0^T\int_0^T [f(\alpha,\beta,x(\alpha,\beta),x_t(\alpha,\beta),x_o(\alpha,\beta))]d\alpha d\beta,$$
$$m(t) = T^{-1}\int_0^T [f(t,\beta,x(t,\beta),x_t(t,\beta),x_s(t,\beta)) - \mu]d\beta,$$
$$n(s) = T^{-1}\int_0^T [f(\alpha,s,x(\alpha,s),x_t(\alpha,s),x_s(\alpha,s)) - \mu]d\alpha. \quad (43.4)$$

In other words, solving the auxiliary equation is equivalent to the problem of finding the doubly T-periodic solutions $x(t,s)$ to the problem

$$x_{ts} = f(t,s,x,x_t,x_s) - m(t) - n(s) - \mu. \quad (43.5)$$

As we shall see below, this is a well posed problem. We may call it the relaxed problem.

Here we have $Px = u(t) + v(s)$, $Qf(\cdots) = m(t) + n(s) + \mu$. The bifurcation equation relative to problem (43.1-2) is now $Qf \equiv 0$, or

$$m(t) \equiv 0, \quad n(t) \equiv 0, \quad \mu = 0. \quad (43.6)$$

Problem (43.1-2) is now equivalent to the system of equations (43.4) and (43.6).

(43.i) Theorem. (An existence theorem for the auxiliary or relaxed problem) (Cesari [14]). If $T > 0$, and N, N_1, N_2, L, b_1, b_2, M_1, M_2, $M_3 \geq 0$ are constants; if A and R are the sets

$$A = [0 \leq t \leq T, \quad 0 \leq s \leq T],$$
$$R = [0 \leq t \leq T, \quad 0 \leq s \leq T, \quad |z| \leq M_1, \quad |p| \leq M_2,$$
$$|q| \leq M_3];$$

if $M_1 \geq N + (N_1 + N_2)T/2 + 3LT^2$, $M_2 \geq N_1 + 3LT$, $M_3 \geq N_2 + 3LT$; if $u_o(t)$, $0 \leq t \leq T$, $v_o(s)$, $0 \leq s \leq T$, are vector functions which are continuous with $u_0'(t)$, $v_0'(s)$; if $f(t,s,z,p,q)$, $(t,s,z,p,q) \in R$ is continuous in R, and

$$u_0(T) = u_0(0), \quad v_0(T) = v_0(0) = 0, \quad u_0'(T) = u_0'(0),$$
$$v_0'(T) = v_0'(0), \quad |u_0(0)| \le N,$$
$$|u_0(t_1) - u_0(t_2)| \le N_1|t_1 - t_2|,$$
$$|v_0(s_1) - v_0(s_2)| \le N_2|s_1 - s_2|,$$
$$f(T,s,z,p,q) = f(0,s,z,p,q), \quad f(t,T,z,p,q) = f(t,0,z,p,q),$$
$$|f(t,s,z,p,q)| \le L,$$
$$|f(t,s,z,p_1,q_1) - f(t,s,z,p_2,q_2)| \le b_1|p_1 - p_2| + b_2|q_1 - q_2|;$$

then for

$$2Tb_1 < 1, \quad 2Tb_2 < 1, \tag{43.7}$$

there exists a vector function $\varphi(t,s)$, $(t,s) \in A$, continuous in A together with φ, φ_s, φ_t, continuous vector functions $m(t)$, $0 \le t \le T$, $n(s), 0 \le s \le T$, and a constant μ, such that

$$\varphi(t,0) = \varphi(t,T) = u_0(t), \quad \varphi_s(t,0) = \varphi_s(t,T),$$
$$\varphi(0,s) = \varphi(T,s) = u_0(0) + v_0(s), \quad \varphi_t(0,s) = \varphi_t(T,s)$$
$$m(0) = m(T), \quad n(0) = n(T), \quad \int_0^T m(\alpha)\,d\alpha = 0, \quad \int_0^T n(\beta)\,d\beta = 0,$$
$$\varphi_{ts} = f(t,s,\varphi(t,s),\varphi_t(t,s),\varphi_s(t,s)) - m(t) - n(s) - \mu,$$
$$\mu = T^{-2}\int_0^T\int_0^T f(\alpha,\beta,\varphi(\alpha,\beta),\varphi_t(\alpha,\beta),\varphi_s(\alpha,\beta))\,d\alpha d\beta,$$
$$m(t) = T^{-1}\int_0^T [f(t,\beta,\varphi(t,\beta),\varphi_t(t,\beta),\varphi_s(t,\beta)) - \mu]d\beta,$$
$$n(s) = T^{-1}\int_0^T [f(\alpha,s,\varphi(\alpha,s),\varphi_t(\alpha,s),\varphi_s(\alpha,s)) - \mu]d\alpha, \tag{43.8}$$

for all $0 \le t \le T$, $0 \le s \le T$. Thus, by extending all functions $\varphi(s,t)$, $m(t)$, $n(s)$, $f(t,s,z,p,q)$ for all $-\infty < t,s < +\infty$, $|z| \le M_1$, $|p| \le M_2$, $|q| \le M_3$ by means of the periodicity of period T in t and s, equation (43.7) is satisfied in the whole ts-plane.

Theorem (43.i) above was proved by Cesari [14], by showing that the transformation $T:x \to y$, where Tx is given by the second number of (43.4) and x varies in a suitable set $\mathcal{S}$ of S, satisfies the hypotheses of Schauder's fixed point theorem in the uniform topology, and thus there is at least one element $x \in \mathcal{S} \subset S$ for which (43.4) holds. (In [128] it was shown that (43.7) can be replaced by a slightly better numerical estimate but the proof is still in the present alternative lines.)

The assumptions of Theorem I do not guarantee uniqueness, as the following example shows: $T = 1$, $f = |z|^{1/2} \sin 2\pi t \sin 2\pi s$. Then, the equation $x_{ts} = |x|^{1/2} \sin 2\pi t \sin 2\pi s$, has the trivial solution $\varphi_1(t,s) = 0$, as well as the solution $\varphi_2(t,s) = (16\pi^4)^{-1} \sin^4 \pi t \sin^4 \pi s$, $0 \leq t,s \leq 1$.

(43.ii) Theorem. (Uniqueness of the solution of the auxiliary or relaxed problem) (Cesari [14]). Under the same hypotheses of (43.i), if $|f(t,s,x_1,p,q) - f(t,s,x_2,p,q| \leq b_0 |x_1 - x_2|$ in R; and $6T(T + 2)(b_0 + b_1 + b_2) < 1$, then the solution $x(t,s)$ to problem (43.4) is unique.

Indeed, under these hypotheses, the transformation T mentioned above is a contraction.

(43.iii) Theorem. (Continuous dependence of the solution of the relaxed problem upon the data) (Cesari [14]). Under the hypotheses of (43.i) and (43.ii), if b_0, b_1, b_2 are sufficiently small, then the solution $x(t,s)$ of problem (43.4) depends continuously on u and v in the uniform topology.

The restriction requested in [14] is actually

$$\delta = (1 - 6Tb_1)(1 - 6Tb_2) - 36T^2 b_1 b_2 > 0,$$

$$k = 6T^2 b_0 + 72T^3 \delta^{-1} b_0 (b_1 + b_2) < 1.$$

For the proof (see [14]) only simple estimates in terms of equations (43.4) are needed.

If we replace f by ϵf in (43.1), then under the hypothesis that f is Lipschitzian in $x, p, q,$ the auxiliary equation (or relaxed problem) has a unique solution which depends continuously on the data and on ϵ, for $|\epsilon|$ sufficiently small [14]. Criteria have been given in [14] for the solution of the bifurcation and hence for the original problem (43.1-2).

The criteria are too detailed to be given here. Some examples will be given below.

By means of the standard transformation $t = \xi + \eta$, $s = \xi - \eta$, problem (43.1-2) is transformed into the problem of doubly periodic solutions of the nonlinear wave equation in R^2,

$$x_{\xi\xi} - x_{\eta\eta} = g(\xi, \eta, x, x_\xi, x_\eta), \quad \text{or} \quad x_{\xi\xi} - x_{\eta\eta} = \epsilon g.$$

The analogous of theorems (43.i-iii) and corresponding criteria hold for this problem.

44. <u>Work of Hale on Doubly Periodic Solutions of the Plane Wave Equation</u>. An essentially equivalent development has been pursued by Hale [62] concerning the problem

$$x_{tt} - x_{ss} = g(t,s,x,x_t,x_s), \quad (t,s) \in R^2, \tag{44.1}$$

$$x(t + T,s) = x(t,s) = x(t,s + T), \tag{44.2}$$

$$x(t,-s) = -x(t,-s), \tag{44.3}$$

for $g: R^5 \to R$ satisfying

$$g(t + T,s,x,p,q) = g(t,s,x,p,q) = g(t,s + 2\pi,x,p,q)$$

$$g(t,s,x,p,q) = -g(t,-s,-x,-p,q).$$

Thus, the linear problem underlying the one above is $x_{tt} - x_{ss} = 0$ with x satisfying (44.2) and (44.3).

Let C denote the linear space of all continuous functions $y(t,s), (t,s) \in R^2$, or $R^2 \to R$, which are T-periodic in t and s, with $y(t,-s) = -y(t,s)$. Let $Q:C \to C$ denote the projection operator defined by $Qy = y_0$ with

$$y_0(t,s) = T^{-1}\int_0^T y(\alpha,t + s - \alpha)d\alpha$$
$$- T^{-1}\int_0^T y(\alpha,t - s - \alpha)d\alpha. \qquad (44.4)$$

Thus, y_0 is a continuous function of t and s of the form $p(t + s) - p(t - s)$ with $p:R^1 \to R^1$ and p is T-periodic. For any $y \in C$ of class C^2, $y_0 = Qy$ is a solution of the underlying linear problem $y_{0tt} - y_{0ss} = 0$ with y_0 satisfying (44.2), (44.3). Let $C_0 = QC$, $C_1 = (I - Q)C$.

Let S denote the linear space of all functions $x(t,s)$, $(t,s) \in R^2$, or $R^2 \to R^1$, which are of class C^2, T-periodic in t and s, with $y(t,-s) = -y(t,s)$. Let $P:S \to S$ denote the same operator $P = Q$ defined by (44.4), and let $S_0 = PS$, $S_1 = (I - P)S$. We take in S the usual uniform norm $\|x\|_2$ as for functions of class C^2.

We note now that for every $\varphi \in C_1$, that is, $\varphi \in C$ with $Q\varphi = 0$, there is one and only one element $x = H\varphi$, $x \in S_1$, (that is, $x \in S$, $PS = 0$), satisfying $x_{tt} - x_{ss} = \varphi$. We have here a linear operator $H:C_1 \to S_1$.

We shall now write problem (44.1-3) in the form $Ex = Nx$ where E is the differential operator defined by $Ex = x_{tt} - x_{ss}$ with $x \in S$, that is, x of class C^2 satisfying conditions (44.1), (44.2). For N we take the operator $Nx = g(t,s,x(t,s),x_t(t,s),x_s(t,s))$ again for $x \in S$.

We are again in the situation of no. 2. The auxiliary equation relative to problem $Ex = Nx$ has now the form

$$x = x^* + H(I - Q)Nx \tag{44.4}$$

where x^* is any element of S_0 and $x \in S$ with $Px = x^*$.

The bifurcation equation is then

$$QNx = 0. \tag{44.5}$$

We consider now with Hale the problem of the perturbation type

$$x_{tt} - x_{ss} = \varepsilon g(t,s,x,x_t,x_s)$$

with x satisfying (44.2), (44.3), and where ε is a real "small" parameter. Then the auxiliary equation becomes

$$x = x^* + \varepsilon H(I - Q)Nx.$$

(44.i) <u>Theorem</u>. Given $R > 0$ let g be continuous in $\Omega(R) = [(t,s,x,p,q) \mid (t,s) \in R^2,\ |x| + |p| + |q| \leq R]$, with Lipschitz first order partial derivatives g_s, g_x, g_p, g_q. Let c, d be arbitrary constants, $0 < c < d < R$. Then there is some $\varepsilon_1 > 0$ such that, for each element $x^* \in S_0$ with $\|x^*\|_2 \leq c$, and for each ε real, $|\varepsilon| \leq \varepsilon_1$, there is a unique function $x \in S$ with $Qx = x^*$, $\|x\|_2 \leq b$, which satisfies (44.4). Thus, x is of class C^2 and satisfies

$$x_{tt} - x_{ss} = \varepsilon(I - Q)g(t,s,x(t,s),x_t(t,s),x_s(t,s)).$$

This theorem has been proved by Hale [62] by showing that the operation $x \to y$ defined by $y = x^* + \varepsilon H(I - Q)Nx$, or $C \to S$ is a contraction in the uniform topologies.

Both in the discussion in [14] and the parallel one in [62], the bifurcation equation is studied by a number of devices, namely, some form of the implicit function theorem in Banach spaces. The difficulty here is that $\ker E$ is an infinite dimensional space.

The details of the many criteria for the solution of the bifurcation equation, and therefore the original problem $x_{ts} = \varepsilon f(t,s,x,x_t,x_s)$, or $x_{tt} - x_{ss} = \varepsilon g(t,s,x,x_t,x_s)$, are too intricated to be presented here. Therefore, we only present a few applications and examples.

1. Consider the equation [62]

$$u_{tt} - u_{xx} = \varepsilon[u_t + bu + cu^3 + f(t,s)] \tag{44.6}$$

where b, c are constants, $b \neq 0$, and f is of class C^2, T-periodic in t and s, odd in s, with Lipschitz first and second derivatives. For $|\varepsilon|$ sufficiently small this equation has a doubly T-periodic solution $u(t,s;\varepsilon)$, odd in s, of class C^2 in t, s, depending continuously on ε.

2. Consider the equation [62]

$$u_{tt} - u_{ss} = \varepsilon[-u_t^3 + f(t,s)] \tag{44.7}$$

where f is as before, and assume that $h(t) = T^{-1}\int_0^T f(\alpha, t - \alpha)\,d\alpha$, is odd in y and not identically zero. Then, for $|\varepsilon|$ sufficiently small, equation (44.7) has a unique doubly T-periodic solution $u(t,s;\varepsilon)$, odd in s, of class C^2 in t, s, depending continuously on ε.

3. Consider the equation [62]

$$x_{ts} = \varepsilon[f(t,s) + Cx + \varepsilon g(t,s,x)] \tag{44.8}$$

where C is a nonzero constant, f and g are T-periodic

t and s, and continuously differentiable in t, s, x. Then, for $|\varepsilon|$ sufficiently small this equation has a doubly T-periodic solution $x(t,s;\varepsilon)$, odd in s, of class C^2 in t, s, depending continuously on ε.

4. Consider the equation

$$x_{ts} = \varepsilon[f(t,s) + Cx + f_1(s) + f_2(t) + \varepsilon g(t,s,x,x_t,x_s)], \tag{44.9}$$

where f, f_1, f_2 are continuous and T-periodic in t and s, where g is T-periodic in t and s, continuous in all its five arguments, and also Lipschitz continuously differentiable in its last three arguments, and C is nonzero constant. Then, for $|\varepsilon|$ sufficiently small, (44.9) has a solution $x(t,s;\varepsilon)$ which depends continuously on ε.

45. <u>Periodicity in only one variable</u>. Cesari discussed in [12,15] the following problem in a strip S,

$$x_{ts} = f(t,s,x,x_t,x_s), \quad (t,s) \in S, \tag{45.1}$$

$$x(t,0) = u(t), \quad x(t+T,s) = x(t,s), \tag{45.2}$$

where $S = [-\infty < t < +\infty, -a \le s \le a]$, $a > 0$ is sufficiently small, and f and u are given functions.

Here $f(t,s,x,p,q)$ is a given Lipschitz continuous function of its five arguments in a region R of the form $[-\infty < t < +\infty, -a_0 \le s \le a_0, |x| \le M_1, |p| \le M_2, |q| \le M_2]$, and u is a given continuous function with Lipschitz continuous derivative u'. Both f and u are T-periodic in t, say

$$|f(t_1,s_1,x_1,p_1,p_2) - f(t_2,s_2,x_2,p_2,q_2)|$$
$$\leq k_1|t_1 - t_2| + k_2|s_1 - s_2| + k_3|x_1 - x_2| + M|p_1 - p_2| + b|q_1 - q_2|,$$
$$|u(t_1) - u(t_2)| \leq N_1|t_1 - t_2|, \quad |u'(t_1) - u'(t_2)| \leq k_4|t_1 - t_2|,$$
$$f(t + T,s,x,p,q) = f(t,s,x,p,q), \quad u(t + T) = u(t).$$

Specifically, it is also assumed that for some real μ the unique solution $q(t)$ of the differential equation $dq/dt = f(t,0,u(t),u'(t),q)$, $0 \leq t \leq T$, with $q(0) = \mu$, satisfies the necessary condition

$$\int_0^T f(t,0,u(t),u'(t),q(t))dt = 0,$$

and moreover

$$\int_0^T f_q(t,0,u(t),u'(t),q(t))dt \neq 0.$$

(45.i) Theorem [15]. If $2Tb < 1$, then there is a constant a, $0 < a < a_0$, and a continuous function $\nu(s)$, $-a \leq s \leq a$, with Lipschitz continuous derivative ν' and with $\nu(0) = 0$, $\nu'(0) = \mu$, and a function $x(t,s)$, $-\infty < t < +\infty$, $-a \leq s \leq a$, T-periodic in t, continuous with continuous derivatives x_t, x_s, x_{ts}, satisfying (45.1), (45.2), and $x_s(t,0) = q(t)$.

This theorem was proved in [15] on the basis of Theorems (43.i-iii) (and analogous ones), and an implicit function theorem previously proved.

As an application, it was shown in [15], that the problem

$$x_{ts} = \sin t + \lambda u_s + sL(t,s,x,x_t,x_s),$$
$$x(t + 2\pi,s) = x(t,s), \quad x(t,0) = u(t),$$

always has a solution $x(t,s)$, 2π-periodic in t, in a strip

S sufficiently narrow, whenever $f = \sin t + \lambda q + sL(t,s,x,p,q)$ and $u(t)$ are given functions satisfying the requirements above.

46. Work of Petrovanu on Periodic Solutions of the Canonic Hyperbolic Equation in R^3 and the Tricomi Problem.

Petrovanu [115] pursued the problem of periodic solutions to relaxed problems associated with the equation

$$x_{\xi\eta\zeta} = f(\xi,\eta,\zeta,x,x_\xi,x_\eta,x_\zeta,x_{\xi\xi},\dots,x_{\eta\eta}) \tag{46.1}$$

where $x = (x_1,\dots,x_n)$, $f = (f_1,\dots,f_n)$. Again, one can seek solutions $x(\xi,\eta,\zeta)$ in R^3 which are T-periodic in ξ,η,ζ. Alternatively, one can seek solutions $x(\xi,\eta,\zeta)$ in a thin slab $[-\infty < \xi,\eta < +\infty,\ |\zeta| \le a]$, for a small, which are T-periodic in ξ and η. Finally, one can seek solutions $x(\xi,\eta,\zeta)$ in a thin cylinder $[-\infty < \zeta < +\infty,\ |\xi|,|\eta| \le a]$, for a small, which are T-periodic in ξ. Petrovanu has taken into consideration these problems in [115]. Thus, in the latter case, the solutions $x(\xi,\eta,\zeta)$ to the relaxed problem are required to satisfy

$$x_{\xi\eta\zeta} = f(\xi,\eta,\zeta,x,\dots,x_{\xi\eta}) - m(\eta,\zeta),$$
$$m(\xi,\eta,\zeta) = T^{-2}\int_0^T f(\alpha,\eta,\zeta,x(\alpha,\eta,\zeta),\dots,x_{\xi\eta}(\alpha,\eta,\zeta))\,d\alpha,$$
$$x(\xi + T,\eta,\zeta) = x(\xi,\eta,\zeta),$$
$$-\infty < \zeta < +\infty, \quad |\eta|,|\xi| \le a.$$

Petrovanu extended Theorems (43.i-iii) to this situation, and he made quantitative improvements leading to better estimates for the constants as discussed in [128].

Petrovanu also studied, in [116], the doubly T-periodic relaxed solutions to the Tricomi problem

$$x(t + T,s) = x(t,s) = x(t,s + T),$$
$$x_t(t,s) = F(t,s,x,y),$$
$$y_s(t,s) = G(t,s,x,y),$$
$$x(0,s) = u(s), \quad y(t,0) = v(t), \qquad (46.2)$$

where $x = (x_1,\ldots,x_m)$, $y = (y_1,\ldots,y_n)$, $F = (F_1,\ldots,F_m)$, $G = (G_1,\ldots,G_n)$; where F and G are functions of t,s,x,y which are T-periodic in t and s; and where u, v are given T-periodic functions of s and t respectively. The corresponding auxiliary, or relaxed problem is now

$$x_t(t,s) = F(t,s,x,y) - m(t), \quad (t,s) \in R^2,$$
$$y_s(t,s) = G(t,s,x,y) - n(s), \quad (t,s) \in R^2,$$
$$x(0,s) = u(s), \quad y(t,0) = v(t),$$
$$m(t) = T^{-1}\int_0^T F(t,\beta,x(t,\beta),y(t,\beta))d\beta,$$
$$n(s) = T^{-1}\int_0^T G(\alpha,s,x(\alpha,s),y(\alpha,s)d\alpha,$$
$$x(t + T,s) = x(t,s) = x(t,s + T).$$

Petrovanu [116] extended Theorems (43.i-iii) to this problem and he made quantitative improvements leading to better estimates of the constants as discussed in [128].

Problem (46.2), without periodicity requirements and in a given interval $|t| \leq a$, $|s| \leq b$ in R^{m+n}, can be reduced to a system of integral equations which had been studied by Tricomi [189], and successively by Santagati [186] and Villari [206]. The results of Petrovanu incorporate this more recent work by requiring only, e.g., in one of the statements, that F and G be continuous in their arguments, and F Lipschitz in x and G Lipschitz in y.

47. Work of Haimovici on Problems Whose Unknowns are Set Functions. Haimovici has extended, in [46-50], much of the work in no. 43 and no. 45 to abstract differential equations where the unknowns are set functions. Precisely Haimovici has taken into consideration equations of the form

$$(d\varphi/d\mu)(P) = f(P, \varphi(E_P)) \qquad (47.1)$$

where P denotes the points of a domain Ω in R^n and $E_P:\Omega \to A$ is an application of Ω into a σ-ring A of subsets of Ω; where $\varphi(E)$, or $\varphi:A \to R$, is a real-valued function defined on the sets E of Ω, $E \in A$; where μ is the Lebesgue measure in R^n; and where f denotes a continuous real-valued function in R^{n+1} which is Lipschitzian in its last variable. Thus, (47.1) is an abstract formulation of first order differential equations. Second order equations can be written in the form

$$(d\varphi/d\mu)(P) = \varphi_1(E_P), \quad (d\varphi_1/d\mu)(P) = f(P, \varphi(E_P). \qquad (47.2)$$

Also, Haimovici has considered a suitable concept of generalized periodicity. In particular, Haimovici has proved the existence of solutions of suitable relaxed (or auxiliary) problems related to (47.1) and (47.2).

The following relevant applications of these ideas are contained in the very recent paper [47] by Haimovici. Let us start with the following problem formulation: To find a mass-distribution on a solid ball S of radius R admitting a density at each point of S, and such that this density at each point P is a function of P and of the mass contained in a certain domain of S depending on P. Precisely, if ρ, θ, φ are the spherical coordinates of P, with respect to the centre of S, let δ be the domain $\rho \leq \gamma \leq R$, $0 \leq t \leq \theta$, $0 \leq s \leq \varphi$, and let u be the mass of δ. Then the density ot P is $\rho^{-2} \sin^{-1}\theta u_{\rho\theta\varphi}$ and the stated problem leads to an equation of the form

$$\rho^{-2} \sin^{-1}\theta u_{\rho\theta\varphi} = f(\rho, \theta, \varphi, u(\rho, \theta, \varphi)).$$

Now for u and f the following periodicity conditions must be expected

$$u(\rho, \theta, \varphi + 2\pi) = u(\rho, \theta, \varphi) + 2\pi\lambda(\rho, \theta),$$

$$f(\rho, \theta, \varphi + 2\pi), \quad u + 2\pi\lambda(\rho, \theta)) = f(\rho, \theta, \varphi, u),$$

where $2\pi\lambda(\rho, \theta)$ is the mass of the domain $\rho \leq \gamma \leq R$, $0 \leq t \leq \varphi$, $0 \leq s < 2\pi$ and where the density is periodic in φ. Thus, the problem here is a periodicity condition with period $\lambda(\rho, \theta)$ which depends on the remaining two coordinates.

Haimovici frames this problem in a more general one with this kind of periodicity and, by bifurcation analysis and the use of implicit function theorems (cf. [15]), proves the existence and uniqueness of the solution under suitable Lipschitz hypotheses on the data.

48. The Work of Hecquet. Hecquet, in a series of papers [74-77], has studied the problem of the periodic relaxed solutions to a number of partial differential equations, such as

$$x_{t_1 \ldots t_n} = f(t_1, \ldots, t_n, x, x_{t_1}, \ldots, x_{t_n}, x_{t_1 t_1}, \ldots, x_{t_2 \ldots t_n}), \tag{48.1}$$

where $t = (t_1, \ldots, t_n) \in R^n$, and periodicity is taken with respect to all variables, or only one of them, or any subset of them. He has done the same for the equation

$$x_{t_1^2 \cdots t_n^2} = f(t_1, \ldots, t_n, x, x_{t_1}, \ldots, x_{t_1^2 \cdots t_{n-1}^2 t_n}), \quad t \in R^n,$$

and for the total differential equation

$$dz = A(t,s,z)dt + B(t,s,z)ds,$$

where $z = (z_1, \ldots, z_n)$, $A = (A_1, \ldots, A_n)$, $B = (B_1, \ldots, B_n)$, and t and s are scalar. In [74] Hecquet has also studied the same problem of the relaxed periodic solutions of (48.1), for $n = 2$, obtaining the existence of relaxed periodic solutions by a method of successive approximations, where the solutions are approximated by polyhedral surfaces $x = x_{mn}(\xi, \eta)$, whose first order partial derivatives are sectionally constants, namely constant on each cell of a suitable partition of the domain.

49. The Work of Hall. In [67] Hall has considered the equation

$$Lx = D_{tt}x(t,s) + (-1)^p D_s^{2p} x(t,s) = \varepsilon f(t,s,u(t,s)), \qquad (49.1)$$

where $p = 1, 2, \ldots$ is any integer, f is 2π-periodic in t and s, and ε is a small parameter. For any $r > 1$, weak solutions $x(t,s)$ are sought which are 2π-periodic in t and s. For any $r > 1$ the class of periodic distributions L_r is first defined, and the space M of all distributions x satisfying $Lx = 0$ is then taken into consideration. The equation $Lx = g$ has a distribution solution x if and only if $g \in M^{\perp}$, and then there is a unique such solution, say $x = Kg$, in $M^{\perp}$, and K is a linear map. In particular, it is proved that, for $g \in M^{\perp} \cap L_{r/(r-1)}$, then $x = Kg \in M^{\perp} \cap L_r$, and K is a linear, bounded, compact map

$K: M^{\perp} \cap L_{r/(r-1)} \to M^{\perp} \cap L_r$. The lengthy analysis of the auxiliary and bifurcation equations cannot be reported here.

The results apply for instance to the equation (49.1) with

$$f = |x|^{r-1} \operatorname{sgn} x + g(t,s), \quad r \geq 2.$$

In [68] Hall has extended his analysis to the solutions $x(t,s)$ which are 2π-periodic in t and s, of the equation

$$\begin{aligned} Lu &= D_{tt}x(t,s) + (-1)^p D_s^{2p} x(t,s) \\ &= \varepsilon f(t,s,x,x_t,x_s,\ldots,D_s^p x), \end{aligned}$$

where f is 2π-periodic in t and s, and ε is a small parameter. Again, the problem is separated into auxiliary and bifurcation equations, the former is solved by Banach fixed point theorem for a contraction map, and the latter by the implicit function theorem.

The final statement being too lengthy to be reported here, we only mention that, for instance, it implies that the equation

$$\begin{aligned} &x_{tt} - x_{ss} = \varepsilon[x_t + ax + bx^3 + g(\ldots)], \\ &x(t,0) = x(t,\pi) = 0, \end{aligned}$$

always has a solution $x(t,s)$ which is 2π-periodic in t whenever a is fixed, b is also fixed and of limited size, and $|\varepsilon|$ sufficiently small.

In [69] Hall has extended his bifurcation type analysis to the equation

$$D_{tt}x(t,s) + \Delta_s^{2p} x = b|x|^{r-1} \operatorname{sgn} x + g(t,s),$$

t scalar, s vector, where Δ is the Laplacian in the s-space, and g is periodic.

By totally different methods, Hall has studied in [149] the solutions periodic in t of some period $2\pi/\omega$ of the equation $x_{tt} - x_{ss} = f(x)$ with $x(t,0) = x(t,\pi) = 0$. In [150] Fink, Hall and Hausrath have studied the periodic solutions of $x_{tt} - x_{ss} + \lambda x = \epsilon(x_t - x_t^3)$ with $x(t,0) = x(t,\pi) = 0$. Also, in [149] Fink, Hall and Khahili have studied the periodic solutions of equations such as $x_{tt} - x_{ss} = x^3$, or $x_{tt} - x_{ss} = \alpha x + \beta x^3$, or $x_{tt} - x_{ss} = -m^2 \sin x$ with $x(t,0) = x(t,\pi) = 0$.

In [70] Hall discusses problems of periodic solutions for abstract equations in Banach spaces by an extended form of the alternative scheme in no. 2 for the case when $Y = X^*$ is the dual of X, weak solutions are sought, and the projection operator $P:X \to X_0$ in X is replaced by the natural map from X to X/X_0, the quotient space.

50. Other Work. Problems of the existence of periodic solutions in t of nonlinear wave equations in R^2 of the perturbation type, such as

$$x_{tt} - x_{ss} = \epsilon f(t,s,x),$$
$$x_{tt} - x_{ss} = \epsilon f(t,s,x,x_t,x_s)$$

are studied by Rabinowitz in [181], [182], [183].

Perturbation type mixed problems, as well as problems of periodic solutions for the nonlinear wave equations, are studied by Vejvoda in [197], [199].

Problems of existence of periodic solutions for the biharmonic wave equation in some bounded domain G in E_N, say

$$x_{tt} + \Delta_s^2 x + cx + x_t + x_t|x_t| = h(t,s)$$

t scalar, s vector, $s \in G \subset E_N$, have been studied by

Havlova in [158]. Also analogous problems for the nonlinear equation of the beam

$$x_{tt} + x_{ssss} = h(t,s) + \varepsilon g(t,s,x,x_t,x_s,x_{ss})$$

under various boundary conditions are studied by Vejvoda and Krylova in [201], [202].

Periodic solutions of abstract differential equations in Banach spaces are studied by Vejvoda and Straskaia in [204].

The nonlinear wave equation in R_2 is also studied for periodic solutions by De Simon and Torelli in [143], [144], [188] with the use of monotone operator theory.

BIBLIOGRAPHY

Selected References on the Alternative Method

1. Bailey, H.R. and R.A. Gambill, "On stability of periodic solutions of weakly nonlinear differential equations," J. Math. Mech. 6 (1957), 655-668.

2. Banfi, C., "Sulla determinazione delle soluzioni periodiche di equazioni nonlineari periodiche," Boll. Un. Mat. Ital. (4) 1 (1968), 608-619.

3. Banfi, C., "Su un metodo di successive approssimazioni per lo studio delle soluzioni periodiche di sistemi debolmente nonlineari," Atti. Accad. Sci. Torino 100 (1965-1966), 471-479.

4. Banfi, C. and G. Casadei, "Calcolo di soluzioni periodiche di equazioni differenziali nonlineari," Calcolo, Vol. 5, Suppl. 1 (1968), 1-10.

5. Bowman, T.T., "Periodic solutions of Liénard systems with symmetries," to appear.

6. Cesari, L., "Functional analysis and periodic solutions of nonlinear differential equations," Contributions to Differential Equations 1, Wiley (1963), 149-187.

7. Cesari, L., "Functional analysis and Galerkin's method," Mich. Math. J. 11 (1964), 385-414.

8. Cesari, L., "Sulla stabilità delle soluzioni di sistemi di equazioni differenziali lineari a coefficienti periodici," Mem. Accad. Italia (6) 11 (1941), 633-695.

9. Cesari, L., <u>Asymptotic Behavior and Stability Problems in Ordinary Differential Equations</u> (particularly pp. 67-72, 123-236), Ergebn. d. Math. No. 16, Springer-Verlag, 1959; 2nd ed., 1963; 3rd ed., 1971. Russian Edition, MIR, Moscow, 1964.

10. Cesari, L., "Existence theorems for periodic solutions of nonlinear Lipschitzian differential systems and fixed point theorems," Contributions to the Theory of Nonlinear Oscillations 5 (1960), 115-172. Annals of Mathematics Study, Princeton, No. 45.

11. Cesari, L., "Existence theorems for periodic solutions of nonlinear differential systems," Symposium on Differential Equations, Mexico City, 1959, Bol. Soc. Mat. Mexicana, 1960, 24-41.

12. Cesari, L., "Periodic solutions of hyperbolic partial differential equations," International Symposium. Nonlinear Differential Equations and Nonlinear Mechanics, Colorado Springs, Proceedings, Academic Press (1963) 33-57. Also: International Symposium Nonlinear Oscillations, Kiev, Ukraine, Proceedings 2 (1964), 440-457.

13. Cesari, L., "Periodic solutions of nonlinear hyperbolic partial differential equations," International Symposium "Les vibrations forcées dans les systèmes non-linèaires," Marseille, Acta (1964), 425-437.

14. Cesari, L., "Existence in the large of periodic solutions of hyperbolic partial differential equations," Arch. Rational Mech. Anal. 20 (1965), 170-190.

15. Cesari, L., "A criterion for the existence in a strip of periodic solutions of hyperbolic partial differential equations," Rend. Circ. Mat. Palermo (2) 14 (1963), 95-118.

16. Cesari, L., "Smoothness properties of periodic solutions in the large of nonlinear hyperbolic differential systems," Funkcial. Ekvac. 9 (1960), 325-338.

17. Cesari, L., "Periodic solutions of nonlinear differential systems," Symposium Active Networks and Feedback Systems. Polytechnic Institute, Brooklyn, Proceedings 10 (1960), 545-560.

18. Cesari, L., "A nonlinear problem in potential theory," Mich. Math. J. 16 (1969), 3-20.

19. Cesari, L., "Functional analysis and differential equations," Studies in Applied Math. 5 SIAM (1969), 143-155.

20. Cesari, L., "Functional analysis and boundary value problems," Analytic Theory of Differential Equations. Lecture Notes No. 183, Springer-Verlag (1971), 178-194.

21. Cesari, L., "Nonlinear Analysis," Lecture Notes C.I.M.E., Centro Internazionale Matematico Estivo, Bressanone, 4-13, giugno 1972. Cremonese, Roma, 1973.

22. Cesari, L., "Alternative methods in nonlinear analysis," International Conference on Differential Equations, Los Angeles, September 1974. Academic Press 1975, 95-148.

23. Cesari, L., "Nonlinear oscillations in the frame of alternative methods," International Conference on Dynamical Systems, Providence, R.I., August (1974). Dynamical Systems, Academic Press 1976, vol. 1, 29-50.

24. Cesari, L., "The alternative method in nonlinear oscillations," Math. Asso. Amer. Studies in Math. (J.K. Hale ed.), to appear.

25. Cesari, L., "Qualitative conditions for nonlinear oscillations," International Conference on Nonlinear Oscillations, East Berlin, September, 1975.

26. Cesari, L., "An abstract existence theorem across a point of resonance," International Symposium on Dynamical Systems, Gainesville, Florida, March 1976. Academic Press.

27. Cesari, L., "Nonlinear oscillations across a point of resonance for nonselfadjoint systems," J. Differential Equations. To appear.

28. Cesari, L., "Nonlinear problems across a point of resonance for nonselfadjoint systems," Nonlinear Analysis, Academic Press. To appear.

29. Cesari, L. and T.T. Bowman, "Some error estimates by the alternative method," Quart. Appl. Math., to appear.

30. Cesari, L. and H.R. Bailey, "Boundedness of solutions of linear differential equations with periodic coefficients," Arch. Rational Mech. Anal. 1 (1958), 246-271.

31. Cesari, L. and J.K. Hale, "Second order linear differential equations with periodic L-integrable coefficients," Riv. Mat. Univ. Parma 5 (1974), 55-61.

32. Cesari, L. and J.K. Hale, "A new sufficient condition for periodic solutions and weakly nonlinear differential systems," Proc. Amer. Math. Soc. 8 (1957), 757-764.

33. Cesari, L., C.A. Borges and D.A. Sanchez, "Functional analysis and the method of harmonic balance," Quart. Appl. Math. 32 (1975), 457-464.

34. Cesari, L. and R. Kannan, "Functional analysis and nonlinear differential equations," Bull. Amer. Math. Soc. 79 (1973), 1216-1219.

35. Cesari, L. and R. Kannan, "Periodic solutions in the large of nonlinear differential equations," Rend. Mat. (2) 8 (1975), 633-654.

36. Cesari, L. and R. Kannan, "Solutions in the large of Liénard systems with forcing terms," Ann. Mat. Pura Appl., to appear.

37. Cesari, L. and R. Kannan, "An abstract existence theorem at resonance," Proc. Amer. Math. Soc., to appear.

38. Christopher, P.A.T., (a) "A new class of subharmonic solutions to Duffing's equation," CO. A. Rep. 195. The College of Aeronautics, Cranfield. Bedford, England 1967; (b) "An extended class of subharmonic solutions to Duffing's

equation," Ibid. 199, 1957; (c) "The response of a second order nonlinear system to a step-function disturbance," Ibid. 205, 1969.

39. Cronin, J., "Periodic solutions of nonautonomous equations," Boll. Un. Mat. Ital. (4) 6 (1972), 45-54.

40. Cronin, J., "Equations with bounded nonlinearities," J. Differential Equations 14 (1973), 581-596.

41. Gambill, R.A., "Stability criteria for linear differential systems with periodic coefficients," Riv. Mat. Univ. Parma 5 (1954), 169-181.

42. Gambill, R.A., "Criteria for parametric instability for linear differential systems with periodic coefficients," Riv. Mat. Univ. Parma 6 (1955), 37-43.

43. Gambill, R.A., "A fundamental system of real solutions for linear differential systems with periodic coefficients," Riv. Mat. Univ. Parma 7 (1956), 311-319.

44. Gustafson, K. and D. Sather, "Large nonlinearities and monotonicity," Arch. Rational Mech. Anal. 48 (1972), 109-122.

45. Gustafson, K. and D. Sather, "Large nonlinearities and closed linear operators," Arch. Rational Mech. Anal. 52 (1973), 10-19.

46. Haimovici, A., "Sur le problème de Dirichlet pour des équations dont les inconnues sont des functions d'ensemble," Boll. Un. Mat. Ital. (4) 3 (1969), 373-392.

47. Haimovici, A., "Periodic solutions of differential equations with set functions as unknowns," Rev. Roumaine Math. Pures Appl. 19 (1974), 161-170.

48. Haimovici, A., "Sur une équation differentielle pour une function d'ensemble," Rev. Roumaine Math. Pures Appl. 9 (1964), 207-210.

49. Haimovici, A., "Ein Dirichletsches Problem für Mengenfunction," Math. Nachr. 33 (1967), 347-353.

50. Haimovici, A., "Sur un système d'equations anx derivées partielles d'une function d'ensemble," Mathematica (8) 31 (1966), 261-265.

51. Hale, J.K., "Evaluations concerning products of exponential and periodic functions," Riv. Mat. Univ. Parma 5 (1954), 63-81.

52. Hale, J.K., "On boundedness of the solutions of linear differential systems with periodic coefficients," Riv. Mat. Univ. Parma 5 (1954), 137-167.

53. Hale, J.K., "Periodic solutions of nonlinear systems of differential equations," Riv. Mat. Univ. Parma 5 (1955), 281-311.

54. Hale, J.K., "On a class of linear differential equations with periodic coefficients," Illinois J. Math. 1 (1957), 98-104.

55. Hale, J.K., "Linear systems of first and second order differential equations with periodic coefficients," Illinois J. Math. 2 (1958), 586-591.

56. Hale, J.K., "Sufficient conditions for the existence of periodic solutions of systems of weakly nonlinear first and second order differential equations," J. Math. Mech. 7 (1958), 163-172.

57. Hale, J.K., "A short proof of a boundedness theorem for linear differential systems with periodic coefficients," Arch. Rational Mech. Anal. 2 (1959), 429-434.

58. Hale, J.K., "On the behavior of the solutions of linear periodic differential systems near resonance points," Contributions to the Theory of Nonlinear Oscillations, Vol. 5, pp. 55-89. Princeton Univ. Press, 1960.

59. Hale, J.K., "On the stability of periodic solutions of weakly nonlinear periodic and autonomous differential systems," Contributions to the Theory of Nonlinear Oscillations, Vol. 5, pp. 91-113, Princeton Univ. Press, 1960.

60. Hale, J.K., "On the characteristic exponents of linear periodic differential systems," Bol. Soc. Mat. Mexicana (2) 5 (1960), 58-66.

61. Hale, J.K., Oscillations in Nonlinear Systems, McGraw-Hill, 1963.

62. Hale, J.K., "Periodic solutions of a class of hyperbolic equations," Arch. Rational Mech. Anal. 23 (1967), 380-398.

63. Hale, J.K., Applications of Alternative Problems, Lecture Notes, Brown University, 1971.

64. Hale, J.K., Ordinary Differential Equations, Interscience, 1969.

65. Hale, J.K., S. Bancroft and D. Sweet, "Alternative problems for nonlinear equations," J. Differential Equations 4 (1968), 40-56.

66. Hale, J.K. and R.A. Gambill, "Subharmonic and ultraharmonic solutions for weakly nonlinear systems," J. Rat. Mech. Anal. 4 (1956), 353-398.

67. Hall, W.S., "On the existence of periodic solutions for the equation $D_{tt}u + (-1)^p D_x^{2p} u = f(\ldots,u)$," J. Differential Equations 7 (1970), 509-526.

68. Hall, W.S., "Periodic solutions of a class of weakly nonlinear evolution equations," Arch. Rational Mech. Anal. 39 (1970), 294-322.

69. Hall, W.S., "The bifurcation of solutions in Banach spaces," Trans. Amer. Math. Soc. 161 (1971), 207-208.

70. Hall, W.S., "On the bifurcation of periodic solutions in Banach spaces, I and II," to appear.

71. Harris, W.A., "Holomorphic solutions of nonlinear differential equations at singular points," SIAM Studies in Applied Mathematics 5 (1969), 184-187.

72. Harris, W.A., Y. Sibuya and L. Weinberg, "Holomorphic solutions of linear differential systems at singular points," Arch. Rational Mech. Anal. 35 (1969), 245-248.

73. Hecquet, G., "Utilization de les approximations d'Euler-Cauchy pour la demonstration d'un theorème de Cesari," C.R. Acad. Sci. Paris, t. 273 (A) (1971), 712-715.

74. Hecquet, G., "Contribution à la recherche des solutions périodiques en x_1 de l'equation $u_{x_1\ldots x_n} = f(x_1,\ldots,x_n, u,u_{x_1},\ldots,u_{x_2\ldots x_n})$. C.R. Acad. Sci. Paris, Ser. A, 276 (1973), 997-1000.

75. Hecquet, G., "Contribution à la recherche des solutions périodiques en x_1 et x_2 de l'equation $u_{x_1\ldots x_n} = f(x_1,\ldots,x_n,u,u_{x_1},\ldots,u_{x_2\ldots x_n})$. C.R. Acad. Sci. Paris, Ser. A., 276 (1973), 1047-1050.

76. Hecquet, G., "Recherche des solutions periodiques en x_1 de l'equation $u_{x_1^2\ldots x_n^2} = f(x_1,\ldots,x_n,u,u_{x_1},\ldots, u_{x_1\ldots x_n})$. Report. Lille, France, 1973.

77. Hecquet, G., "A propos d'un article de Dan Petrovanu sur les solutions periodiques d'un système aux differentielles

totales," Rend. Accad. Naz. Lincei, Cl. Sci. Fis. Mat. Mat. (8) 57 (1974), 40-47.

78. Imaz, C., "Sobre ecuaciones differenciales lineales periodicas con un parametro pequeno," Bol. Soc. Mat. Mexicana (2) 6 (1961), 19-51.

79. Kannan, R., "Periodically disturbed conservative systems," J. Differential Equations 16 (1974), 506-514.

80. Kannan, R., "Existence of solutions of differential equations," Trans. Amer. Math. Soc. 217 (1976), 225-236.

81. Kannan, R., "Existence of solutions of a nonlinear problem in potential theory," Mich. Math. J. 21 (1974), 257-263.

82. Kannan, R. and J. Locker, "Operators J^*J and Hammerstein equations," J. Math. Anal. Appl. 53 (1976), 1-7.

83. Kannan, R. and J. Locker, "Nonlinear boundary value problems and operators TT^*,", to appear.

84. Kannan, R. and J. Locker, "On a class of nonlinear boundary value problems," to appear.

85. Kannan, R. and P.J. McKenna, "An existence theorem by alternative method for semilinear abstract equations," to appear.

86. Kannan, R. and H. Salehi, "Random solutions of differential equations," Boll. Un. Mat. Ital. (4) 12 (1975), 209-213.

87. Kannan, R. and J. Schuur, "Boundary value problems for even order nonlinear ordinary differential equations," Bull. Amer. Math. Soc. 82 (1976), 80-82.

88. Knobloch, H.W., "Remarks on a paper of Cesari on functional analysis and nonlinear differential equations," Mich. Math. J. 10 (1963), 417-430.

89. Knobloch, H.W., "Eine neue Methode zur Approximation von periodischen Lösungen nicht linear Differentialgleichungen zweiter Ordnung," Math. Zeit. 82 (1963), 177-197.

90. Knobloch, H.W., "Comparison theorems for nonlinear second order differential equations," J. Differential Equations 1 (1965), 1-25.

91. Ku, D., Boundary value problems and numerical estimates, Univ. of Michigan thesis, 1976.

92. Landesman, E.M. and A. Lazer, "Nonlinear perturbations of linear elliptic boundary value problems at resonance," J. Math. Mech. 19 (1970), 609-623.

93. Lazer, A.C., "On Schauder's fixed point theorem and forced second order nonlinear oscillations," J. Math. Anal. Appl. 21 (1968), 421-425.

94. Lazer, A.C. and D.E. Leach, "Bounded perturbations of forced harmonic oscillations at resonance," Ann. Mat. Pura Appl. 72 (1969), 49-68.

95. Locker, J., "An existence analysis for nonlinear equations in Hilbert spaces," Trans. Amer. Math. Soc. 128 (1967), 403-413.

96. Locker, J., "An existence analysis for nonlinear boundary value problems," SIAM J. Appl. Math. 19 (1970), 199-207.

97. Mawhin, J., "Application directe de la méthode de Cesari a l'étude des solutions périodiques de systèmes différentiels faiblement non linéaires," Bull. Soc. Roy. Sci. Liège 36 (1967), 193-210.

98. Mawhin, J., "Solutions periodiques de systems differentiels faiblement non lineaires," Bull. Soc. Roy. Sci. Liège 36 (1967), 491-499.

99. Mawhin, J., "Familles de solutions périodiques dans les systèmes différentiels faiblement non lineaires," Bull. Soc. Roy. Sci. Liège 36 (1967), 500-509.

100. Mawhin, J., "Equations integrales et solutions périodiques des systèmes differentiels non linéaires," Bull. Acad. Roy. Belgique (5) 55 (1969), 934-947.

101. Mawhin, J., Le problème des solutions périodiques en mécanique non linéaire, Thèse Univ., Liège, 1969.

102. Mawhin, J., "Degré topologique et solutions périodiques des systèmes différentiels non linéaires," Bull. Soc. Roy. Sci. Liège 38 (1969), 308-398.

103. Mawhin, J., "Periodic solutions of strongly nonlinear differential systems," Fifth International Conference on Nonlinear Oscillations, Kiev 1969, Vol. 1, 380-399.

104. Mawhin, J., "Periodic solutions of nonlinear functional differential equations," J. Differential Equations 10 (1971), 240-261.

105. Mawhin, J., "An extension of a theorem of Lazer on forced nonlinear oscillations," J. Math. Anal. Appl. 40 (1972), 20-29.

106. Mawhin, J., "Equivalence theorems for nonlinear operator equations and coincidence degree theory for some mappings

in locally convex topological vector spaces," J. Differential Equations 12 (1972), 610-636.

107. Mawhin, J., "L_2-estimates and periodic solutions of some nonlinear differential equations," Boll. Un. Mat. Ital. (4) 10 (1974), 341-352.

108. Mawhin, J., "Topology and nonlinear boundary value problems," International Conference Dynamical Systems, Providence, R.I., August 1975. Academic Press, 1976, Vol. 1, 51-82.

109. McKenna, P.J., The alternative method in nonselfadjoint semilinear problems, Univ. of Michigan Thesis, 1976

110. Nagle, K., Boundary value problems for nonlinear ordinary differential equations, Univ. of Michigan Thesis, 1975.

111. Nečas, J., "On the range of nonlinear operators with linear asymptotes which are not invertible," Comment. Math. Univ. Carolinae 14 (1973), 63-72.

112. Osborn, J.E. and D. Sather, "Alternative problems for nonlinear equations," J. Differential Equations 17 (1975), 12-31.

113. Osborn, J.E. and D. Sather, "Alternative problems and monotonicity," J. Differential Equations 18 (1975), 393-410.

114. Perello, C., "Periodic solutions of differential equations with time lag containing a small parameter," J. Differential Equations 4 (1968), 160-175.

115. Petrovanu, D., "Solutions périodiques pour certain equations hyperboliques," Analele Stintifice Iasi 14 (1968), 327-357.

116. Petrovanu, D., "Periodic solutions of the Tricomi problem," Mich. Math. J. 16, 169, 331-348.

117. Rodionov, A.M., "Periodic solutions of nonlinear differential equations with time lag," Trudy Sem. Differential Equations Lumumba Univ. Moscow 2 (1963), 200-207.

118. Sanchez, D.A., "An iteration scheme for Hilbert space boundary value problems," Boll. Un. Mat. Ital. (4) 11 (1975), 1-9.

119. Sather, D., "Branching of solutions of an equation in Hilbert space," Arch. Rational Mech. Anal. 36 (1970), 47-60.

120. Shaw, H.C., Nonlinear elliptic boundary value problems at resonance, Univ. of Michigan Thesis, 1975.

121. Stocking, C.D., Nonlinear boundary value problems in a circle and related questions on Bessel functions, Univ. of Michigan Thesis, 1972.

122. Strygina, S.O. and P.P. Zabreiko, "Periodic solutions of evolutionary equations," Math. Notes of USSR 9 (1971), 378-384; Mat. Zametki 9 (1971), 651-662.

123. Sweet, D., "Periodic solutions for dynamical systems possessing a first integral in the resonance case," J. Differential Equations 14 (1973), 171-183.

124. Williams, S., "A connection between the Cesari and Leray-Schauder methods," Mich. Math. J. 15 (1968), 441-448.

125. Williams, S.A., "A sharp sufficient condition for solution of a nonlinear elliptic boundary value problem," J. Differential Equations 8 (1970), 580-586.

Other References

126. Agmon, S., Lectures on Elliptic Boundary Value Problems, Van Nostrand, Princeton, 1965.

127. Agmon, S., "On the eigenfunctions and on the eigenvalues of general elliptic boundary value problems," Comm. Pure Appl. Math. 15 (1962), 119-147.

128. Azis, A.K., "Periodic solutions of hyperbolic partial differential equations," Proc. Amer. Math. Soc. 17 (1966), 557-566.

129. Azis, A.K. and A.M. Meyers, "Periodic solutions of hyperbolic partial differential equations in a strip,"

130. Antosiewicz, H.A., "Boundary value problems for nonlinear ordinary differential equations," Pacific J. Math. 17 (1966), 191-197.

131. Antosiewicz, H.A., "Un analogue du principe du point fixe de Banach," Ann. Mat. Pura Appl. (4) 74 (1966), 61-64.

132. Bartle, R.G., "Singular points of functional equations," Trans. Amer. Math. Soc. 75 (1953), 366-384.

133. Brézis, H. Operateurs Maximaux Monotones, Am. Elsevier, New York, 1973.

134. Brézis, H., M. Crandall and A. Pazy, "Perturbation of nonlinear maximal monotone sets," Comm. Pure Appl. Math. 23 (1970), 123-144.

135. Cesari, L., "Un problema ai limiti per sistemi di equazioni iperboliche quasilineari alle derivate parziali," Rend. Accad. Naz. Lincei, Cl. Sci. Fis. Mat. Nat. (8) 55 (1974),

136. Cesari, L., "Un problema ai limiti per sistemi di equazioni iperboliche quasilineari nella forma canonica di Schauder," Rend. Accad. Naz. Lincei, Cl. Sci. Fis. Mat. Nat. (8) 55 (1974), 303-307.

137. Cesari, L., "A boundary value problem for quasilinear hyperbolic systems," Riv. Mat. Univ. Parma, to appear.

138. Cesari, L., "A boundary value problem for quasilinear hyperbolic systems in Schauder's canonic form," Ann. Scuola Norm. Sup. Pisa (4) 1 (1974), 311-358.

139. Cesari, L., "Nonlinear oscillations under hyperbolic systems," International Conference Dynamical Systems, Providence, R.I., August 1974. Academic Press, 1976, Vol. 2, 251-261.

140. Coddington, E.A. and N. Levinson, Theory of Ordinary Differential Equations, McGraw-Hill, New York 1955.

141. Cole, R.H., Theory of Ordinary Differential Equations, Appleton-Century-Crofts, New York 1968.

142. Cronin, J., "Fixed points and topological degree in nonlinear analysis," Amer. Math. Soc., Providence 1964.

143. De Simon, L., "Sull'equazione delle onde con termine noto periodico," Rend. Ist. Mat. Univ. Trieste. 1 (1969), 150-162.

144. De Simon, L. and G. Torelli, "Soluzioni periodiche di equazioni alle derivate parziali di tipo iperbolico nonlineare," Rend. Sem. Mat. Univ. Padova 40 (1968), 380-401.

145. De Figueiredo, D.G., (a) "Some remarks on the Dirichlet problem for semilinear elliptic equations," Univ. de Brasilia, Trabalho de Mat. no. 57, 1974; (b) "On the range of nonlinear operators with linear asymptotes which are not invertible," Ibid., no. 59, 1974.

146. De Figueiredo, D.G., "The Dirichlet problem for nonlinear elliptic equations: A Hilbert space approach," Partial Differential Equations and Related Topics (Dold and Eckman ed.), Springer-Verlag Lecture Notes Math., No. 446 (1975), 144-165.

147. Dolph, C.L. and G.J. Minty, "On linear integral equations of the Hammerstein type," Integral Equations, Madison, Univ. of Wisconsin Press (1964), 99-154.

148. Fink, J.P. and W.S. Hall, "Entrainment of frequency in evolution equations," Journ. Differential Equations

149. Fink, J.P., W.S. Hall and S. Khahili, "Perturbation expansions for some nonlinear wave equations," SIAM J. Appl. Math. 24 (1973), 575-595.

150. Fink, J.P., W.S. Hall and A.R. Hausrath, "Discontinuous periodic solutions," to appear.

151. Fink, J.P., W.S. Hall and A.R. Hausrath, "A convergent two-time method for periodic differential equations," J. Differential Equations 15 (1974), 459-498.

152. Friedrichs, K., Special Topics In Analysis, Lecture Notes, New York Univ. 1953-1954.

153. Fučík, S., "Further remarks on a theorem by E. Landesman and A.C. Lazer," Comment. Math. Univ. Carolinae 15 (1974), 259-271.

154. Fučík, S. M. Kučera and J. Nečas, "Ranges of nonlinear asymptotically linear operators," J. Differential Equations 17 (1975), 375-394.

155. Graffi, D., "Forced oscillations for several nonlinear circuits," Ann. of Math. 54 (1951), 262-271.

156. Graffi, D., "Sulle oscillazioni forzate nella meccanica non lineare," Riv. Mat. Univ. Parma 3 (1952), 317-326.

157. Graves, L.M., "Remarks on singular points of functional equations," Trans. Amer. Math. Soc. 79 (1955), 150-157.

158. Havlova, J., "Periodic solutions of a nonlinear telegraph equation," Časopis, Pest. Mat. 90 (1965), 273-289, Prague.

159. Hess, P., "On a theorem by Landesman and Lazer," Indiana Univ. Math. J. 23 (1974), 827-829.

160. Hoffman, K. and R. Kunze, Linear Algebra, Prentice Hall, Englewood Cliffs, New Jersey 1961.

161. Kato, T., "Demicontinuity, hemicontinuity and monotonicity, I and II," Bull. Amer. Math. Soc. 73 (1967), 470-476, 886-889.

162. Krylova, N., "Periodic solutions of hyperbolic partial differential equations with quadratic dissipative terms," Czechoslovak Math. J. 20 (95) (1970), 375-405 Prague.

163. Lazer, A.C., "On the computation of periodic solutions of weakly nonlinear differential equations," SIAM J. Appl. Math. 15 (1967), 1158-1170.

164. Lazer, A.C. and D.A. Sanchez, "On periodically perturbed conservative systems," Mich. Math. J. 16 (1969), 193-200.

165. Leach, D.E., "On Poincaré's perturbation theorem and a theorem of W.S. Loud," J. Differential Equations 7 (1970), 34-53.

166. Lefschetz, S., "Existence of periodic solutions of certain differential equations," Proc. Nat. Acad. Sci. U.S.A. 29 (1943), 29-32.

167. Leray, J. and J. Schauder, "Topologie et équations fonctionnelles," Ann. Sci. École Norm. Sup. 51 (1934), 45-78.

168. Levinson, N., "On the existence of periodic solutions for second order differential equations with a forcing term," J. Math. Phys. 22 (1943), 41-48.

169. Lewis, D.C., "On the role of first integrals in the perturbation of periodic solutions," Ann. Math. 63 (1956), 535-548.

170. Loud, W.S., "Periodic solutions of nonlinear differential equations of Duffing type," Differential and Functional Equations, Benjamin, New York 1967.

171. Lyapunov, A.M., "Sur les figures d'equilibre peu différentes des ellipsoids d'une masse liquide homogène douée d'un mouvement de rotation," Zap. Akad. Nauk St. Petersburg 1 (1906), 1-225.

172. Minty, G.J., "On the maximal domain of a monotone function," Mich. Math. J. 8 (1961), 135-137.

173. Minty, G.J., "Monotone nonlinear operators in Hilbert space," Duke Math. J. 29 (1962), 341-346.

174. Minty, G.J., "On a monotonicity method for the solution of nonlinear equations in Banach spaces," Proc. Nat. Acad. Sci. U.S.A. 50 (1963), 1038-1041.

175. Mizohata, S., "On the existence of periodic solutions for several nonlinear circuits," Mem. Coll. Sci. Univ. Kyoto, Ser. A, Math. 27 (1952), 115-121.

176. Mizohata, S. and M. Yamaguti, Mem. Coll. Sci. Univ. Kyoto, Ser. A, Math. 27 (1951), 109-113.

177. Nagumo, M., "A theory of degree of mappings based on infinitesimal analysis," Amer. J. Math. 73 (1951), 485-496.

178. Nagumo, M., "Degree of mappings in convex linear topological spaces," Amer. J. Math. 73 (1951), 497-511.

179. Nirenberg, L., Functional Analysis, Lecture Notes, New York Univ., 1960-1961.

180. Poincaré, H., Les Méthodes Nouvelles de la Mécanique Céleste, Gauthier-Villars, Paris 1892-1899, Dover, New York 1957.

181. Rabinowitz, P.H., "Periodic solutions of nonlinear hyperbolic partial differential equations, I and II," Comm. Pure Appl. Math. 20 (1967), 145-205; 22 (1969), 15-39.

182. Rabinowitz, P.H., "Time periodic solutions of nonlinear wave equations," Manuscripta Math. 5 (1971), 165-194.

183. Rabinowitz, P.H., "Existence of periodic solutions for some nonlinear hyperbolic partial differential equations," Studies in Appl. Math. 5 (1969), 20-24.

184. Rothe, E., "The theory of topological order in some linear topological spaces," Iowa State College J. Sci. 13 (1939), 373-390.

185. Riesz, F. and B. Sz. Nagy, Functional Analysis, Ungar, New York 1955.

186. Santagati, G., "Su un problema al contorno per una classe di sistemi iperbolici di equazioni alle derivate parziali del primo ordine," Matematiche, Catania 15 (1960), 1-14.

187. Schmidt, E., "Zur Theorie der linearen und nichtlinearen Integralgleichungen und der Verzweigung ihrer Lösungen," Math. Ann. 65 (1908), 370-399.

188. Torelli, G., "Soluzioni periodiche dell'equazione non lineare $u_{tt} - u_{xx} = \epsilon F(x,t,u)$," Rend. Ist. Mat. Univ. Trieste 1 (1969), 123-137.

189. Tricomi, F., Equazioni alle derivate parziali, Roma 1957.

190. Vainberg, M.M., Variational Methods for the Study of Nonlinear Operators, Holden-Day, San Francisco 1964.

191. Vainberg, M.M. and V.A. Trenogin, "The methods of Lyapunov and Schmidt in the theory of nonlinear equations and their further development," Russian Math. Surveys 17 (1962), 1-60.

192. Vainberg, M.M. and P.G. Aizengendler, "The theory and methods of investigation of branch points of solutions," Progress in Mathematics, II, Plenum Press, New York 1968, 1-72.

193. Vainberg, M.M. and I.M. Lavrentiev, "Nonlinear equations of Hammerstein type with potential and monotone operators in Banach spaces," Math. Sbornik USSR 16 (1972), 333-347.

194. Vejvoda, O., "On the periodic solution of a quasi-linear nonautonomous system," Czechoslovak Math. J. 11 (86) (1961), Prague 62-74.

195. Vejvoda, O., "On perturbed nonlinear boundary value problems," Czechoslovak Math. J. 11 (86) (1961), Prague 323-364.

196. Vejvoda, O., "Nonlinear boundary value problems for differential equations," Proc. Conference Differential Equations, Prague (1962), 199-215.

197. Vejvoda, O., "Periodic solutions of a linear and weakly nonlinear wave equation in one dimension," Czechoslovak Math. J. 14 (89) (1964), 341-382, Prague.

198. Vejvoda, O., "Periodic solutions of nonlinear partial differential equations of evolution," Acta. Univ. Carolinae Math. (1967), 293-300.

199. Vejvoda, O., "Periodic solutions of a weakly nonlinear wave equation in E_3 in a spherical symmetrical case," Apl. Mat. 14 (1969), 160-167.

200. Vejvoda, O., "The mixed problem and periodic solutions for a linear and weakly nonlinear wave equation in one dimension," Rozpravy Československé Akad. Věd. Řada Mat. (3) 80 (1970), 1-78.

201. Vejvoda, O. and N. Krylova, "Periodic solutions to partial differential equations especially to a biharmonic wave equation," Symposia Mathematica 7 (1971), 85-96, Istituto di Alta Matematica, Roma.

202. Vejvoda, O. and N. Krylova, "A linear and weakly nonlinear equation of a beam: the boundary-value problem for free extremities and its periodic solutions," Czechoslovak Math. J. 21 (96) (1971), 535-566, Prague.

203. Vejvoda, O. and V. Stasnova, "Periodic solutions of the first boundary value problem for a linear and weakly nonlinear heat equation," Apl. Mat. 13 (1968), 466-477 and 16 (1969), 241.

204. Vejvoda, O. and I. Straskaia, "Periodic solutions to abstract differential equations," Czechoslovak Math. J. 23 (98) (1973), 635-669, Prague.

205. Vejvoda, O. and M. Tvrdy, "Existence of solutions to a linear integro-boundary-differential equation with additional conditions," Ann. Mat. Pura Appl. (4) 89 (1971), 169-216.

206. Villari, G., "Su un problema al contorno per una classe di sistemi di equazioni alle derivate parziali," Boll. Un. Mat. Ital. (3) 13 (1958), 514-521.

207. Yosida, K., Functional Analysis, Springer-Verlag, New York 1971.

APPLICATIONS OF GENERIC BIFURCATION

Shui-Nee Chow*
Michigan State University

John Mallet-Paret**
Brown University

Suppose τ is a family of mappings from one Banach space X into another Banach space Z and suppose there is a norm on the members of τ. Let $T \in \tau$ be given and suppose there is an $x_0 \in X$ such that $Tx_0 = 0$. The operator T is said to be a bifurcation point for τ at x_0 if, for every neighborhood U of T and V of x_0, there is an $S \in U$ and $x_1, x_2 \in V$, $x_1 \neq x_2$, such that $Sx_1 = Sx_2 = 0$.

In many applications, the family τ consists of a two parameter family of mappings $S(\cdot, \lambda, \mu)$, $\lambda \in R$, $\mu \in R$. The parameter λ may represent an eigenvalue parameter and μ may represent an external loading, as in the von Kármán equations for the steady state position of a plate.

If the external loading μ is zero, then normally $x = 0$ is a solution of the equation $S(x, \lambda, 0) = 0$ for all λ. Most of the papers in the literature are concerned with the one

*Supported by the National Science Foundation under GP-28931X3, and GP-43034.

**Supported by the National Science Foundation under GP-28931X3.

parameter family $S(\cdot,\lambda,0)$, $\lambda \in R$. For this family and for certain natural restrictions on $S(\cdot,\lambda,0)$, the point $\lambda = \lambda_0$ being a bifurcation point means there is at least one continuous, nonzero branch of solutions $x(\lambda)$ of the equation $S(x,\lambda,0) = 0$ for λ near λ_0 and $x(\lambda_0) = 0$. Explicit approximations for these branches may be obtained by an application of perturbation methods and Newton's polygonal method. The existence of these branches may be obtained from an application of degree theory.

If the external loading is not zero, then the two parameters enter in the equation in an independent manner and $x = 0$ is generally not a solution of $S(x,\lambda,\mu) = 0$.

A typical situation that occurs in this case (for example, in von Kármán's equations) is that there is a curve $\Gamma: f(\lambda,\mu) = 0$ in (λ,μ)-space, (λ,μ) near $(\lambda_0, 0)$, such that $(\lambda,\mu) \in \Gamma$ implies there is a double root of $S(x,\lambda,\mu) = 0$ near $x = 0$. This curve may also be obtained by proper scaling of λ, μ and perturbation methods.

However perturbation methods seem to be difficult to apply to obtain the complete picture of the number of solutions of $S(x,\lambda,\mu) = 0$ for (λ,μ) in a full neighborhood of $(\lambda_0, 0)$. Degree theory is also not applicable in the case where the family τ depends on more than one parameter.

In [1] we gave a constructive method for discussing all solutions of $S(x,\lambda,\mu) = 0$ in a neighborhood of $(\lambda_0, 0)$ for the case when $S(x,\lambda,\mu) = Ix - \lambda B(x) - \mu C(x)$ and $\dim \ker[I - \lambda_0 B'(0)] = 1 = \dim \operatorname{co} \ker[I - \lambda_0 B'(0)]$. The operators B, C may also depend on λ, μ. The approach uses the Liapunov-Schmidt method together with some elementary ideas from catastrophe theory.

To describe two results more explicitly, suppose ω_0 is a basis for $\ker[I - \lambda_0 B'(0)]$, let z_0 be a basis for $\mathrm{coker}[I - \lambda_0 B'(0)]$, and let π be a projection onto coker $[I - \lambda_0 B'(0)]$.

Theorem 1. Suppose

$$\alpha z_0 = \frac{\partial^2}{\partial a^2} \pi B(a\omega_0)\big|_{a=0}, \quad \alpha \neq 0 \quad \text{and} \quad \pi C(0) \neq 0.$$

Then, in a neighborhood of $(\lambda_0, 0)$, there is a curve Γ in the (λ, μ)-plane with approximate form $(\lambda - \lambda_0)^2 = \gamma\mu$, $\gamma \neq 0$, such that $(I - \lambda B - \mu C)x = 0$ has two, one double, or no zeros depending on whether (λ, u) is "outside" of Γ, on Γ, or "inside" Γ (see Figure 1). Further, if the line $\mu = \bar{\mu}$ intersects Γ, then there are two continuous branches of solutions for varying λ and $\mu = \bar{\mu}$ (see Figures 1 and 2) and if the line $\mu = \bar{\mu}$ does not intersect Γ, then there are two solutions for varying λ and $\mu = \bar{u}$, but these solutions do not lie on a continuous branch (see Figures 1 and 3).

Theorem 2. Suppose $\alpha = 0$ in Theorem 1,

$$\beta z_0 = \frac{\partial^3}{\partial a^3} \pi B(a\omega_0)\big|_{a=0}, \quad \beta \neq 0 \quad \text{and} \quad \pi C(0) \neq 0.$$

Then in a neighborhood of $(\lambda_0, 0)$, there is a curve Γ in the (λ, μ)-plane with approximate form $(\lambda - \lambda_0)^3 = \delta u^2$ such that $(\lambda, \mu) \in \Gamma$, $(\lambda, \mu) \neq (\lambda_0, 0)$ implies there is a double zero of $(I - \lambda B - \mu C)x = 0$. Furthermore, if $\delta < 0$, $(\bar{\lambda}, \bar{\mu}) \in \Gamma$, then there are three solutions for $\lambda < \bar{\lambda}$, $\mu = \bar{\mu}$,

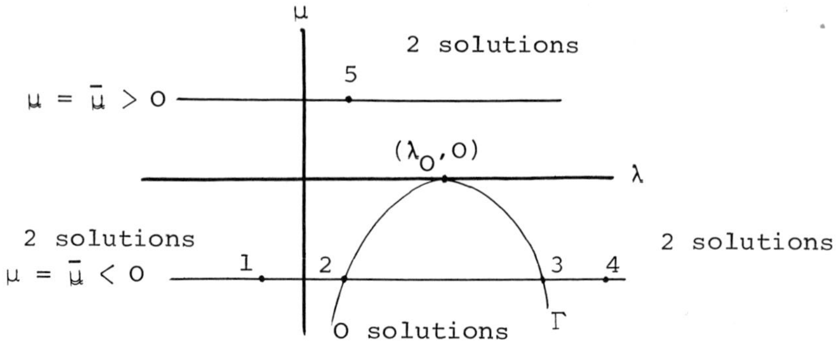

Figure 1

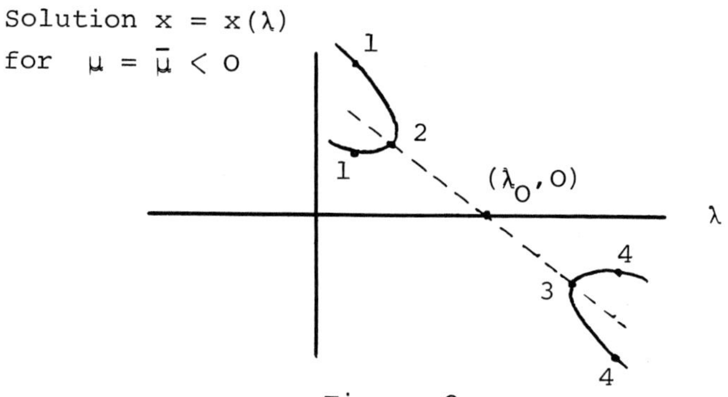

Figure 2

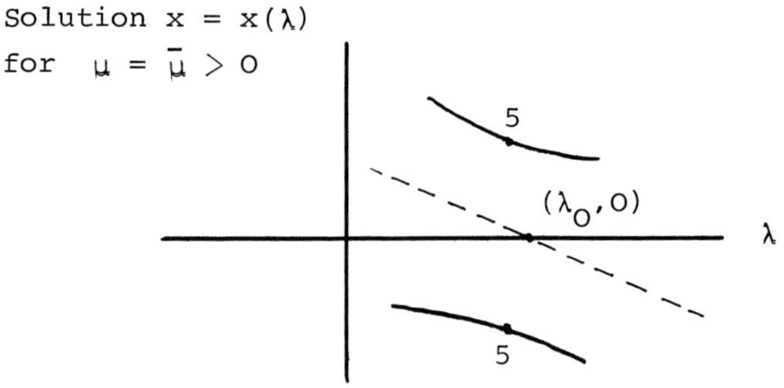

Figure 3

two lying on a continuous branch, and there is only one solution for $\lambda > \bar{\lambda}$ (see Figures 4, 5). If $\delta > 0$, the same conclusions hold with the inequalities for λ reserved.

In this manner, a complete description of the bifurcation phenomena is obtained for simple eigenvalues. A larger number of independent parameters causes no major difficulties with our method.

The case where λ_0^{-1} is a double eigenvalue of $B'(0)$ and the family τ contains two parameters has been discussed in [2] in the same spirit as in [1]. For simplicity we consider the following example. The general result will appear in [2].

Consider the buckling of a simply supported flat rectangular plate

$$D = \{(x,y): 0 < x < \sqrt{2},\ 0 < y < 1\}$$

subject to a compressive thrust applied to $x = 0, \sqrt{2}$, and a normal loading. The equilibrium states of the plate are assumed to be described by the von Kármán euuations

$$\Delta^2 f = -\frac{1}{2}[w,w], \qquad (x,y) \in D,$$

$$\Delta^2 w = [w,f] - \tilde{\lambda}\frac{\partial^2 w}{\partial x^2} - \tilde{\nu} p, \qquad (x,y) \in D,$$

$$f = \Delta f = w = \Delta w = 0, \qquad (x,y) \in \partial D,$$

where

$$[u,v] = \frac{\partial^2 u}{\partial x^2}\frac{\partial^2 v}{\partial y^2} + \frac{\partial^2 u}{\partial y^2}\frac{\partial^2 v}{\partial x^2} - 2\frac{\partial^2 u}{\partial x \partial y}\frac{\partial^2 v}{\partial x \partial y},$$

∂D is the boundary of D,

$\tilde{\lambda}$ is the magnitude of thrust on $x = 0, \sqrt{2}$,

$p(x,y)$ is the normal loading, and

$\tilde{\nu}$ is the magnitude of normal loading.

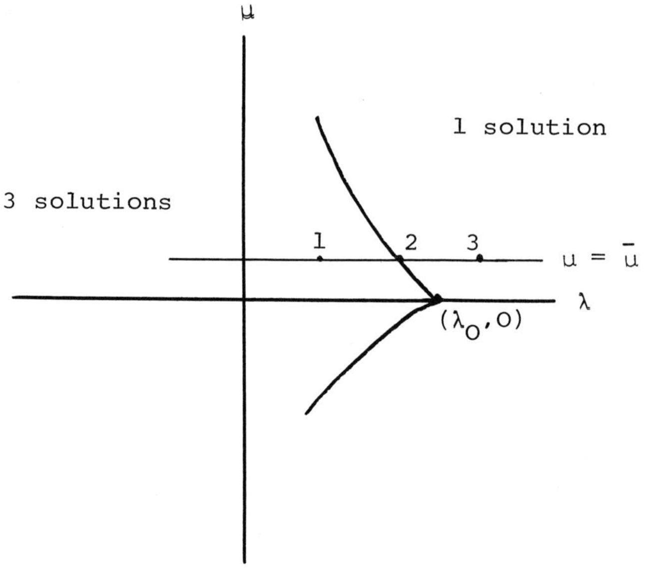

Figure 4

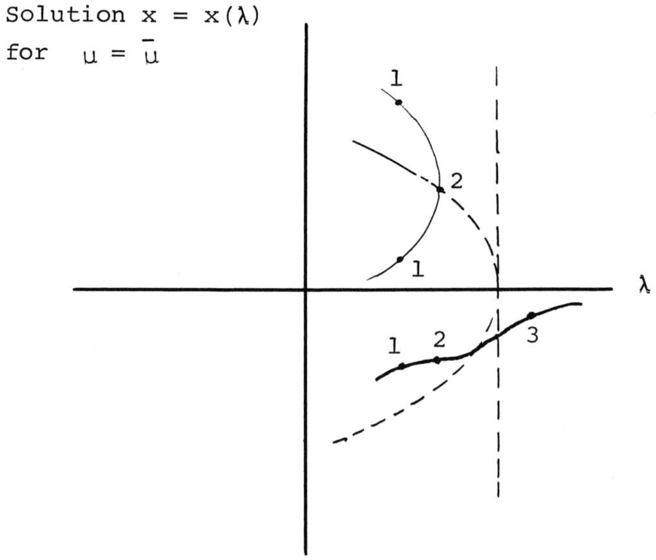

Figure 5

The function $p(x,y)$ is known and fixed, while $\tilde{\lambda}$ and $\tilde{\nu}$ are real parameters. The unknown functions f and w are related, respectively, to the stress and the deflection of the plate. It is known that $\tilde{\lambda} = 9\pi^2/2$ is the first eigenvalue of multiplicity two of the linearized problem

$$\Delta^2 w + \tilde{\lambda}\frac{\partial^2 w}{\partial x^2} = 0, \qquad (x,y) \in D,$$
$$w = \Delta w = 0, \qquad (x,y) \in \partial D.$$

It is also well-known that the above equation may be rewritten in a Hilbert space X (in fact, the Sobolev space $H_0^2(D)$) as a single equation in $w \in X$:

$$(I - \lambda L)w + C(w) - \nu p = 0$$

where $\lambda = 2\tilde{\lambda}/9\pi^2$, $\nu = \tilde{\nu}$, $L:X \to X$ is bounded and linear, and $C:X \to X$ is bounded and trilinear. As we discussed earlier, by applying the Liapunov-Schmidt method we obtain the following bifurcation equations:

$$(1 - \lambda)u_1 + au_1^3 + bu_1u_2^2 - \nu p_1 = \text{h.o.t.}$$
$$(1 - \lambda)u_2 + bu_1^2 + cu_2^3 - \nu p_2 = \text{h.o.t.}$$

where a, b, c, p_1 and p_2 are constants, u_1 and u_2 are scalars, and h.o.t. designates higher order terms and refers to terms of order

$$O(|u|^4 + (1 - \lambda)^2|u| + |1 - \lambda||u|^2 + \nu^2 + |\nu||u|).$$

Note that the third order terms in the bifurcation equations are related to the trilinear map $C:X \to X$.

To characterize the number of zeros of the bifurcation equations near $u = (u_1,u_2) = 0$, $\lambda = 1$ and $\nu = 0$, we studied the intersections of the curves defined by the bifurcation equations as λ and ν vary. The proof is too

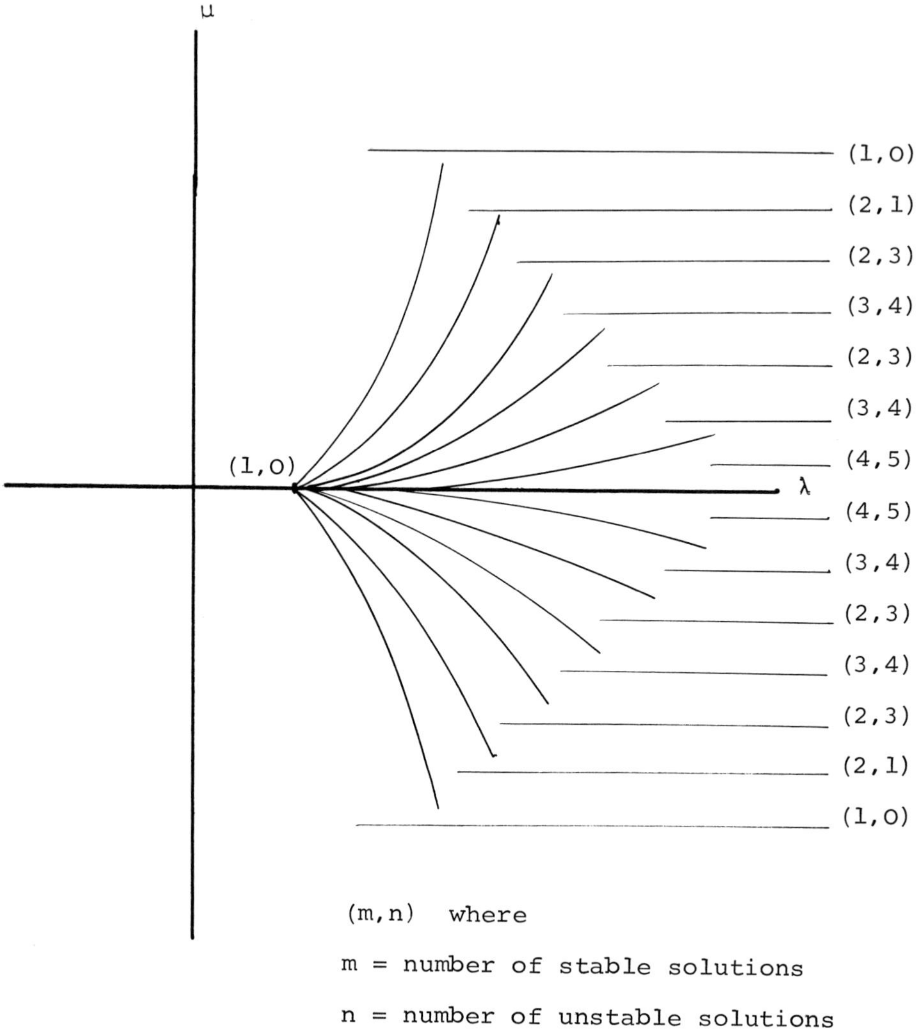
μ
λ
(1,0)
(1,0)
(2,1)
(2,3)
(3,4)
(2,3)
(3,4)
(4,5)
(4,5)
(3,4)
(2,3)
(3,4)
(2,3)
(2,1)
(1,0)
(m,n) where
m = number of stable solutions
n = number of unstable solutions

Figure 6

complicated to give here, but the result may be stated simply as follows: There are twelve arcs of the form

$$\nu = c(\lambda - 1)^{3/2} \qquad c = \text{constant}$$

in the $\lambda\nu$-space (see Figure 6) such that the number of stable and unstable solutions in each region may be determined. We note that as one crosses an arc in the bifurcation diagram (i.e., $\lambda\nu$-space), a secondary bifurcation occurs and the total number of solutions changes by two.

This report is a summary of the following articles written jointly with Professor J.K. Hale, Brown University.

[1] S.N. Chow, J.K. Hale and J. Mallet-Paret, Applications of Generic Bifurcation I, Arch. Rat. Mech. Anal. To appear.

[2] S.N. Chow, J.K. Hale and J. Mallet-Paret, Applications of Generic Bifurcation II. To appear.

PERTURBATION METHODS FOR NONLINEAR PROBLEMS AT RESONANCE

Rangachary Kannan

University of Missouri

1. Introduction. In the recent years there have been several contributions to the general question of existence of solutions of nonlinear perturbations of linear boundary value problems at resonance by using the various techniques of nonlinear functional analysis and the method of "Alternative Problems". The lectures of Cesari in this volume provide an exhaustive survey of these contributions.

We shall study in this paper a method of studying nonlinear problems at resonance without resorting to "Alternative Problems". This survey is essentially motivated by a recent paper of Hess [10]. Hess proved, in a very elegant and concise form the result of Landesman and Lazer [16] on bounded nonlinear perturbations of elliptic problems at resonance. The ideas of Hess were carried further by DeFigureido [7], to the case when the nonlinearity is not required to be bounded but must satisfy certain growth restrictions. DeFigureido obtains, in particular, the results due to Fučik, Kučera and Nečas [8]. Our aim in presenting these results is to point out the salient features and raise some open questions. Roughly speaking the idea in the papers of Hess [10] and

DeFigureido [7] is as follows: perturb the linear part (which is non-invertible) slightly so that the linear part of the perturbed problem is invertible. We consider a sequence of such perturbed problems. After showing that each problem of the sequence is solvable, we show that there exists a subsequence of the sequence of solutions which converges to a solution of the given nonlinear problem at resonance.

In Section 2 of this paper we introduce the general problem of nonlinear perturbations at resonance. After presenting a brief survey of selected preliminary results in Section 3 we state the theorem of Landesman and Lazer in Section 4. An outline of the proof of Landesman and Lazer's result by the method of "Alternative Problems" is given in Section 5. We then present in Section 6 the proof due to Hess [10] of the theorem of Landesman and Lazer. An abstract generalization of the ideas in Hess [10], due to DeFigureido [7], is derived in Section 7. We conclude this paper with some remarks in Section 8.

2. The General Problem of Nonlinear Perturbations at Resonance. The well-known "orthogonality condition" states that the necessary and sufficient condition for the differential equation

$$\begin{aligned} &x'' + m^2 x = p(t) \\ &x(0) = x(2\pi), \quad x'(0) = x'(2\pi) \end{aligned} \tag{1}$$

to have 2π-periodic solutions is that $\int_0^{2\pi} p(t)\cos mt\,dt = \int_0^{2\pi} p(t)\sin mt\,dt = 0$, where m is an integer. In fact, if this condition is not satisfied, no solution of (1) is bounded and this situation is known as the phenomenon of resonance.

Thus, a natural question is to ask for sufficient hypotheses on $h(x)$ (not necessarily linear) so that the boundary value problem

$$\begin{aligned} &x'' + m^2x + h(x) = p(t) \\ &x(0) = x(2\pi), \quad x'(0) = x'(2\pi) \end{aligned} \tag{2}$$

will have 2π-periodic solutions (here $p(t)$ does not necessarily satisfy the orthogonality condition). It must be noted that m^2 is an eigenvalue of the associated linear problem $x'' + \lambda x = 0$, $x(0) = x(2\pi)$ and $x'(0) = x'(2\pi)$.

More generally, let L be a linear, self-adjoint differential operator over a real Hilbert space $S = L^2(\Omega)$, where Ω is a smooth bounded domain in R^n, and let λ be an isolated eigenvalue of the associated linear problem of L. The general problem of nonlinear perturbations at resonance is to find sufficient hypotheses on the nonlinear operator $N:D(N) \to S$ in order that the nonlinear problem

$$Lx - \lambda x + Nx = 0 \tag{3}$$

is solvable.

3. Brief Survey of Preliminary Results. In this section we outline some of the preliminary results regarding the existence of solutions to problems of type (2). Thus, instead of presenting an exhaustive survey of the different results (which may be found in the lectures of Cesari in this volume) we discuss briefly a selected few which lead to the work of Landesman and Lazer [16].

Lazer and Sanchez [20] proved that the nonlinear problem

$$x'' + g(t,x) = p(t)$$
$$x(0) = x(2\pi), \quad x'(0) = x'(2\pi) \tag{4}$$

has at least one 2π-periodic solution if i) $p(t)$ is 2π-periodic and continuous; ii) $g(t,x)$ is continuous and has a continuous derivative with respect to x for $0 < t < 2\pi$ and $-\infty < x < \infty$; and iii) there exists an integer m and real numbers p and q such that $m^2 < q \leq g_x(t,x) \leq p < (m+1)^2$. The above result was also obtained by Ahmad [1] and Kannan [11].

This problem may be rewritten in the form

$$x'' + h(t,x)x = q(t)$$
$$x(0) = x(2\pi), \quad x'(0) = x'(2\pi)$$

where $q(t) = p(t) - g(t,0)$ and $h(t,y) = \int_0^1 g_x(t,sy)ds$. Consequently $n^2 < q \leq h(t,x) < p \leq (n+1)^2$.

In [19] Lazer and Leach proved that

$$x'' + h(t,x,x')x = q(t,x,x')$$
$$x(0) = x(\pi) = 0 \tag{5}$$

has at least one solution provided i) $h(t,x_1,x_2)$, $q(t,x_1,x_2)$ are continuous for $0 \leq t \leq \pi$ and $-\infty < x_1, x_2 < \infty$; ii) there exists an integer n and a $\delta > 0$ such that $(n+\delta)^2 \leq h(t,x_1,x_2) \leq (n+1-\delta)^2$ for all $t \in [0,\pi]$ and $-\infty < x_1, x_2 < \infty$; and iii) q is bounded.

These results concerning problems (4) and (5) have been obtained by a direct method for very general situations by Kannan and Locker [13].

A related problem (5) was studied by Lazer in [17]. Thus, he considers the case of (5) where δ may take the value 0.

In particular, he obtains the following result: The nonlinear problem

$$\begin{aligned} & x'' + cx' + g(x) = p(t) \\ & x(0) = x(2\pi), \quad x'(0) = x'(2\pi) \end{aligned} \tag{6}$$

has at least one solution if i) $xg(x) \geq 0$ for $|x|$ sufficiently large; and ii) $\frac{g(x)}{x} \to 0$ as $|x| \to \infty$.

This result for problem (6) has been extended to very general situations by Cesari and Kannan [5].

The preceding results lead naturally to the question of sufficient hypotheses on $h: R \to R$ such that

$$\begin{aligned} & x'' + n^2 x + h(x) = p(t) \\ & x(0) = x(2\pi), \quad x'(0) = x'(2\pi) \end{aligned} \tag{7}$$

has at least one solution. Lazer and Leach [18] prove the following: Assume i) $p(t)$ is continuous and 2π-periodic; ii) $h(x)$ is continuous, bounded, and nonconstant; and iii) there exists numbers c, d, P, Q such that $h(x) \leq P$ for $x \leq c$ and $h(x) \geq Q$ for $x > d$. Then (7) has at least one solution provided

$$\left(\left[\int_0^{2\pi} p(s)\cos ns\,ds\right]^2 + \left[\int_0^{2\pi} p(s)\sin ns\,ds\right]^2\right)^{1/2} < 2(Q - P).$$

This theorem was then generalized to elliptic partial differential equations by Landesman and Lazer.

4. <u>Landesman and Lazer's Result</u>. Let Ω be a bounded domain in R^n $(n \geq 1)$ and let

$$Lu = \sum_{|\alpha|, |\beta| \leq 1} (-1)^{|\alpha|} D^\alpha a_{\alpha\beta} D^\beta u$$

be a uniformly elliptic, formally self-adjoint differential operator defined on Ω with real valued coefficients

$a_{\alpha\beta} = a_{\beta\alpha} \in L^\infty(\Omega)$. Also let $g:R \to R$ be a bounded continuous function such that the limits $\lim_{s\to+\infty} g(s) = g(\infty)$ and $\lim_{s\to-\infty} g(s) = g(-\infty)$ exist, and suppose that $g(-\infty) < g(s) < g(\infty)$. For a given real-valued $h \in L^2(\Omega)$ we consider the boundary value problem

$$\begin{aligned} (Lu)(x) &= g[u(x)] - h(x), \quad & x \in \Omega \\ u(x) &= 0, & x \in \partial\Omega \end{aligned} \tag{8}$$

for the existence of weak solutions.

Thus if $B(u,v) = \sum_{|\alpha|,|\beta|\le 1} \int_\Omega a_{\alpha\beta} D^\beta u D^\alpha v\,dx$, which is defined and bounded on H_0^1, is the symmetric bilinear form associated with L we are looking for functions $u \in H_0^1$ such that

$$B(u,\varphi) = \int_\Omega \{g[u(x)] - h(x)\}\varphi(x)\,dx$$

for all $\varphi \in H_0^1$. [Note: H_0^1 is the real Hilbert space obtained by completing the space C_0^1 of all real continuously differentiable functions defined on R^n with compact support in Ω with respect to the inner product given by

$$\langle u,v\rangle = \int_\Omega \left[uv + \sum_1^n \left(\frac{\partial u}{\partial x_i}\right)\left(\frac{\partial v}{\partial x_i}\right)\right]dx \quad \text{for } u,v \in C_0^1.]$$

As we stated in Section 2, we shall be considering the general problem of nonlinear perturbations at resonance. Hence we assume that the homogeneous linear problem

$$\begin{aligned} (Lu)(x) &= 0, \quad & x \in \Omega \\ u(x) &= 0, & x \in \partial\Omega \end{aligned} \tag{9}$$

has 0 as an isolated eigenvalue. We assume here that the eigenspace is one dimensional and spanned by $w \in H_0^1$. Thus w is a nontrivial solution of (9), and any other function

$v \in H_0^1$ satisfying $B(v,\varphi) = 0$, for all $\varphi \in H_0^1$, is of the form $v = \lambda w$, $\lambda \in R$.

Landesman and Lazer [16] proved the following theorem.

Theorem 1. Let $\Omega_+ = \{x \in \Omega : w(x) > 0\}$ and $\Omega_- = \{x \in \Omega : w(x) < 0\}$. Then the inequalities

$$g(-\infty)\int_{\Omega_+}|w|dx - g(+\infty)\int_{\Omega_-}|w|dx < \int hwdx < g(+\infty)\int_{\Omega_+}|w|dx - g(-\infty)\int_{\Omega_-}|w|dx \tag{10}$$

are necessary and sufficient for the existence of a weak solution of problem (8).

5. Outline of Proof by "Alternative Problems". The necessity of condition (10) is immediate. So we shall only consider the sufficiency part of the proof. The proof is sketched here, and we refer the reader to the lectures of Cesari in this volume and to the paper of Landesman and Lazer [16] for details.

Let $S_1 = \{v \in L^2(\Omega) \mid \langle v,w\rangle = 0\}$. Then S_1 is a closed linear subspace of $L^2(\Omega)$. The linear functional L_f defined on H_0^1 by $L_f(\varphi) = \langle\varphi, f\rangle$ is continuous and thus, by the Riesz theorem, there exists a unique $Tf \in H_0^1$ such that

$$B(\varphi, Tf) = L_f(\varphi) = \langle\varphi, f\rangle$$

for all $\varphi \in H_0^1$. Thus Tf is the unique weak solution of $Lu = f$, $u = 0$ on $\partial\Omega$. We may thus consider T as a linear operator from $L^2(\Omega)$ into $L^2(\Omega)$. By Rellich's theorem, T is completely continuous from $L^2(\Omega)$ into $L^2(\Omega)$. The nonlinear problem (8) may now be rewritten as the operator equation $u = T[g(u) - h]$.

Clearly, $T(S_1) \subseteq S_1$. Further, the restriction, T_{S_1}, of T to S_1 is a linear operator from S_1 into S_1 and since T is completely continuous, T_{S_1} is also so. Thus there exists a completely continuous linear operator $H: S_1 \to S_1$ which is the inverse of T. Let $P: L^2(\Omega) \to L^2(\Omega)$ be the orthogonal projection of $L^2(\Omega)$ onto the one dimensional subspace of $L^2(\Omega)$ spanned by w. Let $S = L^2(\Omega) \times R$. If, for $(u_1, a_1), (u_2, a_2) \in S$, $\lambda_1, \lambda_2 \in R$, we define scalar multiplication and norm by

$$\lambda_1(u_1, a_1) + \lambda_2(u_2, a_2) = (\lambda_1 u_1 + \lambda_2 u_2, \lambda_1 a_1 + \lambda_2 a_2)$$

and

$$|(u,a)|_S = \|u\|_{L^2(D)} + |a|,$$

then S is a real Banach space.

Let $\beta: S \to S$ be defined by $\beta[u,a] = [u^*, a^*]$ where

$$u^* = aw + H(I - P)[g(u) - h]$$

$$a^* = a - \langle g(u^*) - h, w\rangle.$$

It can be easily seen that if β has a fixed point (u,a), then u is a weak solution of the boundary value problem (8). Using the inequalities (10), Landesman and Lazer prove that β maps a ball of suitable radius into itself. They then show that β is completely continuous and hence, by Schauder's fixed point theorem, β has a fixed point.

6. <u>Hess's Proof of Theorem 1.</u> We shall now discuss the proof of Theorem 1 due to Hess [10].

Since 0 is an isolated eigenvalue of the associated linear problem of L, by an elementary application of Schauder's fixed point theorem we can conclude that there exists a sequence of elements u_n in H_0^1 such that u_n satisfies the perturbed problem

$$B(u_n, \varphi) - \frac{1}{n}\int_\Omega u_n\varphi dx = \int_\Omega (g[u_n] - h)\varphi dx$$

for all $\varphi \in H_0^1$.

We shall now show that the elements u_n remain bounded in H_0^1 as $n \to \infty$. If not, let $\|u_n\|_{H_0^1} \to \infty$ as $n \to \infty$. Also let $v_n = \|u_n\|_{H_0^1}^{-1} u_n$. Then

$$B(v_n, \varphi) - \frac{1}{n}\int_\Omega v_n\varphi dx = \frac{1}{\|u_n\|_{H_0^1}}\int_\Omega [g(u_n) - h]\varphi dx \tag{12}$$

for all $\varphi \in H_0^1$. The boundedness of v_n implies the existence of a subsequence (we shall call it v_n) such that $v_n \rightharpoonup v$ in H_0^1. Passing to the limit in the above equation, $B(v,\varphi) = \lim_n B(v_n,\varphi) = 0$ for all $\varphi \in H_0^1$, since $g: R \to R$ is bounded. By the definition of eigenfunction we can conclude, as observed in Section 4, that $v = \lambda w$, $\lambda \in R$. Also

$$B(v_n - v, v_n - v) = B(v_n, v_n - v) \to 0.$$

By using Gårding's inequality and the compactness of the imbedding of H_0^1 in $L^2(\Omega)$, we conclude that $v_n \to v$ in H_0^1. Thus $v \neq 0$, and hence $\lambda \neq 0$. Hence, from (12), by setting $\varphi = v$ we conclude that $B(v_n, v) = B(v, v_n) = 0$ for all n. Further $\int_\Omega [g(u_n) - h]vdx < 0$ for sufficiently large n. Hence

$$\limsup \int_\Omega g(u_n)vdx \leq \int_\Omega hvdx.$$

Since $u_n = \|u_n\|_{H_0^1} v_n$, $\|u_n\|_{H_0^1} \to \infty$ and since $v_n \to v$ a.e. on Ω, by Lebesgue's theorem we conclude that if $\lambda > 0$

$$g(\infty)\int_{\Omega_+}|w|dx - g(-\infty)\int_{\Omega_-}|w|dx = \lim \int_\Omega g(u_n)wdx$$
$$\leq \int_\Omega hwdx. \tag{13}$$

Similarly if $\lambda < 0$,

$$g(-\infty)\int_{\Omega_+}|w|dx - g(\infty)\int_{\Omega_+}|w|dx = \lim \int_\Omega g(u_n)wdx$$
$$= \lim \lambda^{-1}\int_\Omega g(u_n)vdx$$
$$\geq \lambda^{-1}\int_\Omega hudx = \int_\Omega hwdx. \tag{14}$$

Both (13) and (14) contradict inequalities (10). Hence $\|u_n\|_{H_0^1}$ is bounded and thus for some subsequence $u_n \to v$ in H_0^1. Passing to the limit as $n \to \infty$ in (11) we obtain

$$B(u,\varphi) = \int_\Omega [g(u) - h]\varphi dx, \qquad \varphi \in H_0^1.$$

This proves the theorem.

7. An Abstract Generalization Due to DeFigureido. Let S be a real Hilbert space, $A:S \to S$ be a linear, bounded, self-adjoint operator with closed range $R(A)$ and a finite dimensional null space $N(A)$. Let $N:S \to S$ be a compact nonlinear mapping (i.e., N is continuous and takes bounded sets into relatively compact sets) such that

$$\|Nu\| \leq c\|u\|^\alpha + d \tag{15}$$

for all $u \in S$, where $c > 0$ and $0 \leq \alpha < 1$ are fixed constants. DeFigureido obtains the following theorem [7].

Theorem 2. Let $\{\lambda_n\}$ be a sequence of positive real numbers such that $\lambda_n \to 0$. Also, let the following condition hold:

Given $y \in N(A)$, $\|y\| = 1$, and sequences $t_n \to \infty$, (16) $y_n \to y$, $y_n \in N(A)$, $z_n \in R(A)$, $\|z_n\| \leq K_1$, where K_1 is a constant, we have $(h,y) > \liminf_{n\to\infty} \langle N(t_n y_n + t_n^\alpha z_n), y\rangle$ where h is a given element of S. Then the nonlinear problem

$$Au + Nu = h \tag{17}$$

has a solution $u \in S$.

<u>Proof</u>. We consider the sequence of perturbed problems

$$Au_n - \lambda_n u_n + Nu_n = h. \tag{18}$$

This equation is equivalent to

$$u_n = (A - \lambda_n)^{-1}(h - Nu_n).$$

The operator $T:S \to S$ defined by $Tu = (A - \lambda_n)^{-1}(h - Nu_n)$ is compact, and the growth hypothesis (15) on N enables us to apply Schauder's fixed point theorem. The fact that $\alpha < 1$ implies that T maps a ball of sufficiently large radius into itself, and thus T has a fixed point which is a solution of (18).

We first show that if the sequence $\{u_n\}$ is bounded, then Theorem 2 is proved. Each u_n can be written as $v_n + w_n$ where $v_n \in N(A)$ and $w_n \in R(A)$ since, by virtue of the hypotheses on A, S can be decomposed as $N(A) \oplus R(A)$. By hypothesis N maps bounded sets into relatively compact sets, and thus there exists $g \in S$ and a subsequence of $\{u_n\}$, which will be referred to as $\{u_n\}$ henceforth, such that $Nu_n \to g$. But the boundedness of $\{u_n\}$ implies that $\{v_n\}$ and $\{w_n\}$ are also bounded. Hence we can extract subsequences of $\{v_n\}$ and $\{w_n\}$ such that $v_n \to v$ and $w_n \to w$ where $v,w \in S$. Passing to the limit in (18), we obtain $Aw_n \to h - g$. Since A restricted to $R(A)$ is a linear homeomorphism, we obtain that $w_n \to w$. Let $u = v + w$. It can be then seen that

$$Au + Nu = h,$$

or u is a solution of (17).

We now show that $\{u_n\}$ is bounded. If not, let $\|u_n\| \to \infty$. Writing $u_n = v_n + w_n$ as before and denoting by P the orthogonal projection of S onto $R(A)$ we get, from (18), that

$$Aw_n - \lambda_n w_n + PNu_n = Ph.$$

Since A restricted to $R(A)$ is a linear homeomorphism we get

$$\|w_n\| \leq a\|u_n\|^{\alpha} + b,$$

for sufficiently large n where a and b are fixed constants. Now let $p_n = \|u_n\|^{-1}u_n$, $q_n = \|u_n\|^{-1}v_n$ and $v_n = \|u_n\|^{-1}w_n$ so that $p_n = q_n + v_n$. Clearly, $\{q_n\}$ is bounded and therefore an appropriate subsequence of q_n converges to $q \in N(A)$. In view of (19), $v_n \to 0$. Thus $p_n \to q$ and $\|q\| = 1$ since $\|p_n\| = 1$. From (18) we now obtain that

$$\langle Ap_n, q\rangle - \lambda_n\langle p_n, q\rangle + \|u_n\|^{-1}\langle Nu_n, q\rangle = \|u_n\|^{-1}\langle h, q\rangle.$$

Since A is self-adjoint and $q \in N(A)$ the first term on the left-hand side is 0. Thus we have

$$\lambda_n\|u_n\|\langle p_n, q\rangle = \langle Nu_n - h, q\rangle.$$

Since $p_n \to q$ we have for sufficiently large n,

$$\langle Nu_n - h, q\rangle > 0. \tag{20}$$

But $u_n = v_n + w_n = \|u_n\|q_n + w_n$. And from (19) it follows that $w_n = \|u_n\|^{\alpha}z_n$ where $\{z_n\}$ is bounded. Hence, from (20), we obtain

$$\liminf\langle Nu_n, h\rangle \geq \langle h, y\rangle$$

which contradicts the hypothesis of the theorem.

This completes the proof of Theorem 2.

8. Remarks. a) Hypothesis (15) of Theorem 2 is a natural generalization of the work of Nečas [22]. Nečas, as an abstract generalization of Landesman and Lazer's result, proves the following: Let A be an operator as in Theorem 2 and let $N:S \to S$ satisfy $\|Nu\| \leq c < \infty$ for all $u \in S$. Further let N be compact. Then if i) $l(w) = \lim_{t\to\infty} \langle N(u + tw), w\rangle$ exists, uniformly with respect to bounded sets of u and with respect to $w \in N(A)$, $\|w\| = 1$; and ii) for every $w \in N(A)$, $\|w\| = 1$ we have $\langle w,h\rangle < l(w)$, the equation

$$Au + Nu = h$$

has a solution. Nečas further proves that Theorem 1 follows as a particular case.

b) Theorem 2 also proves the results of Fučik, Kučera and Nečas [8] without resorting to "Alternative Problems". Related results on nonlinear problems at resonance may be seen in Cronin [6], Nirenberg [23] and Berger and Schechter [3].

c) The reader is referred to Mawhin [21] where the connection between coincidence degree and the proof of Theorem 2 is discussed.

d) It would be of interest to generalize the result of Theorem 2 to the case when $a \geq 1$, thereby trying to obtain the linear theory as a particular case. In this connection the following result [12] is of interest: Let L be the linear differential operator defined by

$$D(L) = \{u \in L^2[0,2\pi]\,|\,u(t), u'(t) \text{ are absolutely continuous, } u''(t) \in L^2[0,2\pi] \text{ and } u(0) = u(2\pi),\ u'(0) = u'(2\pi)\}$$

$$Lu = -u''.$$

Further let P be the orthogonal projection of $L^2[0,2\pi]$ onto the subspace of $L^2[0,2\pi]$ generated by the eigenfunctions of L corresponding to the eigenvalues $(m+1)^2$, $(m+2)^2, \ldots$ of the boundary value problem $-x'' - \lambda x = 0$, $x(0) = x(2\pi)$, $x'(0) = x'(2\pi)$. If $M: L^2[0,2\pi] \to L^2[0,2\pi]$ is such that i) M is continuous and bounded; ii) there exists p such that $0 < p < 2m + 1$ and $\langle Mu - Mx, u - v\rangle \geq -p\|u - v\|^2$ for all $u, v \in S$; and iii) there exists $R > 0$ such that $\langle Mu, x^*\rangle \leq 0$ for all $x^* \in (I - P)L^2[0,2\pi]$ satisfying $\|x^*\| = R$ and $u = v + x^*$, where v is the unique solution of $Lv + P[Mv - m^2 v] = 0$, then the nonlinear problem $Lu - m^2 u + Mu = 0$ has at least one solution.

This result was then generalized in Kannan and Locker [15] to the case when $D(M) \subset S$. It must be noted that condition (iii) is very closely related to (16) of Theorem 2. In fact condition (ii) of the theorem of Nečas (see (a)) implies condition (iii) above. Generalization of the results of Theorem 2 may be seen in Kannan and Locker [14].

e) Of course, one of the chief features of the method of "Alternative Problems" is illustrated in Cesari [4] where he handles nonlinearities which are "large". It would be of interest to see whether the methods of this paper would go over to the case of large nonlinearities. (See also Gustafson and Sather [9].)

f) It would be of interest to see how the degree of the maps $I + T_n$, where $T_n = (A - \lambda_n)^{-1}(h - Nu)$, is related to the degree of the map W introduced by Williams [24]. In fact Williams [24] shows that the existence of a solution to problems at resonance is equivalent to the nonvanishing nature of the degree of a single map which is closely related to the coupled map β of Section 5. A similar idea is present in the work of Bebernes [2]. Thus Bebernes considers the problem

$$\begin{aligned} & x'' = f(t,x,x') \\ & x(0) = x(1), \quad x'(0) = x'(1). \end{aligned} \tag{21}$$

For each $\lambda \in [0,1]$ he considers the perturbed problem

$$\begin{aligned} & x'' = \lambda f(t,x,x') + (1 - \lambda)x \\ & x(0) = x(1), \quad x'(0) = x'(1) \end{aligned}$$

which may be written as

$$\begin{aligned} & x'' - x = \lambda f(t,x,x') - \lambda x \\ & x(0) = x(1), \quad x'(0) = x'(1). \end{aligned} \tag{22}$$

This perturbed problem has a linear part which is invertible and $x(t)$ is a solution of (22) if and only if $x(t)$ is a solution of

$$x(t) = \lambda \int_0^1 G(t,s)[f(s,x(s),x'(s)) - x(s)]ds$$

where $G(t,x)$ is the unique Green's function of $x'' - x = 0$, $x(0) = x(1)$ and $x'(0) = x'(1)$. If we now know that $x(t)$ is a solution of (22) implies that either $(t,x(t),x'(t)) \in D$, a bounded open set in $[0,1] \times R \times R$ or there exists $T \in [0,1]$ such that $[T,x(T),x'(T)] \notin \bar{D}$ then by an easy application of Leray-Schauder theory we can establish the existence of a solution of (21). The similarity in ideas between the work of Bebernes and the papers of Hess [10] and DeFigureido [7] is obvious.

REFERENCES

1. Ahmad, S., "An existence theorem for periodically perturbed conservative systems," Michigan Math. J. 20 (1973), 385-392.

2. Bebernes, J.W., "A simple alternative problem for finding periodic solutions of second order ordinary differential systems," Proc. Amer. Math. Soc. 42 (1974), 121-127.

3. Berger, M.S. and M. Schechter, "On the solvability of semilinear operator equations and elliptic boundary value problems," Bull. Amer. Math. Soc. 78 (1972), 741-745.

4. Cesari, L., "Functional analysis and periodic solutions of nonlinear differential equations," Contributions to Differential Equations 1 (1963), 149-167.

5. Cesari, L. and R. Kannan, "Periodic solutions in the large of nonlinear ordinary differential equations," Rend. di Math. 8 (1975), 633-654.

6. Cronin, J., "Equations with bounded nonlinearities," J. Differential Equations 14 (1973), 581-586.

7. DeFigureido, D.G., "On the range of nonlinear operators with linear asymptotes which are not invertible," Comment. Math. Univ. Carolinae 15 (1974), 415-428.

8. Fučik, S., M. Kučera and J. Nečas, "Ranges of nonlinear asymptotically linear operators," J. Differential Equations 17 (1975), 375-394.

9. Gustafson, K. and D. Sather, "Large nonlinearities and monotonicity," Arch. Rat. Mech. Anal. 48 (1972), 109-122.

10. Hess, P., "On a theorem by Landesman and Lazer," Indiana Univ. Math. J. 23 (1974), 827-829.

11. Kannan, R., "Periodically perturbed conservative systems," J. Differential Equations 16 (1974), 506-514.

12. Kannan, R., "Existence of periodic solutions of nonlinear differential equations," Trans. Amer. Math. Soc. 217 (1976), 225-236.

13. Kannan, R. and J. Locker, "On a class of nonlinear boundary value problems," (to appear).

14. Kannan, R. and J. Locker, "Perturbation methods and alternative problems," (to appear).

15. Kannan, R. and J. Locker, "Nonlinear boundary value problems and operators TT*," (to appear).

16. Landesman, E.M. and A.C. Lazer, "Nonlinear perturbations of linear elliptic boundary value problems at resonance," Indiana Univ. Math. J. 19 (1970), 609-623.

17. Lazer, A.C., "On Schauder's fixed point theorem and forced second order nonlinear oscillations," J. Math. Anal. Appl. 21 (1968), 421-425.

18. Lazer, A.C. and D.E. Leach, "Bounded perturbations of forced harmonic oscillators at resonance," Annali di Mat. Pura. Appl. 82 (1969), 49-68.

19. Lazer, A.C. and D.E. Leach, "On a nonlinear two-point boundary value problem," J. Math. Anal. Appl. 26 (1969), 20-27.

20. Lazer, A.C. and D.A. Sanchez, "On periodically perturbed conservative systems," Michigan Math. J. 16 (1969), 193-200.

21. Mawhin, J., "Nonlinear perturbations of Fredholm mappings in normed spaces and applications to differential equations," Univ. de Brasilia, Trabalho de Mat. no. 61, 1974.

22. Nečas, J., "On the range of nonlinear operators with linear asymptotes which are not invertible," Comment. Math. Univ. Carolinae 14 (1973), 63-72.

23. Nirenberg, L., "An application of generalized degree to a class of nonlinear operators," Trosieme Coll. Analyse Fonctionelle, Vander Louvain 1971, 57-74.

24. Williams, S.A., "A connection between the Cesari and Leray-Schauder methods," Michigan Math. J. 15 (1968), 441-448.

MEASURABILITY OF SOLUTIONS OF NONLINEAR EQUATIONS

Rangachary Kannan
University of Missouri, St. Louis

Habib Salehi
Michigan State University

1. Introduction. In an earlier paper [13] the authors consider random nonlinear differential equations of the type

$$Lu + Nu = 0 \tag{1}$$

where L is a random, linear, self-adjoint, positive, differential operator and N is a random nonlinear operator, for the existence of measurable solutions. The main idea is to utilize the existence results from the deterministic case and then apply the results from the theory of random operators to prove the measurability of the solutions. Thus, in [13] we consider equations of type (1) where 0 may or may not be an eigenvalue of L and N is a monotone operator. When 0 is not an eigenvalue of the associated linear problem of L we rewrite (1) as a nonlinear Hammerstein equation of the type $u + KNu = 0$ and then, applying the results on the existence of solutions of nonlinear Hammerstein equations in the deterministic case and the randomness properties of the operator $(I + KN)^{-1}$, we prove the measurability of solutions. When 0 is an eigenvalue, we apply the method of "Alternative Problems" and rewrite the equations into an equivalent system of two operator equations ([5], [6]). We then draw upon the iterative methods that are available, since N is monotone, to establish the randomness of the solution.

In this paper we extend the above ideas to the case when N is not monotone. The Leray-Schauder degree theory or the Schauder fixed point theorem are two of the principal tools in many problems in the deterministic theory of nonlinear differential equations. However, a major difference of this approach as compared with monotonic nonlinearities is that iterative techniques are not available. Thus, we have to resort to proving the measurability of solutions by more direct methods.

We consider here several models of random nonlinear differential equations where the Leray-Schauder theory or the Schauder fixed point theorem is used in the corresponding deterministic cases. The different models selected here illustrate the different situations that may arise. In each case we reduce the measurability question to one of proving the measurability of the graph of the solution function.

In Section 2 of the paper we introduce all the preliminary concepts and results that will be needed in this paper. A theorem on joint measurability of a function of two variables is presented in Section 3. Then in Section 4 we consider several models in random equations.

2. Preliminaries. In this section we state some results and definitions from nonlinear functional analysis and random operator theory as they relate to this paper.

For a real Banach space $\mathcal{X}$, let $\mathcal{X}^*$ be its dual, $\|\cdot\|$ be the norm and $\langle\ ,\ \rangle$ denote the pairing between $\mathcal{X}$ and $\mathcal{X}^*$. The symbol $(\ ,\ )$ will be used to denote inner product over a real Hilbert space. As usual for a linear operator T between two Banach spaces, T^* will denote the adjoint of T between the dual spaces.

2.1. A real Banach space $\mathcal{X}$ is said to be in <u>normal position</u> [1] with respect to a real Hilbert space $\mathcal{J}$ if

(i) $\mathcal{X}^* \subseteq \mathcal{J} \subseteq \mathcal{X}$

(ii) $(u,v)_{\mathcal{J}} = \langle u,v\rangle$, where $u \in \mathcal{X}^*$, $v \in \mathcal{J}$ and

(iii) $\mathcal{J}$ is dense in $\mathcal{X}$.

2.2. A nonlinear operator N from $\mathcal{X}$ into $\mathcal{X}^*$ is said to be <u>monotone</u> if

$\langle Nx - Ny, x - y\rangle \geq 0$ for all $x,y \in \mathcal{D}(N)$ where $\mathcal{D}(N)$ indicates the domain of N.

2.3. A monotone operator N from $\mathcal{X}$ into $\mathcal{X}^*$ is said to be <u>maximal monotone</u> if it is maximal in the sense of ordering by inclusion of graphs.

We note that any monotone operator which is continuous over the entire real Banach space $\mathcal{X}$ is maximal monotone.

2.4. An operator N from a real Banach space $\mathcal{X}$ into $\mathcal{X}^*$ is said to be <u>hemicontinuous</u> if for all $u \in \mathcal{D}(N)$, $v \in \mathcal{X}$, $u + t_n v \in \mathcal{D}(N)$, where $t_n \geq 0$, $t_n \to 0$, implies $N(u + t_n v)$ converges weakly to Nu. N is said to be <u>demicontinuous</u> if $u_n \in \mathcal{D}(N)$ and $u_n \to u$ in $\in \mathcal{D}(N)$ implies Nu_n converges weakly to Nu.

In Theorems I, II, III below, $\mathcal{X}$ is assumed to be reflexive.

2.5. <u>Theorem I</u> [2], [14]. Let $N:\mathcal{X} \to \mathcal{X}^*$ be monotone; then N is <u>maximal monotone</u> if and only if range of $I + N$, $\mathcal{R}(I + N)$, is $\mathcal{X}^*$, i.e., $I + N$ is surjective.

2.6. <u>Theorem II</u> [4]. Let $N:\mathcal{X} \to \mathcal{X}^*$ be maximal monotone and coercive, i.e. $\langle Nu,u\rangle/\|u\| \to \infty$. Then $\mathcal{R}(N) = \mathcal{X}^*$.

2.7. <u>Theorem III</u> [4]. Let $N:\mathcal{X} \to \mathcal{X}^*$ be hemicontinuous, monotone and coercive. Then N is maximal monotone and $\mathcal{R}(N) = \mathcal{X}^*$.

2.8. <u>Theorem IV</u> [11]. Let N be a monotone operator from $\mathcal{X}$ into $\mathcal{X}^*$. If $\mathcal{D}(N)$ is open, then hemicontinuity of N is equivalent to its demicontinuity.

In the following we summarize some results from the theory of random operators.

2.9. Let Ω be a measure space and $\mathcal{B}$ be a σ-algebra of subsets of Ω. For any metric space $\mathcal{Y}$, $\mathcal{B}_{\mathcal{Y}}$ denotes the σ-algebra of subsets of $\mathcal{Y}$ generated by closed subsets of $\mathcal{Y}$. A correspondence g from Ω into $2^{\mathcal{Y}}$ (subsets of $\mathcal{Y}$) is called a $\mathcal{Y}$-valued random correspondence if for each $B \in \mathcal{B}_{\mathcal{Y}}$, the set $g^{-1}(B) = \{\omega \in \Omega : g(\omega) \cap B \neq \emptyset\}$ is in $\mathcal{B}$.

In case g is a function we will, as usual, use the term random variable for g.

2.10. Let $(\Omega,\mathcal{B})$ be a measurable space, $\mathcal{X}$ and $\mathcal{Y}$ be metric spaces. A mapping T (single-valued) from $\Omega \times \mathcal{X}$ into $\mathcal{Y}$ is called a <u>random operator</u> if for each fixed $x \in \mathcal{X}$, the function $T(\cdot,x)$ is a $\mathcal{Y}$-valued random variable. A random operator T is said to be continuous, hemicontinuous, demicontinuous if for each $\omega \in \Omega$, $T(\omega,\cdot)$ is, respectively, continuous, hemicontinuous, or demicontinuous.

2.11. Let T be a random operator from $\Omega \times \mathcal{X}$ into $\mathcal{Y}$. An equation of the type

$$T(\cdot, x(\cdot)) = y(\cdot)$$

where y is a given random variable with values in $\mathcal{Y}$, is called a random operator equation.

Examples of random differential and integral operator equations may be seen in [3] and [9].

2.12. Any $\mathcal{X}$-valued random correspondence $x(\omega)$ which satisfies

$$T(\omega, x(\omega)) = y(\omega)$$

where the second argument of T is any value in the range of $x(\omega)$, is said to be a random solution of the equation

$$T(\omega)x(\omega) = y(\omega)$$

($T(\omega)x$ stands for $T(\omega, x)$).

It must be noted that not every solution of a random equation is a random solution, as can be seen from examples in [3] or [9].

2.13. Theorem V [9] and [15]. Let $(\Omega, \mathcal{B}, \mu)$ be a complete probability space, $\mathcal{X}$ be a separable metric space, and $\mathcal{Y}$ be a metric space. Let T be a separable random operator [15] from $\Omega \times \mathcal{X}$ and $\mathcal{Y}$ such that $T(\omega, \cdot)$ is invertible and $T^{-1}(\omega, \cdot)$ is continuous. Then T^{-1} is also a random operator.

Outline of Proof. Let $S(\cdot, r)$ and $\bar{S}(\cdot, r)$ denote, respectively, the open and closed balls of radius r around

To prove the measurability of T^{-1} it suffices to show that for an arbitrary $y \in \mathcal{Y}$ and a closed ball $\bar{S}(x', r)$ the event $\{\omega : T^{-1}(\omega, y) \in \bar{S}(x', r)\}$ is in $\mathcal{B}$. It can be easily shown that

$$\{\omega : T^{-1}(\omega, y) \in \bar{S}(x', r)\}$$
$$= \bigcap_{n=1}^{\infty} \bigcup_{S(x', r+1/n)} \{\omega : T(\omega, x') \in S(y, 1/n)\}$$

Because T is measurable and separable, it follows from this relation that T^{-1} is a random operator.

2.14. <u>Theorem VI</u> [13]. Let $(\Omega, \mathcal{B}, \mu)$ be a complete probability space and $\mathcal{X}$ and $\mathcal{Y}$ be separable Banach spaces. Assume that T is a random operator from $\mathcal{X}$ into $\mathcal{Y}$ such that for each $\omega \in \Omega$, $T(\omega)$ is demicontinuous. Then for any $\mathcal{X}$-valued random variable x, the function $T(\omega, x(\omega))$ is a $\mathcal{Y}$-valued random variable.

2.15. <u>Theorem VII</u> [19]. Let g be a correspondence of $(\Omega, \mathcal{B})$ into a complete separable metric space $\mathcal{X}$. If the graph of g belongs to the σ-algebra generated by $\mathcal{B} \times \mathcal{B}_{\mathcal{X}}$, then g is measurable.

3. <u>Joint Measurability</u>. In several areas of mathematics such as stochastic processes and control theory, it is of primary interest to impose minimal conditions on a function of two variables, regarding each variable separately, so that measurability with respect to both variables jointly is guaranteed. One of the earlier papers in this area is the work of Scorza-Dragoni [18]. Recently the interest in this problem has been revived, as can be seen from the paper [8]. Most of these results deal with a situation where the given function is real-valued and both the arguments belong to topological spaces. An extension of this result to the case where the function can have values in a metric space, one argument varying in a metric space and the other argument belonging to

an arbitrary measurable space, was obtained recently [16]. In view of the mathematical models provided in Section 4 of this paper it is evident that a theorem in this general setting will play an important role in the theory of random operators. Its chief advantage is that no restriction concerning the measure space is made (no topology is required). Because of this, and for the sake of completeness, we state the theorem below and provide a sketch of the proof.

3.1. Theorem [16]. Let $(\Omega, \mathcal{B}, \mu)$ be a measure space, $\mathcal{X}$ be a separable metric space, and $\mathcal{Y}$ be a metric space. Assume that $T: \Omega \times \mathcal{X} \to \mathcal{Y}$ is a continuous random operator. Then T is a measurable map from $\Omega \times \mathcal{X}$ into $\mathcal{Y}$.

Proof. Let $x_1, x_2, \ldots, x_n, \ldots$ be a dense subset of $\mathcal{X}$ and let k range through positive integers. Let

$$\Omega_n^k = \{x : x \in \mathcal{X},\ d(x, x_n) \leq 1/k\} \setminus \bigcup_{j=1}^{n-1} \Omega_j^k$$

It can be easily seen that (i) $\bigcup_n \Omega_n^k = \mathcal{X}$; and (ii) $\Omega_{n_1}^k \cap \Omega_{n_2}^k = \emptyset$, if $n_1 \neq n_2$. For each k define the function $T_k(\omega, x): \Omega \times \mathcal{X} \to \mathcal{Y}$ as follows: Fix $\omega \in \Omega$ and $x \in \mathcal{X}$. There exists exactly one Ω_n^k such that $x \in \Omega_n^k$. Set

$$T_k(\omega, x) = T(\omega, x_n)$$

Clearly, for each k the function $T_k(\omega, x)$ is countably valued and hence it is measurable (jointly). Let $\omega \in \Omega$, $x \in \mathcal{X}$ and $\epsilon > 0$. Because of the continuity of $T(\omega, \cdot)$ in x for each ω one can show the existence of a $k_o(\omega, x, \epsilon)$ such that for all $k \geq k_o$, $T(\omega, x)$ and $T_k(\omega, x)$ are not farther apart than ϵ. This shows that the sequence of $T_k(\omega, x)$ convergence to $T(\omega, x)$ pointwise. Hence $T(\omega, x)$ is measurable.

4. Stochastic Models. We will now consider several situations in which we apply the theorem of the previous section to obtain the randomness of the solutions of nonlinear differential equations. As mentioned in the introduction, we consider random problems of the type

$$Lu + Nu = 0 \tag{1}$$

where L is a random, positive, linear, self-adjoint, differential operator and 0 is an eigenvalue, with finite multiplicity, of the associated linear problem of L. Also N is a nonlinear random operator. The case when 0 is not an eigenvalue would then be a particular case.

Equation (1) is first rewritten into an equivalent system of two operator equations as in [6]. In this paper we will only outline the method by which this system is solved in the deterministic case. We then prove, by applying Theorem 3.1, the randomness of the solution. The details of the deterministic case may be seen in [5], [6], and [7].

Let $\mathcal{S}$ be a real Hilbert space such that the eigenfunctions corresponding to the distinct eigenvalues of L form a complete orthonormal system in $\mathcal{S}$. Also let $L: \mathcal{D}(L) \subset \mathcal{S} \to \mathcal{S}$ be a random linear differential operator of the type mentioned above. Let $\mathcal{S}_0(\omega)$ be the null space of L corresponding to a fixed ω and let $P(\omega): \mathcal{S} \to \mathcal{S}_0(\omega)$ be the projection operator. Let $H(\omega) = [L(\omega)|\mathcal{D}(L) \cap N(L)^{\perp}]^{-1}$ be the generalized inverse of L. Then (1) can be rewritten ([5]) in the equivalent form

$$x + H(I - P)Nx = Px \tag{2}$$

$$PNx = 0. \tag{3}$$

Model I. Our first example is the following random nonlinear equation

$$\begin{aligned} x'' + \text{grad } G(x, \omega) &= p(t, \omega) \\ x(0) = x(2\pi), \quad x'(0) &= x'(2\pi) \end{aligned} \tag{4}$$

where $G \in C^2(R,R)$ in x and for every $a \in R$

$$M^2 < q \leq \left(\frac{\partial^2 G(a,\omega)}{\partial x_i \partial x_j}\right) \leq p \leq (M + 1)^2, \quad M \text{ an integer.}$$

Also $p(t, \omega)$ is continuous in t so that the operator $N(x, \omega) = -\text{grad } G(\omega, x) + p(t, \omega)$ is a random operator. We will show that under the above hypotheses (4) has a unique random 2π-periodic solution.

For each fixed $\omega \in \Omega$, (4) is of the type (1). Hence we have the equivalent system of equations (2) and (3). Let $x^* \in \mathscr{A}_0(\omega)$ be an arbitrary element. In this model, L is a deterministic operator and thus $\mathscr{A}_0(\omega) = \mathscr{A}_0 = \mathcal{N}(L)$ = set of all 2π-periodic functions.

Equation (2) can then be rewritten into another nonlinear equation by the following process. Let $\mu = \frac{p + q}{2}$. Then (2) is equivalent to

$$x - \mu H(I - P)x + \mu H(I - P)x + H(I - P)Nx = x^*$$

or

$$\begin{aligned} x = &-[I - \mu H(I - P)]^{-1}\{H(I - P)[\mu x + Nx]\} \\ &+ [I - \mu H(I - P)]^{-1}x^*. \end{aligned}$$

This equation is of the type

$$x = Tx,$$

where T is a strict contraction [6]. By the Banach theorem, T has a unique fixed point and thus (2) has a unique solution $x \in \mathscr{A}$ for each $x^* \in \mathscr{A}_0$. It can be easily proved that $[I + H(I - P)N]^{-1}$ is continuous from $\mathscr{A}_0$ into $\mathscr{A}$.

We now rewrite (3) in the form

$$PN[I + H(I - P)N]^{-1}x^* = 0. \tag{5}$$

By applying the Leary-Schauder theory we show that this equation has at least one solution. Also the boundary conditions force this solution to be unique.

Randomness of the Solution. Equation (5) can be written as

$$\int_0^{2\pi} N[I + H(I - P)N]^{-1}x^* dt = 0$$

This equation is of the type

$$f(\omega, x^*) = 0$$

where $f(\omega, x^*) : \Omega \times \mathscr{A}_0 \to \mathscr{A}_0$. Since $\mathscr{A}_0$ is the space of constant functions, f can be treated as a map from $\Omega \times \mathcal{R} \to \mathcal{R}$. Further f is single-valued. The randomness of N and the fact that H and $I - P$ are deterministic implies that $H(I - P)N$ is random and separable. By the theorem of Nashed and Salehi [15] and Theorem 2.14, $[I + H(I - P)N]^{-1}$ is random (since it is continuous in x^* for each fixed ω). Hence f is measurable in ω. Also the continuity of $[I + H(I - P)N]^{-1}$ and N implies that f is continuous in x^*. Hence, by Theorem 3.1 on joint measurability, f is jointly measurable. Hence the set of couples $(\omega, x^*) \in \Omega \times \mathcal{R}$ such that

$$f(\omega, x^*) = 0$$

is a measurable set.

But this set is precisely the graph of the function $\varphi: \Omega \to \mathcal{R}$ defined by $\varphi(\omega) = x^*$, where x^* is the solution of (5). By Theorem 2.15, $\varphi(\omega)$ is measurable in ω. Now we apply the theorem of Nashed and Salehi [15] again to conclude that the randomness of $[I + H(I - P)N]^{-1}$ implies the measurability of $[I + H(I - P)N]^{-1}\varphi(\omega)$ which establishes the randomness of the solution of (5).

Model II. We now consider an example where the random differential equation does not have a unique solution for each $\omega \in \Omega$, but by virtue of the properties of the nonlinearity one can extract a single-valued measurable selection. In [13] we prove the following theorem.

Theorem. Let $L: \mathcal{X}^* \to \mathcal{H}$ be a random self-adjoint, differential operator with eigenvalues $\{\lambda_i(\omega)\}$, satisfying $\lambda_i(\omega) \geq 0$, $\lambda_i \to \infty$; with $\mathcal{X}^*$ a separable, reflexive Banach space; and with $\mathcal{H}$ a real Hilbert space such that $(\mathcal{X}^*, \mathcal{H}, \mathcal{X})$ is in normal position. Let $N: \mathcal{X}^* \to \mathcal{H}$ be a random, hemicontinuous, monotone, coercive operator. Then the equation

$$Lx + Nx = 0$$

has a random solution.

The proof in [13] used an iterative technique. We present here a different proof which uses Theorem 3.1.

Proof. Proceeding as in the earlier example, the random problem can be rewritten as

$$x + H(I - P)Nx = x^*$$

$$PN[I + H(I - P)N]^{-1}x^* = 0.$$

Since $H(I - P)$ is maximal monotone and N is monotone hemicontinuous the auxiliary equation is solvable for each $x^*(\omega) \in \mathcal{H}_0(\omega)$. The bifurcation equation is then shown to be of the type $Tx^* = 0$ where $T: \mathcal{H}_0(\omega) \to \mathcal{H}_0(\omega)$ is maximal monotone and coercive and thus $Tx^* = 0$ is solvable for each

ω. By the theory of maximal monotone operators we know that the solution set is closed and convex for each ω.

The chief feature of this example distinguishing it from the previous situation is that P is a random operator. However, since $x^*(\omega) \in \mathcal{R}(P(\omega))$ for each ω, we can rewrite the bifurcation equation as

$$PN[I + H(I - P)N]^{-1}Px = 0$$

And this problem can now be treated as one of the type

$$f(\omega, x) = 0$$

where $x \in \mathcal{S}$, a separable metric space. Repeating the arguments of the earlier section we can show that the (possibly) multivalued solution $x^*(\omega)$ is measurable in ω and thus the original random problem has a multivalued measurable solution. In particular since the set of solutions of the bifurcation equation is closed and convex and $\mathcal{S}$ is a real Hilbert space, it might be possible to apply arguments similar to the ones used in [17] to show that we could take the single valued measurable selection $x^*(\omega)$ which associates with each ω the element of minimum norm associated with the solution set. If this is possible, then since $[I + H(I - P)N]^{-1}$ is random [13], $x(\omega) = [I + H(I - P)N]^{-1}x^*$ is a measurable selection.

Model III. We now present a situation where the bifurcation equation has a unique solution $x^* = 0$ and thus solving the nonlinear problem reduces to solving the auxiliary equation which is over an infinite dimensional space.

$$x'' + \frac{d}{dt}\left(\frac{dG(\omega,x)}{dx}\right) + Ax = p(t,\omega)$$
$$x(0) = x(2\pi), \quad x'(0) = x'(2\pi). \tag{6}$$

This equation corresponds to the random version of the Liénard type of equations. Here G is a C^2 function in x, $\int_0^{2\pi} p(t,\omega)\,dt = 0$, and we assume that the nonlinear operator $Nx = -\frac{d}{dt}\left(\frac{dG(\omega,x)}{dx}\right) - Ax + p(t,\omega)$ is a random operator from $\Omega \times H^1[0,2\pi]$ into $L^2(0,2\pi]$. Further A is a constant > -1. Also L is the linear differential operator defined by

$$Lx = -x''$$

$$\mathcal{D}(L) = \{x \in H^2[0,2\pi] \mid x(0) = x(2\pi),\ x'(0) = x'(2\pi)\}.$$

Here H^1 and H^2 denote Sobolev spaces.

Proceeding as before the random nonlinear problem is equivalent to the system of equations

$$x + H(I - P)Nx = x^*$$

$$PNx = 0.$$

<u>Outline of Proof of the Deterministic Case</u> [7]. The bifurcation equation, when rewritten as an integral equation (note $Px = \int_0^{2\pi} x(t,\omega)\,dt$), implies that $Px = x^* = 0$. Thus we have to solve the equation

$$x + H(I - P)Nx = 0. \tag{7}$$

The operator L can be written as TT^* where

$$Tx = -x' \quad x \in \mathcal{D}(T) = \{H^1[0,2\pi],\ x(0) = x(2\pi)\}.$$

and

$$T^*x = x' \quad x \in \mathcal{D}(T^*) = \{H^1[0,2\pi],\ x(0) = x(2\pi)\}$$

This induces a natural decomposition of $H(I - P)$ in the form J^*J where $J^* = (I - P)\{T^* | \mathcal{D}(T^*) \cap \mathcal{N}(T^*)^{\perp}\}^{-1}$ and $J = \{T | \mathcal{D}(T) \cap \mathcal{N}(T)^{\perp}\}^{-1}(I - P)$. Equation (2) can then be rewritten as

$$y + JN[J^* y] = 0$$

where $x = J^* y$.

The operator JNJ^* can then be proved to be compact from $L_2[0,2\pi]$ into itself because of the appropriate continuity and boundedness hypotheses on N and compactness of J and J^*. We now apply the Leray-Schauder theory to prove the existence of at least one solution.

Proof of the Randomness. The randomness of N from $\Omega \times H^1[0,2\pi]$ into $L^2(0,2\pi]$ implies that JNJ^* is a random operator from $\Omega \times L^2[0,2\pi]$ into $L^2[0,2\pi]$. The auxiliary equation is of the type

$$f(\omega,x) = 0$$

where f is measurable in ω, because JNJ^* is random, and f is continuous in x, since J, N, and J^* are so. Thus $f(\omega,x)$ satisfies the conditions of Theorem 3.1. In fact this was the main motivation to prove Theorem 3.1 from a separable metric space into a separable metric space. In our model $L_2[0,2\pi]$ has a complete orthonormal system generated by $\sin nt$ and $\cos nt$. The randomness of the solution now follows as in the earlier arguments. Hence by [10, p.54] a choice of a measurable selection is possible.

Model IV. Finally we give an example where neither the auxiliary nor the bifurcation equation has a unique solution. In the model we show the full force of Theorem 3.1. Thus we consider the more general model of (1):

$$x'' + \frac{d}{dt}\left(\frac{dG}{dx}\right) + g(x,\omega) = p(t,\omega)$$

where

i) $\int_0^{2\pi} p(t,\omega)\,dt = 0$

ii) $\frac{g(x,\omega)}{|x|} \to 0$ as $|x| \to \infty$

iii) $x(\omega)g(\omega,x) \geq 0$ for $|x| \geq b$

iv) $G(\omega,x) \in C^2(R,R)$ in x

v) the operator $Nx = -\frac{d}{dt}(\frac{dG}{dx}) - g(x,\omega) + p(t,\omega)$

is a random operator from $\Omega \times H^1[0,2\pi] \to L^2[0,2\pi]$.

Proceeding as in the earlier cases the random problem is equivalent to the system

$$x + H(I - P)Nx = x^*$$

$$PNx = 0.$$

We now rewrite the auxiliary equation, as in the previous case, by using the decomposition of $H(I - P)$ in the form J^*J. Thus any $x(t,\omega) = J^*y(t,\omega) + x^*$ is a solution to the random problem where $y(t,\omega)$ is a solution of

$$y + JN(J^*y + x^*) = 0$$

$$PN(J^*y + x^*) = 0.$$

Outline of Proof in the Deterministic Case (7). We consider, for each ω, the operator T from $\mathscr{S}_1 \oplus \mathscr{S}_0$ into $\mathscr{S}_1 \oplus \mathscr{S}_0$, where $\mathscr{S}_1 + \mathscr{S}_0 = L_2[0,2\pi]$, defined by

$$Tz = T(y,x^*) = (JN[J^*y + x^*],\ PN[J^*y + x^*] - x^*).$$

Then solving the nonlinear problem (for a fixed ω) is equivalent to solving

$$(y,x^*) + T(y,x^*) = 0. \tag{8}$$

Clearly T is compact from $\mathscr{S}_1 \oplus \mathscr{S}_0$ into itself. We show that by the hypotheses of the theorem there exists R_1 and R_2 such that the equation

$$(y,x^*) + \lambda T(y,x^*) = 0 \qquad 0 \leq \lambda < 1$$

has no solution with $\|y\| = R_1$, $\|x^*\| = R_2$. Then by the Leray-Schauder theory there exists at least one solution of (3).

<u>Randomness of the Solution</u>. We apply our Theorem 3.1 to the separable metric space $\mathscr{A}_1 \oplus \mathscr{A}_0$ and thus solving the non-linear problem is equivalent to solving

$$(y,x^*) + T(y,x^*) = 0$$

which is of the type $f(\omega,z) = 0$ where $z = (y,x^*)$. Clearly f is measurable because N is random and the other operators involved are deterministic. Also f is continuous in (y,x^*) because of the appropriate continuity of J, N and J^*. We now repeat our arguments showing that $\varphi(\omega) = [y(\omega), x^*(\omega)]$ is measurable in ω and $x(\omega) = J^* y(\omega) + x^*(\omega)$ is also so. According to [10, p.54] a choice of a measurable selection is possible.

REFERENCES

1. Amann, H., "Existence theorems for equations of Hammerstein type," Applicable Anal. 1 (1972), 385-397.

2. Brézis, H., Operateurs Maximaux Monotones, North Holland, American Elsvier, 1973.

3. Bharucha-Reid, A.T., Random Integral Equations, Academic Press, 1972.

4. Browder, F.E., "Nonlinear maximal monotone operators in Banach spaces," Math. Ann. 175 (1968), 89-113.

5. Cesari, L., "Functional analysis and Galerkin's method," Michigan Math. J. 11 (1964), 385-414.

6. Cesari, L. and R. Kannan, "Functional analysis and non-linear differential equations," Bull. Amer. Math. Soc. 79 (1973), 1216-1219.

7. Cesari, L. and R. Kannan, "Periodic solutions in the large, of nonlinear ordinary differential equations," Rend. Di. Math. 8 (1975), 633-654.

8. Davies, R.O., "Separate approximate continuity implies measurability," Proc. Camb. Phil. Soc. 73 (1973), 461-465.

9. Hanš, O., "Random operator equations," Proc. 4th Berkeley Symposium on Math. Statistics and Prob., Vol. II, Univ. of California Press, Berkeley, Calif. 1961, 185-202.

10. Hildenbrand, W., Core and Equilibribria of a Large Economy, Princeton University Press, 1974.

11. Kato, T., "Demicontinuity, hemicontinuity and mono-tonicity," Bull. Amer. Math. Soc. 70 (1964), 548-550, 73 (1967), 886-889.

12. Kannan, R., "Periodically perturbed conservative systems," Tech. Report No. 49, Department of Mathematics, The University of Michigan 1974.

13. Kannan, R. and H. Salehi, "Random nonlinear equations and monotonic nonlinearities," to appear in J. Math. Anal. Appl.

14. Minty, G.J., "On a "monotonicity" method for the solution of nonlinear equations in Banach spaces," Proc. Nat. Acad. Sci., U.S.A. 50 (1963), 1038-1041.

15. Nashed, M.Z. and H. Salehi, "Measurability of generalized inverses of random linear operators," SIAM J. Appl. Math. 25 (1973), 681-692.

16. Neubrunn, T., "Measurability of certain functions on Cartesian products," Mat.-Fyz. Casopis Sloven. Akad. Vied 10 (1960), 216-221; MR 25 #3140.

17. Rockafellar, R.T., "Measurable dependence of convex sets and functions on parameters," J. Math. Anal. Appl. 28 (1969), 4-25.

18. Scorza-Dragoni, G., "Un teorema sulle funzioni continue rispetto ad una misurabile rispetto ad un'altra variabile," Rend. Sem. Mat. Padova 17 (1948), 102-106.

19. Debreu, G., "Integration of correspondences," Proc. Fifth Berkeley Symp. Math. Stat. and Probability, II, Part 1, 351-372, University of California Press, Berkeley.

SOME MATHEMATICAL PROBLEMS
FROM PLATE AND SHELL THEORY

George H. Knightly*

University of Massachusetts

1. Introduction. In this article we consider aspects of three related problems arising in the nonlinear theory of thin elastic plates and shells. Each of the problems is governed by a coupled system of two nonlinear fourth order partial differential equations, together with appropriate homogeneous boundary conditions, and can be recast as a single nonlinear equation in a suitable Hilbert space H,

$$L_\lambda w + \alpha Q(w) + C(w) = P, \quad w \text{ in } H. \tag{1.1}$$

Here w measures deflection of the plate or shell from its unstressed state, L_λ, Q and C are, respectively, linear, quadratic and cubic operators while λ is a loading parameter, α is a constant depending on the geometry and P represents a given external load.

The problems to be studied involve the bending $(P \neq 0)$ and buckling $(P = 0)$ of (i) a clamped plate of general shape (see, e.g., [15], [5], [2], [19], [20], [28]), (ii) a simply supported rectangular plate (see e.g., [1], [21]) and (iii) a simply supported cylindrical panel (see [23] for a related problem). Other plate and shell problems involving ordinary

*Supported in part by NSF Grant No. MPS75-07579.

or partial differential equations models can also be investigated within the framework of equation (1.1) (see, e.g., [4], [6], [13], [14], [22]). Although a variety of methods have been used to discuss such problems (see, e.g., [1], [10], [11], [17], [26], [38]), a Hilbert space setting seems to provide for a unified approach, at least for some questions.

In §2, classical formulations of the three problems are given, the derivation of equation (1.1) is sketched and properties of the operators are listed. The existence question is addressed in §3 via a priori estimates and a fixed point theorem. A local theory of branching for equation (1.1) in the case $P = 0$ is developed in §4. In §5, the earlier work is applied in discussing branching questions for the three given problems.

Most of the results presented here are not new. However, we include in §5 a synopsis of a part of some forthcoming joint work with Professor Duane Sather on the buckling problem for the cylindrical panel.

2. Formulation of the operator equation. We now show how equation (1.1) can be obtained from the classical equations of equilibrium in a representative case, that of an axially compressed cylindrical panel subjected to radial loading. This derivation follows ideas expressed by Berger and Fife in [2, 5,6] (see also Fichera [9], in the linear case).

Suppose that the (right circular) cylindrical panel in the unbuckled state has radius R and middle surface given by $\Omega = \{(x,y): 0 < x < a, 0 < y < b\}$, where the axial coordinate, x, measures distance parallel to the axis from one end of the panel and the circumferential coordinate, y, measures distance along circular arcs $x = \text{const}$ from one edge of the panel. The governing equations (the von Kármán-Donnell equations) are

$$\Delta^2 f = -\frac{1}{2}[w,w] + \alpha w_{xx} \tag{2.1a}$$

$$\Delta^2 w = [w,f] - \lambda w_{xx} - \alpha f_{xx} + p, \text{ in } \Omega, \tag{2.1b}$$

where Δ^2 denotes the biharmonic operator and

$$[w,f] = w_{xx}f_{yy} + w_{yy}f_{xx} - 2w_{xy}f_{xy}. \tag{2.2}$$

Here, after suitable scaling of the parameters and dependent variables to suppress several physical constants, w represents the radial component of deflection of the panel, f is a certain stress function, λ measures the uniform compressive axial load on the panel (negative values of λ correspond to uniform tension on the panel), α is proportional to R^{-1} and $p = p(x,y)$ corresponds to a given, radically directed, external force on the panel. In addition, we shall suppose that the ends and edges of the panel are simply supported so that the conditions at the boundary, $\partial\Omega$, are

$$w = f = \Delta w = \Delta f = 0 \quad \text{on} \quad \partial\Omega. \tag{2.3}$$

Definition 1. A classical solution of (2.1), (2.3) is a pair of functions f, w in $C^4(\Omega) \cap C^2(\bar{\Omega})$ that satisfy (2.1) and (2.3) pointwise.

For the problem (2.1), (2.3) it will be convenient to introduce the Hilbert space H_s defined in [21] as the closure W, in the norm $\|\cdot\|_{2,2}$ of the Sobolev space $W_{2,2}(\Omega)$, of the set of smooth functions defined on $\bar{\Omega}$ and vanishing on $\partial\Omega$. Here we use an equivalent norm on H_s

given by

$$\|u\|^2 = \int_\Omega (\Delta u)^2,$$

with corresponding inner product $(\cdot,\cdot)$.

If equations (2.1a) and (2.1b) are multiplied by a smooth function φ in H_s, then an integration by parts leads to the generalized equations

$$(f,\varphi) = -\frac{1}{2}(B(w,w),\varphi) + \alpha(Aw,\varphi) \tag{2.4a}$$

$$(w,\varphi) = (B(w,f) - \lambda Aw - \alpha Af + P,\varphi), \tag{2.4b}$$

where

$$(B(w,f),\varphi) = \int_\Omega [(w_{xy}f_y - w_{yy}f_x)\varphi_x + (w_{xy}f_x - w_{xx}f_y)\varphi_y], \tag{2.5}$$

$$(Aw,\varphi) = \int_\Omega w_x\varphi_x \tag{2.6}$$

and

$$(P,\varphi) = \int_\Omega p\varphi. \tag{2.7}$$

As the notation suggests, it can be shown that the right hand sides of (2.5), (2.6) and (2.7) are bounded linear functionals of φ in H_s for fixed w, f, in H_s and p in $L_2(\Omega)$, so that $B(w,f)$, Aw and P are elements of H_s according to the Riesz representation theorem. Since (2.4) holds for arbitrary φ in H_s we obtain the equations

$$f = -\frac{1}{2}B(w,w) - \alpha Aw \tag{2.8a}$$

$$w = B(w,f) + \lambda Aw + \alpha Af + P, \quad (w,f \text{ in } H_s). \tag{2.8b}$$

Now (1.1) is obtained if we use (2.8a) to eliminate f from (2.8b) and define

$$L_\lambda w = w - \lambda Aw + \alpha^2 A^2 w,$$

$$Q(w) = B(w,Aw) + \frac{1}{2}AB(w,w),$$

$$C(w) = \frac{1}{2}B(w,B(w,w)).$$

Definition 2. A generalized solution of (2.1), (2.3) is a pair of functions f, w in H_s satisfying (1.1) and (2.8a).

We have sketched a proof of the first part of the following theorem. The more difficult second part, the regularity, can be established by simple modifications in the proof of Theorem B.1 in [21]. Similar results for more general cases are proved in [6]. Here $\partial'\Omega$ denotes $\partial\Omega$ with the corners deleted.

Theorem 1. Every classical solution of (2.1), (2.3) is a generalized solution. Every generalized solution is a classical solution in Ω and up to $\partial'\Omega$.

In addition to the problem just described for the cylindrical panel, we shall consider two problems involving flat plates: the simply supported rectangular plate under arbitrary normal loading and uniform compression on a pair of opposite edges, and the clamped plate of general shape subjected to general edge and normal loading.

The model for the simply supported rectangular plate is identical to (2.1), (2.3) with $\alpha = 0$ (see e.g., [1], [21]). In this case, the rectangle $\Omega = \{(x,y): 0 < x < a,\ 0 < y < b\}$ corresponds to the mid plane of the undeflected plate and the compressive force acts normal to the edges $x = 0$ and $x = a$. The Hilbert space H_s and operators A, B, C are defined in the same way as in the panel problem, while L_λ reduces to $I - \lambda A$. It is easily seen that the operator equation satisfied by the deflection w is

$$w - \lambda Aw + C(w) = P, \qquad (w \text{ in } H_s)$$

which is just (1.1) when $\alpha = 0$.

For the problem of the clamped plate we suppose that Ω is a bounded open set in the x,y-plane constituting the mid plane of the undeflected plate. We take as model the von Kármán equations ([15], [5], [2]) in the form

$$\Delta^2 f = -\frac{1}{2}[w,w]$$

$$\Delta^2 w = [w,f] + \lambda[w,F] + p, \quad \text{in} \quad \Omega,$$

together with the boundary conditions $f = f_x = f_y = w = w_x = w_y = 0$ on $\partial\Omega$. Again w represents normal deflection of the plate, f is an excess stress function and p is the normal load. The function F determines the form of the pre-stressing or edge loading and, for example, may provide for tension on some parts of the edge and compression on other parts; λ measures the magnitude of this loading.

The Hilbert space H_c for the clamped plate problem again uses the inner product $(u,v) = \int_\Omega \Delta u \Delta v$, but H_c is the completion in this norm (or, equivalently, in the norm $\|\cdot\|_{2,2}$ of the Sobolev space $W_{2,2}(\Omega)$) of $C_0^\infty(\Omega)$, the functions of compact support in Ω having continuous derivatives of all orders. The operators B and C are formally the same as above, but now A is given by

$$(Aw, \varphi) = \int_\Omega [F,w]\varphi$$

for w, φ in H_c. Note that F is not assumed to lie in H_c and that A need not be positive. Once again (1.1) takes the form $w - \lambda Aw + C(w) = P$, now for w in H_c.

Results similar to Theorem 1 are valid for the above plate problems, [2], [6], [21].

The next lemma summarizes properties common to the three problems under consideration; similar lemmas are proved in

[2], [6], [20] and [21]. We use the symbol H for the Hilbert space whenever a statement is valid for both of the spaces H_s and H_c. We speak of a <u>characteristic value</u>, μ, of A when $u = \mu Au$ for some nontrivial u in H and an <u>eigenvalue</u>, λ, of L_λ when $L_\lambda u = 0$ for nontrivial u in H (of course, these notions merge when $\alpha = 0$). In either case we call u an <u>eigenfunction</u>.

<u>Lemma 1</u>. (i) the operator A is bounded, linear, compact, self-adjoint and, in the problems of the cylindrical panel and rectangular plate, positive.

(ii) L_λ is bounded, linear and self-adjoint for each real λ. It can be factored, $L_\lambda = (I - \mu_+ A)(I - \mu_- A)$, where $\mu_\pm = \frac{1}{2}(\lambda \pm \sqrt{\lambda^2 - 4\alpha^2})$, and λ is an eigenvalue of L_λ if and only if at least one of the numbers $\mu_\pm$ is a characteristic value of A, i.e., if and only if

$$\lambda = \mu + \alpha^2\mu^{-1} \tag{2.9}$$

for some characteristic value, μ, of A. If λ is an eigenvalue of L_λ then the eigenfunctions of A corresponding to characteristic values μ satisfying (2.9) are eigenfunctions of L_λ and span its null space $\mathfrak{N} = \mathfrak{N}(L_\lambda)$.

(iii) If u, v, w lie in H, then the form $(B(u,v),w)$ is symmetric in u, v, w and linear in each entry.

(iv) Q and C are compact, homogeneous, continuous, polynomial operators. Q is quadratic, the gradient of the real-valued functional

$$\tau(w) = \frac{1}{3}(Q(w),w), \quad w \text{ in } H_s, \tag{2.10}$$

and has at each w in H_s a Fréchet differential $Q'(w)$, which satisfies

$$Q'(w)u = B(w,Au) + B(u,Aw) + AB(w,u), \quad u \text{ in } H_s, \tag{2.11}$$

and is Lipschitz continuous in w. C is cubic, the gradient of the real-valued functional

$$\sigma(w) = \frac{1}{4}(C(w),w), \quad w \text{ in } H, \tag{2.12}$$

and has at each w in H a Fréchet differential $C'(w)$, which satisfies

$$C'(w)u = \frac{1}{2} B(u,B(w,w)) + B(w,B(u,w)), \quad u \text{ in } H, \tag{2.13}$$

and is Lipschitz continuous in w. Furthermore, $\sigma(w) = 0$ with w in $N(L_\lambda)$ for some eigenvalue λ if and only if $w = 0$.

3. Existence. Several approaches to existence questions have been taken in various plate and shell problems. For example, in [10] Fife develops Schauder estimates in working with the von Kármán equations. Morosov [26] uses a representation in terms of a biharmonic Green's function when solving a clamped plate problem in the case of pure bending ($\lambda = 0$ in (1.1)). Methods dealing with formulations of the type (1.1) are given in [13], [14], [19], [28] and are demonstrated in this section.

If T is a compact operator on H and the set of all possible solutions, w in H, of the equation

$$w = tTw \tag{3.1}$$

lies in a fixed sphere in H as t ranges over $0 \leq t \leq 1$, then the Schaefer version [35] of the Schauder fixed point theorem ensures the existence of at least one solution of

(3.1) at $t = 1$. Equation (1.1) takes the form (3.1) with $t = 1$ if we set

$$Tw = \lambda Aw - \alpha^2 Aw - \alpha Q(w) - C(w) + P. \tag{3.2}$$

That T is compact follows easily from Lemma 1. Thus the key to existence in this approach lies in deriving an a priori estimate for (3.1) with T given by (3.2).

One can always determine such an a priori bound for restricted values of the parameters. For example, since $(Q(u),u) = \frac{3}{2}(B(u,u),Au))$ and $(C(u),u) = \frac{1}{2}\|B(u,u)\|^2$ for u in H, we find, upon taking the inner product of (3.1) with any solution w in H and integrating by parts,

$$\|w\|^2 = t\{\lambda(Aw,w) + \frac{\alpha^2}{8}\|Aw\|^2 - \frac{1}{2}\|\frac{3\alpha}{2} Aw + B(w,w)\|^2 + (P,w)\}. \tag{3.3}$$

If μ_1 is the smallest (in magnitude) characteristic value of A then $\|A\| = |\mu_1|^{-1}$ so that (3.3) implies

$$\|w\| \leq \|P\|[1 - \frac{|\lambda|}{|\mu_1|} - \frac{\alpha^2}{8|\mu_1|^2}]^{-1}. \tag{3.4}$$

If A is positive and $\lambda \leq 0$, then

$$\|w\| \leq \|P\|[1 - \frac{\alpha^2}{8\mu_1^2}]^{-1}. \tag{3.5}$$

Theorem 2. (i) If $8|\lambda\mu_1| + \alpha^2 < 8\mu_1^2$, then (1.1) has at least one solution, w, in H and every solution satisfies (3.4).

(ii) If A is a positive operator, if $\lambda \leq 0$ and $0 \leq \alpha < 2\sqrt{2}\mu_1$, then (1.1) has at least one solution, w, in H and every solution possesses the bound (3.5).

Corollary. Let $P = 0$ in equation (1.1) and suppose that the hypotheses of either part (i) or part (ii) in Theorem 2 are satisfied. Then (1.1) has no nontrivial solution.

If the operator A is positive, then when $\lambda < 0$ the plate or shell is undergoing forces of tension on its edges, while the edge forces are compressive when $\lambda > 0$. Thus arbitrary forces of tension, but only sufficiently small forces of compression are allowed in Theorem 2.

Remark 1. If $\|P\|$ is sufficiently small one can show that the solution given by Theorem 2 is unique. The argument uses the relevant bound (3.4) or (3.5) and proceeds along the lines of the derivation of those estimates (see also [20]).

Results similar to Theorem 2 are obtained for the flat plate, under more general edge conditions, by Hlaváček and Naumann [13, 14]. The boundary conditions in [13] and in the present work exclude rigid motion solutions w. In the paper [14] rigid motions are admitted. In fact under conditions similar to those in Theorem 2 the existence of families of solutions is shown such that two members of a family differ by a rigid motion. In the current context a principal step in the existence proof of [13, 14] lies in showing that

$$\lim_{\|u\|\to\infty} \frac{(u-T_0(u),u)}{\|u\|} = +\infty, \tag{3.6}$$

where T_0 is the operator T in (3.2) with $P = 0$. An abstract theorem (see, e.g., [8]) using compactness of T_0 then implies that $I - T_0$ is an onto mapping. To prove (3.6) we may compute as in (3.3) to see that

$$\begin{aligned}(u-T_0(u),u) &= \|u\|^2 - \lambda(Au,u) - \frac{\alpha^2}{8}\|Au\|^2 \\ &\quad + \frac{1}{2}\left\|\frac{3\alpha}{2} Au + B(u,u)\right\|^2 \geq \gamma(\lambda,\alpha)\|u\|^2,\end{aligned}$$

where $\gamma(\lambda,\alpha) = 1 - (8\lambda\mu_1 + \alpha^2)(8\mu_1^2)^{-1}$ if $\lambda \geq 0$ and

$\gamma(\lambda,\alpha) = 1 - \frac{\alpha^2}{8\mu_1^2}$ if $\lambda \leq 0$. Thus $\gamma(\lambda,\alpha) > 0$ and (3.6) is satisfied under the hypotheses of Theorem 2.

More complete theorems have been proved for the problem of the clamped plate, [19], [28] (the latter treatment includes clamped shells). Equation (1.1) becomes

$$w - \lambda Aw + C(w) = P, \quad w \text{ in } H_c. \tag{3.7}$$

Theorem 3. The problem (3.7) for the clamped plate has at least one solution w in H_c for arbitrary loading, $-\infty < \lambda < \infty$, P in H_c. Moreover, there is a number $K = K(\lambda, P, F, \Omega)$, increasing with $|\lambda|$ and $\|P\|$, such that every solution of (3.7) satisfies $\|w\| \leq K$.

The proof given by Rabinowitz [28] contains a clean, indirect derivation of the essential estimate $\|w\| \leq K$. A somewhat more technical, direct derivation of an explicit form for K is contained in [19].

Similar results for other plate and shell problems have been obtained by Mark Harris in his doctoral dissertation at the University of Massachusetts.

4. Branching.

In this section we take $P = 0$ and consider a slightly more general version of (1.1):

$$T_\lambda(w) \equiv L_\lambda w + N(w) = 0, \quad w \text{ in } H, \tag{4.1}$$

where $N = N_1 + N_2$ is a nonlinear operator defined on H such that N_1 and N_2 are continuous polynomial operators (see, e.g. [12, Ch. XXVI]), homogeneous of degree k_1 and k_2, respectively, with $k_2 > k_1 \geq 2$. In addition, we suppose that N_1 is the (strong) gradient of a real-valued functional $\nu(w) = (k_1 + 1)^{-1}(N_1(w), w)$ on H. The development given below is closely related to several instances of the Lyapunov-Schmidt approach, among them are [18], [20], [21], [30], [31], [32], [34]. Of many other works containing wider discussions of branching questions we list just a few: [16], [24], [27], [33], [36], [37].

We shall discuss the branching only at positive eigenvalues of L_λ; results analogous to Theorem 5, 6 and 7 are also valid at negative eigenvalues. Of course, when A is positive this is no restriction.

We are particularly interested in the detailed behavior of solutions of (4.1) having small norm (however, some global results are discussed in §5). We say that a real number λ_0 is a branch point or bifurcation point of (4.1) if there is a family of real numbers $\{\lambda_\delta\}$ and a family $\{w_\delta\} \subset H$ such that $\|w_\delta\| > 0$, $T_{\lambda_\delta}(w_\delta) = 0$, $\lim_{\delta\to 0} \lambda_\delta = \lambda_0$ and $\lim_{\delta\to 0} w_\delta = 0$. Using the properties listed in Lemma 1 for L_λ and A and the properties given above for N it is easy to show that every branch point of (4.1) is an eigenvalue of L_λ.

Now suppose that λ_0 is a positive eigenvalue (when $\lambda_0 < 0$ a similar procedure may be followed) of L_λ of multiplicity $n \geq 1$ so that the null space $\mathfrak{N}$ of L_{λ_0} is n-dimensional. If S denotes the orthogonal projection of H onto $\mathfrak{N}$ and $\mathfrak{N}^\perp$ denotes the orthogonal complement of $\mathfrak{N}$ in H, then each w in H may be resolved into components, $w = v + V$, with $v = Sw$ in $\mathfrak{N}$ and $V = (I - S)w$ in $\mathfrak{N}^\perp$. A basis $v_1, \ldots, v_n$ for $\mathfrak{N}$ may be introduced such that (δ_{ij} is the Kronecker delta symbol)

$$(Av_i, v_j) = \delta_{ij} \qquad (i,j = 1,2,\ldots,n). \tag{4.2}$$

For arbitrary v in $\mathfrak{N}$, represented by $v = \xi_1 v_1 + \xi_2 v_2 + \cdots + \xi_n v_n$, equation (4.1) is decomposed into the <u>auxiliary equation</u> $(I - S)T_\lambda(v + V) = 0$, and the <u>branching equation</u> or <u>bifurcation equation</u>, $ST_\lambda(v + V) = 0$.

Since $(I - S)L_{\lambda_0}$ has a bounded inverse $[(I - S)L_{\lambda_0}]^{-1}$ on $\mathfrak{N}^\perp$ and $L_\lambda v$ is in $\mathfrak{N}$, it follows that the auxiliary equation may be written as

$$0 = V + \tilde{N}(\xi,\eta,V) \equiv F(\xi,\eta,V), \tag{4.3}$$

where $\tilde{N}(\xi,\eta,V) = [(I - S)L_{\lambda_0}]^{-1}\{-\eta AV + (I - S)N(v + V)\}$, $\xi = (\xi_1,\ldots,\xi_n)$, and $\eta = \lambda - \lambda_0$. Now $F(\xi,\eta,V)$ is analytic in ξ, η and V at $(0,0,0)$ and has Fréchet derivative $\partial F/\partial V = I$ at $(0,0,0)$. For sufficiently small $\rho_0 > 0$, $\eta_0 > 0$ it follows that (4.3) has a unique small solution $V = V(\xi,\eta)$ analytic in ξ and η and satisfying for some constant $K > 0$,

$$\|V(\xi,\eta)\| \leq K|\xi|^{k_1}, \quad \text{for } |\xi| < \rho_0,\ |\eta| < \eta_0. \tag{4.4}$$

The branching equation may be rewritten as a system of n nonlinear equations in the unknowns $\xi_1,\ldots,\xi_n$

$$0 = -\eta\xi_i + (N(\sum_{j=1}^{n} \xi_j v_j + V(\xi,\eta)), v_i), \quad (i = 1,\ldots,n). \tag{4.5}$$

If in (4.5) we now set

$$z = (z_1,\ldots,z_n) = \eta^{-h}(\xi_1,\ldots,\xi_n), \tag{4.6}$$

where $h = (k_1 - 1)^{-1}$, then after cancelling a factor $\eta^{k_1 h}$ we obtain

$$0 = -z_i + (N_1(\sum_{j=1}^{n} z_j v_j + \eta^{-h}V(\eta^h z,\eta)) + \eta^{(k_2-k_1)h} N_2(\sum_{j=1}^{n} z_j v_j + \eta^{-h}V(\eta^h z,\eta)), v_i, \quad (i = 1,\ldots,n) \tag{4.7}$$

Formally setting $\eta = 0$ we are led, in view of $k_2 > k_1$ and (4.4), to the system of polynomial equations in $\mathbb{R}^n$

$$0 = -z_i + (N_1(z_1v_1 + z_2v_2 + \cdots + z_nv_n), v_i) \equiv F_i(z),$$
$$(i = 1, \ldots, n). \tag{4.8}$$

Suppose that $z = \beta \neq 0$ is a solution of (4.8) at which the Jacobian $J = \partial(F_1, \ldots, F_n)/\partial(z_1, \ldots, z_n)$ is different from zero. We conclude from the implicit function theorem that, for $|\eta|$ sufficiently small, the system (4.7) has a nontrivial solution $z(\eta) = \beta + Z(\eta)$, where $Z(\eta)$ is analytic in η^h and satisfies $\lim_{\eta \to 0} Z(\eta) = 0$. Then $\xi(\eta) = \eta^h z(\eta)$ satisfies (4.5) and in turn generates a solution, w, of (4.1), so that we have the following theorem.

<u>Theorem 4</u>. Let λ_0 be a positive eigenvalue of L_λ for which the null space, N, is n-dimensional with basis $v_1, \ldots, v_n$ satisfying (4.3). For each nontrivial solution $z = \beta$ of (4.8) at which the Jacobian J is not zero there exists a $\delta > 0$ and a nontrivial solution of (4.1) having the form

$$w(\lambda) = (\lambda - \lambda_0)^h(\beta_1v_1 + \cdots + \beta_nv_n) + W, \tag{4.9}$$

where W is analytic in $(\lambda - \lambda_0)^h$ for $|\lambda - \lambda_0| < \delta$ and satisfies $\lim_{\lambda \to \lambda_0} (\lambda - \lambda_0)^{-h}W = 0$. The solution (4.9) is valid for $|\lambda - \lambda_0| < \delta$ if k_1 is even and for $\lambda_0 \leq \lambda < \lambda_0 + \delta$ if k_1 is odd.

A similar result may be proved when k_1 is odd and $\lambda < \lambda_0$ by setting $\eta = \lambda_0 - \lambda$ in the above discussion.

Theorem 4 provides a way of finding solutions of (4.1). The next result, whose proof is similar to that of Theorem

4.3 in [33] shows that all continuous branches from a bifurcation point may be determined this way in some cases.

Theorem 5. In Theorem 4 suppose also that $J \neq 0$ at every nontrivial solution of (4.8) and the system

$$0 = (N_1(z_1v_1 + z_2v_2 + \cdots + z_nv_n), v_i), \quad (i = 1, \ldots, n) \tag{4.10}$$

has no nontrivial solution. Then every solution $\xi(\eta)$ of (4.5), continuous near $\eta = 0$ with $\xi(0) = 0$, is obtained by means of the substitution (4.6) from a solution $z(\eta)$ of (4.7) continuous near $\eta = 0$ with $\beta = z(0) \neq (0,0,\ldots,0)$.

Theorems 4 and 5 represent the nondegenerate situation in which complete information on the number of small solutions and their dominant terms can be obtained from a study of the systems (4.8) and (4.10). Application of these results to plate and shell models will be made in §5.

The final theorem in this section is essentially a special case of work of Sather [30, 31]. If $n = 2$, it enables one to find continuous, small solutions of (4.1) in some cases when the Jacobian condition of Theorem 4 cannot be verified. The result relies on the facts that the polynomial operator N_1 is the gradient of the functional $\nu(w) = (k_1 + 1)^{-1}(N_1(w), w)$ and that the system (4.8) can be interpreted as the usual necessary condition for $\nu(w)$ restricted to the "ellipse" $E = \{u \in \mathfrak{N}: (Au, u) = 1\}$ to have an extremum, together with a general implicit function theorem due to Macmillan [25] and Bliss [7]. We recall that $h = (k_1 - 1)^{-1}$.

Theorem 6. Let λ_0 be a positive eigenvalue of L_λ of multiplicity two and suppose that the functional

$\nu(w) = (k_1 + 1)^{-1}(N_1(w), w)$ is not constant on E. Let the restriction of ν to E have a positive relative extremum at v^* in E and set $r = (k_1 + 1)\nu(v^*)$. Then there is a positive constant δ such that equation (4.1) has a nontrivial solution of the form

$$w(\lambda) = r^{-1}(\lambda - \lambda_0)^h v^* + W, \tag{4.11}$$

where W is continuous for $|\lambda - \lambda_0| < \delta$ and satisfies $\lim_{\lambda \to \lambda_0} (\lambda - \lambda_0)^{-h} W = 0$. The solution (4.11) is valid for $|\lambda - \lambda_0| < \delta$ if k_1 is even and for $\lambda_0 \leq \lambda < \lambda_0 + \delta$ if k_1 is odd.

Similar results hold for the negative relative extrema of ν on E (see [31]).

5. Branching in the Plate and Shell Problems. Here we examine the branching question for the plate and shell problems being considered. In this discussion we take $P = 0$ in (1.1), however the case $P \neq 0$ can be studied using similar methods (see, e.g., [20] and the paper of Professors Mallet-Paret and Chow in these proceedings).

The clamped plate. A great deal of mathematical work on plate and shell problems in recent years was stimulated by the papers of Berger and Fife [5,6] and Berger [2]. A central result in those papers is given now for the problem of the clamped plate; similar theorems are valid for a variety of plate and shell problems (see also [3], [4]). The proof utilizes variational methods for infinite dimensional spaces. For this problem (1.1) is

$$w - \lambda Aw + C(w) = 0, \quad w \text{ in } H_c. \tag{5.1}$$

Theorem 7. Every characteristic value of A is a branch point of (5.1) for the problem of the clamped plate. In fact, for each positive (resp., negative) characteristic value, λ_i, of A there exists a family $\{w_\lambda^i\}$ of nontrivial solutions of (5.1) for $\lambda > \lambda_i$ (resp., $\lambda < \lambda_i$) such that $w_\lambda^i \to 0$ $\lambda \downarrow \lambda_i$ (resp., $\lambda \uparrow \lambda_i$). Furthermore, each family has a representative on every one of the infinite dimensional manifolds $\partial A_R = \{u : u \in H_C \text{ and } 2\|u\|^2 + (C(u),u) = 4R\}$.

Remark 2. It should be noted that if w is a solution of (5.1), then so is $-w$, since the operator C is homogeneous of degree three.

Taken together, the results of Theorems 3 and 7 and the Corollary to Theorem 2 give the picture of a pair of nontrivial solutions branches leaving $w = 0$ at each characteristic value λ_o and extending to $w = \infty$ as $\lambda \to +\infty$, if $\lambda_o > 0$ (as $\lambda \to -\infty$, if $\lambda_o < 0$). This is, of course, in distinct contrast with the situation for linear eigenvalue problems, in which solutions of arbitrary magnitude exist at each characteristic value.

Theorem 7 gives no indication of the smoothness of the branches. From the work [27] of Rabinowitz, it follows that at each characteristic value of odd multiplicity a connected branch of nontrivial solutions of (5.1) bifurcates from $w = 0$ and extends either to $w = \infty$ or to $w = 0$ at some other characteristic value.

More detailed information concerning both the smoothness of branches and the number of solutions can be obtained using the results of §4 if the search is restricted to small solutions near a positive characteristic value of A. (Analogous results are valid near the negative characteristic values.) Comparing (5.1) with (4.1) we see that, in the notation of §4, $N_1(w) = C(w)$, $k_1 = 3$, $N_2 = 0$ and $\nu(w) = \sigma(w)$ (see Lemma 1). The system (4.10) becomes

$$0 = (C(z_1 v_1 + \cdots + z_n v_n), v_i) = 0, \quad (i = 1, \ldots, n). \tag{5.2}$$

If (5.2) has a solution z, it follows that $0 = \sigma(v)$, where $v = z_1 v_1 + \cdots + z_n v_n$, and, since the v_i are smooth, Lemma 3.2 of [2] shows that $v = 0$, i.e. $z = 0$. A similar estimate shows that when $\lambda < \lambda_0$ the counterpart of system (4.8) has no nontrivial solution, so that no branch emanates "to the left" from $(\lambda, w) = (\lambda_0, 0)$ when $\lambda_0 > 0$. Thus, <u>all continuous solutions branching from $w = 0$ at a positive characteristic value of A can be determined by studying the system (4.8) if that system has nonvanishing Jacobian at each of its zeros</u>. For example, when λ_0 is a simple characteristic value (4.8) becomes $0 = -z + rz^3$, $r = (C(v_1), v_1) > 0$, with nontrivial solutions $z = \pm\sqrt{r}$ at which the Jacobian is 2. Theorem 4 with Remark 2 yields a pair of nontrivial solutions of (5.1), $\pm w(\lambda)$, analytic in $(\lambda - \lambda_0)^{1/2}$, where $w(\lambda) = (\lambda - \lambda)^{1/2} r^{-1/2} v_1 + o(|\lambda - \lambda_0|^{1/2})$, $\lambda_0 < \lambda < \lambda_0 + \delta$. According to Theorem 5 there are no other continuous branches from $w = 0$ at $\lambda = \lambda_0$.

If λ_0 is a positive characteristic value of A of multiplicity two and if $\sigma(w)$ is not constant on the ellipse $E = \{v : v \in \mathfrak{N}(I - \lambda_0 A), \ (Av, v) = 1\}$, then σ has at least two distinct positive relative extrema on E (a maximum and a minimum). According to Theorem 6 and Remark 2 each of these extrema yields a pair of solutions of (5.1) of the form (4.11) with $h = 1/2$.

The Simply Supported Rectangular Plate. Again the governing equation is (5.1), but now the Hilbert space is H_S and the operator A is positive. Accordingly, much of the discussion concerning branching for the clamped plate applies to this case as well and we shall not repeat those statements. For this problem, however, the linear equation is completely solved, having characteristic values

$$\mu_{pq} = \left(\frac{\pi}{a}\right)^2 \left(p + \frac{q^2a^2}{pb^2}\right)^2, \qquad (p,q = 1,2,\ldots), \tag{5.3}$$

and corresponding eigenfunctions

$$w_{pq} = c_{pq} \sin \frac{p\pi x}{a} \sin \frac{q\pi y}{b}, \tag{5.4}$$

with constants c_{pq} chosen so that $(Aw_{pq}, w_{pq}) = 1$.

The characteristic value of greatest interest is the smallest one (the "buckling load" of the plate). It is simple except when $a^2/b^2 = m(m + 1)$ for some positive integer m, in which case the multiplicity, n, is two. Branching at a simple characteristic value proceeds as for the clamped plate. Next we look at a case in which $n = 2$ (see also [21]).

We take $a = \sqrt{2}$, $b = 1$ so that the first characteristic value is $\mu_1 \equiv \mu_{11} = \mu_{21} = 9\pi^2/2$ having two dimensional null space $\mathfrak{N}$ spanned by $v_1 = w_{11}$ and $v_2 = w_{21}$. With this basis the system (4.8) becomes

$$\begin{aligned} 0 &= -z_1 + cz_1^3 + dz_1z_2^2 \\ 0 &= -z_2 + dz_1^2z_2 + ez_2^3, \end{aligned} \tag{5.5}$$

where $c = (C(v_1), v_1)$, $d = (B(v_1,v_1), B(v_2,v_2))/2 + \|B(v_1,v_1)\|^2$ and $e = (C(v_2), v_2)$. A tedious calculation and estimation process reveals that $d > c > e > 0$, so that the solutions of (5.5) are the intersections of the line $z_1 = 0$ and ellipse $cz_1^2 + dz_2^2 = 1$ with the line $z_2 = 0$ and ellipse $dz_1^2 + ez_2^2 = 1$. There are exactly eight such intersections, all nondegenerate, so that the Jacobian of (5.5) at each is not zero. Equation (4.10) in this case has no nontrivial solutions. According to Theorems 4 and 5 and Remark 2, exactly eight nontrivial continuous solution branches of (5.1) bifurcate from $w = 0$ at $\lambda = \mu_1$. Furthermore, of the eight solutions of (5.5) the pair $\pm(0, e^{-1/2})$ minimizes $\sigma(w)$ on E. The corresponding solutions of (5.1) provide, among the eight solutions determined this way, a minimum for the potential energy when $\lambda - \mu_1$ is small. (Further discussion of such stability matters is given by D. Sather [34].)

The Cylindrical Panel. Here we report on part of some forthcoming joint work with Professor Duane Sather.

For the panel problem the discussion of local branching in §4 applies with $L_\lambda = I - \lambda A + \alpha^2 A^2$, $\alpha > 0$, $N_1(w) = \alpha Q(w)$, $N_2(w) = C(w)$, $k_1 = 2$, $k_2 = 3$ and $\nu(w) = \tau(w)$. The Hilbert space is H_s and A is the same operator as in the rectangular plate example; its characteristic values and eigenfunctions are given by (5.3) and (5.4).

If all characteristic values, μ, of A satisfy $\mu \geq \alpha$, then (2.9) implies that the mth eigenvalue, λ_m, of L_λ is obtained from the mth characteristic value, μ_m, of A by means of (2.9). Furthermore, part (ii) of Lemma 1 shows that the multiplicity of λ_m equals that of μ_m. From (5.3) one obtains a lower bound for $\mu_{pq} \geq \mu_{p1} \geq (2\pi/b)^2$. Thus

$\mu_{pq} \geq \alpha$, $(p, q = 1, 2, \ldots)$ if $(2\pi/b)^2 > \alpha$, i.e., if

$$\frac{b\alpha^{1/2}}{2\pi} \leq 1. \tag{5.6}$$

Condition (5.6) is equivalent to one given by Koiter [23] to define a <u>narrow panel</u>. Thus, for a narrow panel the first eigenvalue, λ_1, of L_λ (the buckling load) comes from the first characteristic value, μ_1, of A. It follows that λ_1 is simple unless $(a/b)^2 = m(m+1)$ for a positive integer m, in which case the multiplicity is two. We discuss only the latter case.

To fix the ideas, suppose that (5.6) holds and m is odd with $(a/b)^2 = m(m+1)$ so that $\mu_1 = \mu_{m1} = \mu_{m+11}$ and $\lambda_1 = \mu_1 + \alpha^2\mu_1^{-1}$. The null space $\mathfrak{N}$ has basis $v_1 = w_{m1}$, $v_2 = w_{m+11}$, in terms of which the system (4.8) becomes

$$\begin{aligned} 0 &= -z_1 + cz_1^2 + dz_2^2 \\ 0 &= -z_2 + 2dz_1z_2, \end{aligned} \tag{5.7}$$

where $c = \alpha(Q(v_1), v_1)$ and $d = \alpha(Q(v_2), v_1)$. A calculation shows that $0 < c < 2d$ so that (5.7) has exactly three distinct, nontrivial solutions. The Jacobian of (5.7) is nonzero at each of these solutions and (4.10) in this case has only the trivial solution. Thus Theorems 4 and 5 show that <u>exactly three nontrivial continuous solution branches pass through</u> $w = 0$ <u>at</u> $\lambda = \lambda_1$.

REFERENCES

1. Bauer, L. and E.L. Reiss, "Nonlinear buckling of rectangular plates," SIAM J. Appl. Math. 13 (1965), 603-626.

2. Berger, M.S., "On von Kármán's equations and the buckling of a thin elastic plate, I. The clamped plate," Comm. Pure Appl. Math. 22 (1967), 687-719.

3. Berger, M.S., "An application of the calculus of variations in the large to the equations of nonlinear elasticity," Bull. Amer. Math. Soc. 73 (1967), 520-525.

4. Berger, M.S., "The existence of equilibrium states of thin elastic shells, (I)," Indiana Univ. Math. J. 20 (1971), 591-602.

5. Berger, M.S. and P.C. Fife, "On von Kármán's equations and the buckling of a thin elastic plate," Bull. Amer. Math. Soc. 72 (1966), 1006-1011.

6. Berger, M.S. and P.C. Fife, "von Kármán's equations and the buckling of a thin elastic plate II. Plate with general edge conditions," Comm. Pure Appl. Math. 21 (1968), 227-241.

7. Bliss, G., "A generalization of Weierstrass' preparation theorem for a power series in several variables," Trans. Amer. Math. Soc. 13 (1912), 133-145.

8. Brezis, H., "Equations et inéquations non-linéaires dans les espaces vectoriels, en dualité," Ann. Inst. Fourier, Grenoble 18 (1968), 115-176.

9. Fichera, G., Linear Elliptic Differential Systems and Eigenvalue Problems, Springer, Berlin, 1965.

10. Fife, P.C., "Nonlinear deflection of thin elastic plates under tension," Comm. Pure Appl. Math. 14 (1961), 81-112.

11. Friedrichs, K.O. and J.J. Stoker, "The nonlinear boundary value problem of the buckled plate," Amer. J. Math. 63 (1941), 839-888.

12. Hille, E. and R.S. Phillips, Functional Analysis and Semigroups, Amer. Math. Soc., Providence, R.I., 1957.

13. Hlaváček, I. and J. Naumann, "Inhomogeneous boundary value problems for the von Kármán equations, I," Apl. Mat. 19 (1974), 253-269.

14. Hlaváček, I. and J. Naumann, "Inhomogeneous boundary value problems for the von Kármán equations, II," Apl. Mat. 20 (1975), 280-297.

15. Kármán, T.v., "Festigkeitsproblem im Maschinenbau," Encyk. der Math. Wissenschaften IV^4 (1910), 348-352. Leipzig 1907-1914.

16. Keller, J.B. and S. Antman, Bifurcation Theory and Nonlinear Eigenvalue Problems, Benjamin, New York, 1969.

17. Keller, H.B., J.B. Keller and E.L. Reiss, "Buckled states of circular plates," Quart. Appl. Math. 20 (1962/63), 55-65.

18. Kirchgassner, K., "Multiple eigenvalue bifurcation for holomorphic mappings," in Contributions to Nonlinear Functional Analysis, E.H. Zarantonello, ed., Academic Press, New York, 1971.

19. Knightly, G.H., "An existence theorem for von Kármán's equations," Arch. Rational Mech. Anal. 27 (1967), 233-242.

20. Knightly, G.H. and D. Sather, "On nonuniqueness of solutions of the von Kármán equations," Arch. Rational Mech. Anal. 36 (1970), 65-78.

21. Knightly, G.H. and D. Sather, "Nonlinear buckled states of rectangular plates," Arch. Rational Mech. Anal. 54 (1974), 356-372.

22. Knightly, G.H. and D. Sather, "Nonlinear axisymmetric buckled states of shallow spherical caps," SIAM J. Math. Anal. 6 (1975), 913-924.

23. Koiter, W.T., "Buckling and postbuckling of a cylindrical panel under axial compression," NLR Report S476, Amsterdam, 1956, see also Reports and Transactions, National Aero. Res. Inst. 20.

24. Krasnoselski, M.A., Topological Methods in the Theory of Nonlinear Integral Equations, Macmillan, New York, 1964.

25. MacMillan, W., "A reduction of a system of power series to an equivalent system of polynomials," Math. Ann. 72 (1912), 157-179.

26. Morosov, N.F., "On the nonlinear theory of thin plates," Dokl. Akad. Nauk SSSR 114 (1957), 968-971.

27. Rabinowitz, P.H., "A global theorem for nonlinear eigenvalue problems and applications," in Contributions to Nonlinear Functional Analysis, E.H. Zarantonello, ed., Academic Press, New York, 1971.

28. Rabinowitz, P.H., "A note on topological degree for potential operators," J. Math. Anal. Appl. 51 (1975), 483-492.

29. Reiss, E.L., "Bifurcation buckling of spherical caps," Comm. Pure Appl. Math. 18 (1965), 65-82.

30. Sather, D., "Branching of solutions of an equation in Hilbert space," Arch. Rational Mech. Anal. 36 (1970), 47-64.

31. Sather, D., "Nonlinear gradient operators and the method of Lyapunov-Schmidt," Arch. Rational Mech. Anal. 43 (1971), 222-244.

32. Sather, D., "Branching of solutions of a class of nonlinear equations," Math. Z. 123 (1971), 105-112.

33. Sather, D., "Branching of solutions of nonlinear equations," Rocky Mtn. J. Math. 3 (1973), 203-250.

34. Sather, D., "Branching and stability for nonlinear shells," to appear in Proceedings IUTAM/IMU Symposium on Applications and Methods of Functional Analysis to Problems of Mechanics, Marseille, Sept. 1-6, 1975.

35. Schaefer, H., "Über die Methode der a-priori Schranken," Math. Ann. 129 (1955), 415-416.

36. Stakgold, I., "Branching of solutions of nonlinear equations," SIAM Rev. 13 (1971), 289-332.

37. Vainberg, M.M. and V.A. Trenogin, "The Lyapunov and Schmidt methods in the theory of non-linear equations and their subsequent development," Russian Math. Surveys 17 (1962), 1-60.

38. Wolkowisky, J.H., "Existence of buckled states of circular plates," Comm. Pure Appl. Math. 20 (1967), 549-560.

SOME RESONANCE PROBLEMS FOR ELLIPTIC BOUNDARY VALUE PROBLEMS

Alan C. Lazer

University of Cincinnati

1. <u>Introduction</u>. In the theory of ordinary differential equations the term "resonance" is sometimes used to describe the following phenomenon: Consider the response of a simple harmonic oscillator subject to an impressed periodic force. If, for simplicity, we assume that the force has period 2π the corresponding differential equation has the form

$$L[x] = x'' + k^2 x = F(t) = F(t + 2\pi).$$

If k is an integer then there are no 2π-periodic solutions, or even bounded solutions, unless F satisfies the two conditions

$$\int_0^{2\pi} F(t) \sin k\, t\, dt = \int_0^{2\pi} F(t) \cos k\, t\, dt = 0.$$

The phenomenon of resonance is a simple instance of the so called Fredholm alternative for certain "well-behaved" boundary value problems for linear elliptic differential systems. As it is difficult to make the notion of "well-behaved" precise, let us try to formulate abstractly the minimal properties that should be enjoyed by such systems. Let B be a Banach space, D a subspace of B, and $L:D \to B$ a linear (in general unbounded) operator. Assume that L has closed range M and that there exists a compact linear operator $T: M \to B$ such that $T(M) \subseteq D$ and $LTy = y$ for $y \in M$. Typically, one thinks of B as a function space, D a subspace of B whose elements satisfy certain smoothness requirements, and perhaps other conditions which are linear and homogeneous, and L a linear differential operator. In the example given above we can take B to be the space of continuous 2π-periodic functions, D the subspace whose members have a continuous second derivative, and $L[x] = x'' + k^2x$.

For the general situation it is natural to consider the equation

$$Lx = Nx \tag{1.1}$$

where N is a nonlinear function on B. If $M = B$, then (1.1) is clearly equivalent to the operator equation

$$x = y + TN(x) \tag{1.2}$$

where $y \in \text{Kernel } L$ is arbitrary. Fixing $y \in \text{Kernel } L$, there are two natural conditions one can place on N to assure solvability of this equation. First, if N is bounded and continuous, then the mapping $x \to y + TN(x)$ is compact, continuous, and has its range contained in some closed ball. In this case we can invoke Schauder's fixed point theorem to ascertain the existence of a solution. Second, if N satisfies a Lipshitz condition with sufficiently small Lipshitz constant, then the contraction mapping principle can be used to solve (1.2).

The case where $\text{Range } L = M \neq B$ is definitely the more interesting. By a resonance problem we shall understand an equation of the form (1.1) where $M \neq B$ and N is, in some sense, small compared with L. This seems to be a natural restriction if one hopes to obtain solvability criteria which somehow relate to the fact that Range $L \neq B$. Otherwise the linear part of (1.1) ceases to be the dominant part and the question of whether or not L is onto would seem to be unimportant.

We should point out that problems of the form (1.1) have been the subject of a long series of investigations by Professor Cesari and his co-workers. (See for example [2], [6], and [7].) One of the most striking successes of these investigations has been in dealing with problems in which the nonlinearity is large, such as in the paper [6]. It was the work of Cesari which gave the author and D.E. Leach the impetus to consider a resonance problem in [13] for the simple harmonic oscillator operator considered above.

In this article we shall deal mainly with a rather restricted problem of the above type. Our treatment will be entirely expository except for the fifth section which is new in its present form.

2. An Introductory Example. The term "resonance", as applied to a nonlinear partial differential of elliptic type, was first used in the paper [12]. In this section we shall illustrate the principal result of [12] by considering the simplest possible case: namely, the case of an ordinary differential equation of second order. By assuming that the nonlinearity satisfies a local Lipshitz condition we can derive the main result by simple continuity arguments and

thereby dispense with the ponderous functional analytic methods used in [12].

Let $g(x)$ be a locally Lipshitzian function defined for $x \in (-\infty, \infty)$ and such that the quantities $\lim_{x \to \infty} g(x) = g(\infty)$ and $\lim_{x \to -\infty} g(x) = g(-\infty)$ exist and are finite. Moreover, we assume that

$$g(-\infty) < g(x) < g(\infty), \quad x \in (-\infty, \infty) \tag{2.1}$$

although g need not be monotone. We consider the boundary value problem

$$x'' + x + g(x) = h(t) \tag{2.2}$$

$$x(0) = x(\pi) = 0 \tag{2.3}$$

where h denotes a function continuous on the interval $[0, \pi]$. As g is bounded and satisfies a local Lipshitz condition there exists a unique solution $x(t,a)$ of (2.2), defined on $[0, \pi]$, such that $x(0,a) = a$, and, by a standard result of differential equations, $x(0,a)$ <u>is continuous in a</u>. It follows that there exists a solution of the boundary value problem given by (2.2) and (2.3) if and only if there exists a number $\bar{a}$ such that $x(\pi, \bar{a}) = 0$.

Now according to the well-known "variation of parameters formula" we have the identity

$$x(t,a) = a \sin t + \int_0^t \sin(t - s)[h(s) - g(x(s,a))]ds.$$

From this equation and the boundedness of g we infer the existence of a number L such that

$$|x(t,a) - a \sin t| \leq L. \tag{2.4}$$

Furthermore, we see that the solvability condition $x(\pi, \bar{a}) = 0$ is equivalent to solving

$$\int_0^\pi h(s) \sin s \, ds = \int_0^\pi g(x(s, \bar{a})) \sin s \, ds. \tag{2.5}$$

Assuming the existence of a number $\bar{a}$ satisfying (2.5), we note from (2.1) that

$$g(-\infty) < g(x(t,\bar{a})) < g(\infty)$$

for all $t \in [0,\pi]$. Hence the inequalities

$$g(-\infty)\int_0^\pi \sin s\, ds < \int_0^\pi h(s) \sin s\, ds < g(\infty)\int_0^\pi \sin s\, ds \qquad (2.6)$$

are necessary for the solvability of the problem given by (2.2) and (2.3).

Conversely, let us assume that the inequalities (2.6) are satisfied and consider the continuous function

$$F(a) = \int_0^\pi g(x(s,a)) \sin s\, ds - \int_0^\pi h(s) \sin s\, ds.$$

From (2.4) we see that if $0 < t < \pi$, then

$$\lim_{a\to\infty} x(t,a) = +\infty, \qquad \lim_{a\to-\infty} x(t,a) = -\infty$$

and so

$$\lim_{a\to\infty} F(a) = g(\infty)\int_0^\pi \sin s\, ds - \int_0^\pi h(s) \sin s\, ds > 0$$

and

$$\lim_{a\to-\infty} F(a) = g(-\infty)\int_0^\pi \sin s\, ds - \int_0^\pi h(s) \sin s\, ds < 0.$$

Thus we infer the existence of a number $\bar{a}$ such that $F(\bar{a}) = 0$ which is equivalent to (2.5). We have therefore shown that the inequalities (2.6) are both necessary and sufficient for the solvability of the problem given by (2.2) and (2.3).

3. A Functional Approach and the Role of the Alternative Method.

Again let us assume that g is continuous and the inequalities (2.1) hold. If we now drop the assumption that g satisfies a local Lipshitz condition, then the above existence proof is no longer applicable. We can still show that the inequalities (2.6) are both necessary and sufficient for solvability by considering a sequence of functions $\{g_n(t)\}_1^\infty$ such that each function $g_n(t)$ satisfies the same hypotheses as $g(t)$, each function $g_n(t)$ is locally Lipshitzian, $\lim_{n\to\infty} g_n(t) = g(t)$ uniformly on $(-\infty, \infty)$, and such that the inequalities (2.6) are satisfied where g is replaced by g_n. If x_n denotes the solution of the problem given by (2.2) and (2.3) with g replaced by g_n, then it is not difficult to show that some subsequence of the sequence $\{x_n\}_1^\infty$ converges to a solution of the original problem.

We shall now indicate a slightly more involved approach which has the advantage that with suitable modifications it can be used to treat an analogous problem for a partial differential equation of elliptic type.

This method will be based on the following elementary result:

Lemma. Let M denote the set of functions, continuous on $[0,\pi]$, such that if $f \in M$, then

$$\int_0^\pi f(t) \sin t \, dt = 0. \tag{3.1}$$

If $f \in M$, then there exists a unique solution $Tf \in M$ which is a solution of

$$x'' + x = f \quad x(0) = x(\pi) = 0. \tag{3.2}$$

Moreover the linear map $T:M \to M$ is compact with respect to the uniform norm on $[0,\pi]$.

The lemma will follow from Ascoli's lemma and a closed form representation of T which will be exhibited below.

Let $w(t) = \sqrt{2/\pi}\,\sin t$ and, for any $f \in C[0,\pi]$, let $\langle f,w\rangle_0 = \int_0^\pi f(t)w(t)dt$ $Pf = \langle f,w\rangle_0 w$. We note that if $I:C[0,\pi] \to C[0,\pi]$ denotes the identity map on $C[0,\pi]$, then for any $f \in C[0,\pi]$, $(I - P)f \in M$. Using this notation it is easy to check that if $f \in M$, then

$$(Tf)(t) = (I - P)\int_0^t \sin(t - s)f(s)ds.$$

Furthermore, for any $f \in M$ the general solution of the boundary value problem (3.2) is given by

$$x(t) = aw(t) + (Tf)(t).$$

Returning to the boundary value problem given by (2.2) and (2.3), we see from the above that if $x'' + x + g(x) = h(t)$, $x(0) = x(\pi) = 0$, then it is necessary and sufficient that $P(h - g(x)) = 0$ and that $x = aw + T(h - g(x))$ for some constant a. These conditions can obviously be rewritten in the following canonical form corresponding to the Cesari-type alternative problem.

$$x = aw + T(I - P)(h - g(x)) \tag{3.3}$$

$$P[h - g(x)] = 0. \tag{3.4}$$

The most natural approach in attempting to solve (3.3) and (3.4) would be to first fix a and solve the <u>auxillary equation</u> (3.3) (using some fixed point principle). Calling such a solution x_a the next step would be to try to solve the <u>determining equation</u> $F(a) = P[h - g(x_a)] = 0$.

The two obstacles with this approach are:

1. Assuming the existence of x_a for arbitrary <u>a</u>, it is not clear that x_a is a <u>single valued function of a</u>, i.e. x_a may not be unique.

2. It is not clear that one can choose a "branch" of x_a which makes the function F continuous.

However, Howard Shaw has shown in his lecture that for the natural extention of this problem to elliptic partial differential equations the homological properties of the graph of the multi-valued function x_a imply the existence of a number a and corresponding x_a satisfying (3.3) and (3.4).

To overcome the problems described above a different method of attack was used in [12]. As $P[h - g(x)] = \langle w, h - g(x)\rangle_0 w$, any pair $(x,a) \in M\times(-\infty,\infty)$ which satisfies (3.3) and (3.4) is a fixed point of the mapping $G: M \times (-\infty,\infty) \to M \times (-\infty,\infty)$ defined by $G(x,a) = (x^*, a^*)$ where

$$x^*(x,a) = aw + T(I - P)[h - g(x)]$$

$$a^*(x,a) = a + \langle w, h - g(x^*(x,a)\rangle_0 .$$

As will be seen below, it is important that $x^*(x,a)$ be used to "update" the second equation. From the boundedness of g,

$$|x^*(x,a)(t) - aw(t)| = |x^*(x,a)(t) - a\sqrt{\tfrac{2}{\pi}}\sin t| \leq K$$

for $t \in [0,\pi]$, where K is independent of $x \in M$. Therefore, if (2.6) holds, then

$$\lim_{a\to\infty} \langle w, h - g(x^*(x,a))\rangle_0 = \sqrt{\tfrac{2}{\pi}}\Big(\int_0^\pi h(s)\sin s\, ds - g(\infty)\int_0^\pi \sin s\, ds\Big) < 0,$$

$$\lim_{a\to-\infty} \langle w, h - g(x^*(x,a))\rangle_0 = \Big(\int_0^\pi h(s)\sin s\, ds - g(-\infty)\int_0^\pi \sin s\, ds\Big) > 0$$

where the limits are uniform with respect to $x \in M$. It follows from the definition of $a^*(x,a)$ that there exists $r_1 > 0$, independent of $x \in M$, such that $|a^*(x,a)| \leq r_1$ if $|a| \leq r_1$. Consequently we infer the existence of an $r_2 > 0$ such that $|x^*(x,a)| \leq r_2$ if $|a| \leq r_1$.

One can easily show that G is a continuous compact mapping on the Banach space $M \times (-\infty,\infty)$ (with norm $|(x,a)| = \max_{[0,\pi]} |x(t)| + |a|$ and obvious linear structure).

As the set $A = \{(x,a) \in M \times (-\infty,\infty) \mid |x| \leq r_2, |a| \leq r_1\}$ is closed and convex and $G(A) \subseteq A$, it follows from a well-known result of Schauder that there exists a fixed point of G. This proves that the result of the previous section still holds when g is only assumed to be continuous.

4. <u>An Elliptic Boundary Value Problem</u>. We now consider a generalization of the previous result to higher dimensions. Let D be a bounded domain in R^n $(n \geq 1)$ and suppose that $Lu = \sum_{|i|,|j|\leq 1} (-1)^{|i|} D^i a_{ij}(x) D^j u$ is a uniformly elliptic, formally selfadjoint, linear differential pperator defined on D, with real-valued coefficients $a_{ij} = a_{ji} \in L^\infty(D)$. Let $g: R \to R$ be a bounded continuous function such that limits $\lim_{x\to\infty} g(x) = g(\infty)$ and $\lim_{x\to-\infty} g(x) = g(-\infty)$ exist, and suppose that $g(-\infty) < g(x) < g(\infty)$. Let $h \in L^2(D)$ be real valued. We consider the boundary value problem

$$\begin{aligned} (Lu)(x) &= g(u(x)) - h(x), \quad x \in D \\ u(x) &= 0 \quad\quad\quad\quad\quad\;\; , \quad x \in \partial D. \end{aligned} \tag{4.1}$$

Let

$$B(u,v) = \sum_{|i|,|j|\leq 1} \int_D a_{ij} D^i u D^j v dx$$

be the symmetric bilinear form associated with L, which is defined and bounded on $\overset{o}{H}_1 \times \overset{o}{H}_1$, where $\overset{o}{H}_1$ denotes the usual Sobolev space of real functions defined on D, provided with the norm $|\ |_1$. (See [11], [12].)

By a <u>weak solution</u> of (4.1) we mean a function $u \in \overset{o}{H}_1$ such that

$$B(u,\varphi) = \int_D (g(u(x)) - h(x))\varphi(x) dx$$

for all $\varphi \in \overset{o}{H}_1$.

With regard to this problem we have the following

Theorem 4.1. *Let the homogeneous linear problem*

$$(Lu)(x) = 0, \quad x \in D$$
$$u(x) = 0 \quad , \quad x \in \partial D \tag{4.2}$$

have a nontrivial weak solution $w \in \overset{o}{H}_1$ *and suppose that if* v *is any other solution of* (4.1), *then* $v = \lambda w$, $\lambda \in R$. *Let* $D^+(D^-) = \{x \in D : w(x) > 0 \ (w(x) < 0)\}$. *Then the inequalities*

$$g(-\infty)\int_{D^+}|w|dx - g(\infty)\int_{D^-}|w|dx < \int_D h\, w\, dx$$
$$< g(\infty)\int_{D^+}|w|dx$$
$$- g(-\infty)\int_{D^-}|w|dx]$$

are both necessary and sufficient for the existence of a weak solution of (4.1).

Theorem 4.1 was first proved in [12] where it was assumed that the operator L had a slightly less general form than that given above. The above formulation is due to Hess [11]. The proof of the necessity of the above inequalities is straightforward. The proof of the sufficiency, as given in [12], makes use of the alternative method in much the same way that it was used in the simple example of the previous section. Subsequently, Hess gave a much shorter proof in [11].

This result has been extended by several authors. We shall give a rough description of one strong extension due to Nirenberg [17], [18].

Let D be a sufficiently regular bounded domain in R^n and let L be a sufficiently regular elliptic operator of order $m = 2r$. Let L operate on scalar functions and satisfy coercive boundary conditions $Bu = 0$ on ∂D and assume that these boundary conditions are expressed in terms of differential operators of order $\leq r$. We emphasize that L and the boundary conditions need not be selfadjoint. It is known that $d = \dim(\text{Kernel } L) < \infty$ and $d^* =$ codimension (Range L) $< \infty$ in a suitable Sobolev space. Let us first assume $d = d^* > 0$. Furthermore, assume that $w = 0$ is the only function in the kernel of L which vanishes on a set of positive measure in D. We consider the boundary value problem

$$\begin{aligned} (Lu)(x) &= g(x,u(x)), \quad x \in D \\ (Bu)(x) &= 0 \quad\quad\quad\;\, , \quad x \in \partial D \end{aligned} \tag{4.3}$$

where $g(x,s)$ is a smooth function, bounded for $(x,s) \in \bar{D} \times R$. (Nirenberg actually allows g to depend on the derivatives of u up to order $m - 1$.) Let us suppose that

$$\lim_{s\to\infty} g(x,s) = h_+(x) \quad \text{and}$$

$$\lim_{s\to-\infty} g(x,s) = h_-(x)$$

exist, uniformly for $x \in D$. Let $w_1, w_2, \ldots, w_d$ be a basis for Kernel L and let $w_1^*, w_2^*, \ldots, w_d^*$ be a basis for (Range $L)^\perp$ in $L^2(D)$. Let $\bar{S}^{d-1}$ denote the unit sphere in R^d. If $w = (w_1, \ldots, w_d)$ and $a = (a_1, \ldots, a_d) \in \bar{S}^{d-1}$, let $a \cdot w = \sum_{j=1}^{d} a_j w_j$ and define $\varphi: \bar{S}^{d-1} \to R^d$ by $\varphi(a) = (\varphi_1(a), \ldots, \varphi_d(a))$ where

$$\varphi_i(a) = \int_{a\cdot w > 0} h_+ w_i^* dx + \int_{a \cdot w < 0} h_- w_i^* dx$$

for $i = 1, 2, \ldots, d$. It can be shows that φ is a continuous mapping under the above assumptions. Finally, let us assume that $\varphi(a) \neq 0$ for all $a \in \bar{S}^{d-1}$ and define $\Psi(a): \bar{S}^{d-1} \to \bar{S}^{d-1}$ by $\Psi(a) = \varphi(a)/|\varphi(a)|$.

Theorem 4.2. *If the topological degree of* Ψ *is different from zero, then the problem* (4.3) *is solvable.*

To show that this result extends Theorem 4.1 in the smooth case we note that if L is selfadjoint with Dirichlet boundary conditions and $d = 1$, we may take $w_1 = w_1^* = w$. For the special problem (4.1) we obviously have

$$h_+(x) = g(\infty) - h(x), \quad h_-(x) = g(-\infty) - h(x)$$

Hence $\varphi: \underline{\bar{S}}^0 \to R$ satisfies

$$\varphi(-1) = g(-\infty)\int_{w>0} w dx + g(\infty)\int_{w<0} w dx - \int_D hw dx$$

and

$$\varphi(1) = g(\infty)\int_{w>0} w dx + g(-\infty)\int_{w<0} w dx - \int_D hw dx.$$

Under the assumption of Theorem 4.1, $\varphi(-1) < 0$ and $\varphi(1) > 0$ so $\Psi: \underline{\bar{S}}^0 \to \underline{\bar{S}}^0$ is the identity map with degree equal to one.

Let us consider the problem

$$\begin{aligned} (Lu)(x) &= g(u(x)) - h(x), & x \in D \\ (D^\alpha u)(x) &= 0 \quad |\alpha| < m, & x \in \partial D \end{aligned} \qquad (4.1)'$$

where L is a real, selfadjoint strongly elliptic operator of order $2m$ with smooth coefficients; h and g are smooth functions; and it is again assumed that $-\infty < g(-\infty) < g(s) < g(\infty) < +\infty$ and that $d > 0$. For this problem we may take $w_k = w_k^*$, $k = 1,\ldots,d$. In this case we again have $h_+(x) = g(\infty) - h(x)$ and $h_-(x) = g(-\infty) - h(x)$. Therefore if $a \in \underline{\bar{S}}^{d-1}$, then $a\cdot\varphi(a) = g(\infty)\int_{a\cdot w>0} a\cdot w \, dx + g(-\infty)\int_{a\cdot w<0} a\cdot w \, dx - \int_D a\cdot w \, h \, dx$. If $a\cdot\Psi(a) > 0$ for all $a \in \underline{\bar{S}}^{d-1}$, it is not difficult to show that $\Psi: \underline{\bar{S}}^{d-1} \to \underline{\bar{S}}^{d-1}$ is homotopic to the identity mapping and hence has degree equal to one. We therefore have the following:

<u>Corollary 4.3</u>. <u>If, for all</u> $a \in \underline{\bar{S}}^{d-1}$,

$$\int_D a\cdot w \, h \, dx < g(\infty)\int_{a\cdot w>0} a\cdot w \, dx + g(-\infty)\int_{a\cdot w<0} a\cdot w \, dx,$$

<u>then the problem</u> (4.1)′ <u>is solvable</u>.

This result was also discovered independently by Williams [23] using the alternative method. For the case where $d^* \neq d$ the assertion of Theorem 4.2 still holds if it is assumed that the stable homotopy class of the map $\Psi:\underline{\bar{S}}^{d-1} \to \underline{\bar{S}}^{d^*-1}$, defined as above, is nontrivial. This requirement coincides with the stipulation that Ψ has nonzero degree if $d = d^*$. Nirenberg's results are applications of an extension of the notion of Leray-Shauder degree.

In the next section we shall consider an extension of the sufficiency part due to S. Ahmad, J. Paul and the author [1] which is based on ideas from the calculus of variations.

5. <u>A Critical Point Principle.</u> In this section we indicate how the results of the previous section can be generalized by means of a certain finite dimensional critical point principle. A less abstract and more detailed exposition will appear in [1]. will appear in [1].

Let H be a real, separable, Hilbert space and $E:H \to H$ a bounded, linear, selfadjoint operator on H. We assume that $H = H^+ \oplus H^0 \oplus H^-$ where each summand is closed; $E(H^+) \subseteq H^+$, $E(H^-) \subseteq H^-$; and H^0 = Kernel E. Moreover we assume that $H^+ \perp H^-$, $H^0 \perp H^+$, $H^0 \perp H^-$, and that there exist constants $m_1 > 0$ and $m_2 > 0$ such that

$$\langle Ex,x\rangle \geq m_1 |x|^2 \quad \text{if} \quad x \in H^+, \tag{5.1}$$

and

$$\langle Ex,x\rangle \leq -m_2 |x|^2 \quad \text{if} \quad x \in H^-. \tag{5.2}$$

It is not difficult to show that such a decomposition of H exists if and only if either 0 is an isolated point of the spectrum of E or E^{-1} exists, in which case $H^0 = \{0\}$. Let

$N:H \to H$ be a <u>bounded</u> operator, continuous with respect to weak convergence, i.e. if $\{x_n\}_1^\infty$ converges weakly to $x \in H$ then $\{N(x_n)\}_1^\infty$ converges <u>strongly</u> to $N(x)$. We further assume the existence of a continuous functional $\eta:H \to R$ such that $N(x) = \text{grad. } \eta(x)$, i.e. if $x \in H$ and $w \in H$, then

$$\lim_{t \to 0+} \frac{\eta(x + tw) - \eta(x)}{t} = \langle N(x), w\rangle.$$

The continuity of N implies that N is a strong gradient ([22, p.41]).

We consider the solvability of the nonlinear operator equation

$$Ex = N(x). \tag{5.3}$$

Theorem 5.1. If $\lim_{\substack{|x_o| \to \infty \\ x_o \in H^o}} \eta(x_o) = +\infty$, *then* (5.3) *is solvable.*

If $H^o = \{0\}$, so that E^{-1} exists, then the condition of Theorem 5.1 holds vacuously and (5.3) is solvable. This case can also be handled through a straightforward application of Leray-Schauder theory.

The proof of Theorem 5.1 is based on the following finite dimensional result:

Lemma 5.2. Let $f:R^m \times R^n \to R$ *be of class* C^1. *Suppose there exist numbers* $r > 0$ *and* $c > 0$ *such that*

$$f(x,y) \leq f(0,y^*) \tag{5.4}$$

whenever $|x| = r$ *and* $|y|, |y^*| \leq c$, *and*

$$\langle\frac{\partial f}{\partial y}(x,y), y\rangle = \sum_{k=1}^{n} \frac{\partial f}{\partial y_k}(x,y)y_k > 0 \tag{5.5}$$

whenever $|x| \le r$ *and* $|y| = c$. *If* $B = \{(x,y) \in R^m \times R^n: |x| \le r, |y| \le c\}$ *then there exists* $(x_o, y_o) \in B$ *such that* $\nabla f(x_o, y_o) = 0$. Here $|\cdot|$ is the Euclidean norm.

We sketch the proof of the lemma which is motivated by the well-known method of Ważewski in ordinary differential equations. (See [1] or [5, p.179]). By standard approximation arguments it suffices to assume that $f \in C^2$. Let $\varphi(t,x,y)$ denote the solution of the initial value problem

$$\varphi'(t,x,y) = \nabla f(\varphi(t,x,y)), \quad \varphi(0,x,y) = (x,y).$$

If the lemma is false, there exists $a > 0$ such that $|\nabla f(x,y)| \ge a$ if $(x,y) \in B$ and hence

$$\frac{d}{dt} f(\varphi(t,x,y)) = |\nabla f(\varphi(t,x,y))|^2 \ge a^2. \tag{5.6}$$

Let $D = \{y | y \in R^n, |y| \le c\}$. As f is bounded on B, it follows that for all $y \in D$, $\varphi(t,0,y)$ must leave B in a finite time. Hence there exists a *largest* number $T(y)$ with $0 \le T(y) < \infty$ such that $\varphi(t,0,y) \in B$ for $0 \le t \le T(y)$. Clearly $\varphi(T(y),0,y) \in \partial B$. If $\varphi(T(y),0,y) = (\bar{x},\bar{y})$, with $|\bar{x}| = r$ and $|\bar{y}| \le c$, then by (5.6), $f(\bar{x},\bar{y}) = f(\varphi(T(y),0,y)) > f(\varphi(0,0,y)) = f(0,y)$ which contradicts (5.4). Consequently $\varphi(T(y),0,y) = (\bar{x},\bar{y})$ with $|\bar{x}| < r$, $|\bar{y}| = c$.

By (5.5) the vector field $\nabla f(x,y)$ is pointing outward on the cylinder $\{(x,y) | |x| < r, |y| = c\}$ so it follows from an implicit function argument that $T:D \to R$ is continuous. Moreover if $|y| = c$, then $(0,y) \in \partial B$ so $T(y) = 0$. If $P:R^m \times R^n \to R^n$ is the projection defined $P(x,y) = y$, we have a continuous mapping $r:D \to \partial D$ given by $r(y) = P(\varphi(T(y),0,y))$. But if $y \in \partial D$, $|y| = c$ so $r(y) =$

$P(\varphi(0,0,y)) = P((0,y)) = y$. As ∂D is not a retract of D, we have a contradiction and the lemma is proved.

To derive the theorem from the lemma one considers the function $f: H \to R$ defined by $f(z) = \frac{1}{2}\langle Ez, z\rangle - \eta(z)$. We have

$$\frac{d}{dt} f(z + tw)\big|_{t=0} = \langle \nabla f(z, w\rangle = \langle Ez, w\rangle - \langle N(z), w\rangle.$$

If $z = x_0 + x_1 + y$, with $x_0 \in H^0$, $x_1 \in H^-$ and $y \in H^+$, then the conditions on E, given before the statement of Theorem 5.1, imply that

$$\langle \nabla f(z), y\rangle = \langle Ey, y\rangle - \langle N(z), y\rangle.$$

Consequently, if K is a number such that $|N(z)| \leq K$ for all $z \in H$, then according to (5.1)

$$\langle \nabla f(z), y\rangle \geq m_1 |y|^2 - K|y|.$$

Hence

$$\langle f(z), y\rangle > 0 \quad \text{if} \quad |y| \geq c \tag{5.7}$$

where c is any fixed number with $c > K/m_1$.

We may assume $\eta(0) = 0$ so if $y \in H^+$, then $f(y) = \frac{1}{2}\langle Ey, y\rangle - \eta(y) \geq \frac{1}{2} m_1 |y|^2 - K|y|$. Hence $f(y) \geq -\frac{1}{2} K^2/m_1$ for all $y \in H^+$. Moreover if $y \in H^+$ with $|y| \leq c$, $x_0 \in H^0$, and $x_1 \in H^-$, then by (5.2)

$$\begin{aligned} f(x_0 + x_1 + y) &= \tfrac{1}{2}\langle Ey, y\rangle + \tfrac{1}{2}\langle Ex_1, x_1\rangle \\ &\quad - [\eta(x_0 + x_1 + y) - \eta(x_0)] - \eta(x_0) \\ &\leq \tfrac{1}{2}|E|c^2 - \tfrac{1}{2} m_2 |x_1|^2 + K(|x_1| + c) - \eta(x_0). \end{aligned}$$

Since $\eta(x_0) \to \infty$, there exists $r > 0$ such that

$$f(x_0 + x_1 + y) \leq -\tfrac{1}{2} K^2/m_1 \leq f(y^*),$$

$$\text{if} \quad |x_0|^2 + |x_1|^2 = r^2, \ |y|, \ |y^*| \leq c \ \text{and} \ y^* \in H^+ \tag{5.8}$$

We cannot apply the lemma directly as H is not finite dimensional. However H is separable and thus there exist nondecreasing sequences of finite dimensional subspaces

$\{H_n^0\}_1^\infty$, $\{H_1^-\}_1^\infty$, and $\{H_n^+\}_1^\infty$ with $H_n^0 \subseteq H^0$, $H_n^+ \subseteq H^+$ and $H_n^- \subseteq H^-$ such that if $H_n = H_n^- \oplus H_n^0 \oplus H_n^+$, then $\bigcup_{n=1}^\infty H_n$ is dense in H. We apply the lemma to the restriction of f to H_n. Using (5.7) and (5.8) for each $n = 1,2,\ldots,$ there exists, by the lemma, $u_n \in H_n$ such that $0 = \langle \nabla f(u_n), w\rangle = \langle Eu_n, w\rangle - \langle N(u_n), w\rangle$ for all $w \in H_n$ and such that if r and c are as above $|u_n|^2 \leq r^2 + c^2$. This implies the existence of a subsequence $\{u_{n_k}\}_1^\infty$ of $\{u_n\}_1^\infty$ which converges weakly to $u \in H$ and therefore, by the conditions on N, we have $\{N(u_{n_k})\}_1^\infty$ converges strongly to $N(u)$. Fixing $w \in \bigcup_{n=1}^\infty H_n$ we have

$$\langle Eu, w\rangle - \langle N(u), w\rangle = \lim_{n_k \to \infty} (\langle Eu_{n_k}, w\rangle - \langle N(u_{n_k}), w\rangle) = 0.$$

As $\bigcup_{n=1}^\infty H_n$ is dense in H, this implies that $Eu = N(u)$.

Let us again consider the boundary value problem

$$\begin{aligned} (Lu)(x) &= g(x, u(x)), \quad && x \in D \\ u(x) &= 0 \quad , && x \in \partial D \end{aligned} \tag{5.9}$$

where, for simplicity, we assume that L is a second order, strongly elliptic, selfadjoint operator with real coefficients which are measurable and bounded on D. Let us assume that g is continuous and bounded on $\bar{D} \times R$. Let $H = H_1^0$, $\frac{\partial G}{\partial u} = g$, and $\langle\ ,\ \rangle$ denote the usual inner product on H_1^0. If we define $\eta: H \to R$ by $\eta(v) = \int_D G(x, v(x))dx$, then for $v, w \in H$

$$\lim_{t \to 0+} \frac{\eta(v + tw) - \eta(v)}{t} = \int_D g(x, v(x))w(x)dx. \tag{5.10}$$

For fixed $v \in H$ the right hand side of (5.10) defines a continuous linear functional on H so we may write

$$\int_D g(x, v(x))w(x)dx = \langle N(v), w\rangle \tag{5.11}$$

for some unique $N(v) \in H$. The boundedness of g implies that $N:H \to H$ is bounded and the continuity of N follows from an elementary measure theoretic argument. Thus η is continuously differentiable and we may write $\operatorname{grad} \eta = N$.

If B is the continuous symmetric bilinear form on $H \times H$ which was defined in the previous section, then u is a weak solution of (5.9) if and only if

$$B(u,w) = \int_D g(x,u(x))w(x)\,dx, \qquad \text{for all } w \in H.$$

Now the symmetry of B implies the existence of a bounded selfadjoint operator $E:H \to H$ such that $B(v,w) = \langle Ev,w\rangle$, for all $v, w \in H$. Thus by (5.1), the boundary value problem (5.9) is equivalent to the operator equation $Eu = Nu$. The kernel of E consists of all $w \in H = \overset{\circ}{H}_1$ such that $B(w,v) = 0$ for all $v \in H_1$. This is clearly equivalent to condition that w is a weak solution of the linear homogeneous problem

$$(Lu)(x) = 0, \quad x \in D \quad \text{and} \quad u(x) = 0, \quad x \in \partial D. \tag{5.12}$$

Thus from Theorem 5.1 we obtain

<u>Theorem 5.3</u>. <u>If</u> $w_1,\ldots,w_r$ <u>is a basis of the solution space of</u> (5.12) <u>then</u> (5.9) <u>is solvable if</u>

$$\int_D G(x, s_1w_1 + \cdots + s_rw_r)\,dx \to +\infty$$

<u>as</u> $s_1^2 + \cdots + s_r^2 \to +\infty$.

It is a simple matter to derive Theorem 4.1 and Corollary 4.3 from this result.

The above result depends strongly on the fact that L is selfadjoint.

6. <u>Further Related Results</u>.

In addition to the work of Nirenberg, Hess, and Williams, cited above, there have recently appeared a number of papers which are in some way related to resonance type problems. We mention the works of

Berger and Podolak [3], Brezis [4], Cronin [8], DeFigueiredo [9], Fučik, Kučera, and Nečas [10], Mahwin [14], [15], Nečas [16], Rabinowitz [19], Schechter [20], Shaw [21], and Tromba [24].

Most of the above works require that the nonlinearities appearing in the equations be bounded. However, DeFigueiredo and Mawhin have obtained results for the solvability of problems of the type (4.3) where the kernel of L is nontrivial and the nonlinearity satisfies a growth restriction of the type

$$|g(x,u)| \leq K|u|^s + h(x)$$

where $h \in L^2(D)$ and K and s are sufficiently small.

REFERENCES

1. Ahmad, S., A. Lazer and J. Paul, "Elementary critical point theory and perturbations of elliptic boundary value problems at resonance," Indiana Univ. Math. J., to appear.

2. Bancroft, S., J.K. Hale and D. Sweet, "Alternative problems for nonlinear functional equations," J. Differential Equations 4 (1968), 40-56.

3. Berger, M.S. and E. Podolak, "On nonlinear Fredholm operator equations," Bull. Amer. Math. Soc. (80) 1974, 861-864.

4. Brezis, H., "Une équation elliptique nonlineaire a la resonance," Séminaire Lions, Schwartz, 1971.

5. Cesari, L., <u>Asymptotic Behavior and Stability Problems in Ordinary Differential Equations</u>, Ergebn. d. Math. # 16 (2nd edition), Springer-Verlag, New York, 1963.

6. Cesari, L., "Functional analysis and periodic solutions of nonlinear differential equations," Contributions to Differential Equations 1 (1963), 149-187.

7. Cesari, L., "Functional analysis and Galerkin's method," Mich. Math. J. 11 (1964), 385-414.

8. Cronin, J., "Equations with bounded nonlinearities," J. Differential Equations 14 (1973), 581-596.

9. DeFigueiredo, D.G., "On the range of nonlinear operators with linear asymptotes which are not invertible," Univ. de Brasilia, Trabalho de Math., no. 59, 1974.

10. Fučik, S., M. Kučera and J. Nečas, "Ranges of nonlinear asymptotically linear operators," J. Differential Equations 17 (1975), 375-394.

11. Hess, P., "On a theorem of Landesman and Lazer," Indiana Univ. Math. J. 23 (1974), 827-829.

12. Landesman, E.M. and A.C. Lazer, "Nonlinear perturbations of linear elliptic boundary value problems at resonance," J. Math. Mech. 19 (1970), 609-623.

13. Lazer, A.C. and D.E. Leach, "Bounded perturbations of forced harmonic oscillators at resonance," Ann. Mat. Pura Appl. 82 (1969), 49-68.

14. Mawhin, J., "Nonlinear perturbations of Fredholm mappings in normed spaces and applications to differential equations," Univ. de Brasilia, Trabalho de Mat., no. 61, May 1974.

15. Mawhin, J., "Topology and nonlinear boundary value problems," Univ. de Louvain, Sém. Math. Appl. Mec.; no. 75, September 1974.

16. Nečas, J., "On the range of nonlinear operators with linear asymptotes which are not invertible," Comm. Math. Carolinae 14 (1973), 63-72.

17. Nirenberg, L., "Generalized degree and nonlinear problems," Contributions to Nonlinear Functional Analysis (edited by E. Zarantonello), Academic Press, New York, London 1971, 1-9.

18. Nirenberg, L., "An application of generalized degree to a class of nonlinear problems," Trosieme Coll. B.R.M. cl'Analyse Functionnelle, Vander, Louvain, 1971, 57-74.

19. Rabinowitz, P., "A note on a nonlinear elliptic equation," Indiana Univ. Math. J. 22 (1972), 43-49.

20. Schechter, M., "A nonlinear elliptic boundary value problem," Ann. Scuola Norm. Sup. Pisa. 27 (1973), 707-716.

21. Shaw, H.C., "A note on a theorem by Landesman and Lazer," Notices Amer. Math. Soc. (22) (1975), p.A-150.

22. Vainberg, M.M., Variational Methods for the Study of Nonlinear Operators, Holday-Day, San Francisco 1964.

23. Williams, S.A., "A sharp sufficient condition for solution of a nonlinear elliptic boundary value problem," J. Differential Equations 8 (1970), 580-586.

24. Tromba, S., "Stable homotopy groups and nonlinear partial differential equations," to appear.

EXPOSITORY INTRODUCTION TO SOME ASPECTS OF DEGREE THEORY

Erich H. Rothe

University of Michigan

1. Introduction. In this conference monotone mappings play an important role. Therefore, in this lecture, those aspects of degree theory will be emphasized which lead to an existence theorem for monotone mappings. This application of theory was made possible by the investigations of Petryshyn and Browder-Petryshyn ([13], [4]). The classical Leray-Schauder theory [9] does not cover this topic since it is restricted to mappings differing from the identity map by a completely continuous map.

It is hoped that this expository introduction will be comprehensible to readers unfamiliar with the concept of the degree of a mapping.

The Leray-Schauder theory, as well as that of Browder-Petryshyn, presupposes the finite dimensional degree theory introduced by L.E.J. Brouwer in 1911 [5]. Section 2 of this paper treats this theory in the form given it by M. Nagumo [10]. As indicated at the end of that section a different approach, based on intersection theory, is used by Alexandroff and Hopf in their topology book [1].

In Section 3, an exposition of the Browder-Petryshyn theory of the generalized degree in Banach spaces is given. However this exposition does not cover the theory in its full generality, but only a special part which will be sufficient for the purpose of this paper. For a full account the reader is referred to [3], [4], [14]. The Leray-Schauder theory is not treated in this paper. Here the reader is referred to [6], [8], [11], [12].

In all degree theories one of the most important properties is the invariance under general homotopy. But even the invariance under the special linear convex homotopy (4.1) has a number of nearly immediate, very basic consequences. Section 4 deals with this subject for the generalized degree.

Finally, in Section 5, an application to monotone mappings is given.

2. The Finite Dimensional Case. Let E^n denote n-dimensional, real Euclidean space. Let G be an open bounded subset of E^n. The boundary and closure of G will be denoted by ∂G and $\bar{G}$, respectively. We consider the equation

$$f(x) = y_0 \tag{2.1}$$

where f is a continuous map from $\bar{G}$ into E^n and y_0 is a given point of E^n. We want to define a number $A(f,G,y_0)$ which, in a way to be made precise later on, counts the number of solutions $x \in G$ of equation (2.1). This definition will be attained in three steps under the basic restriction

$$y_0 \notin f(\partial G). \tag{2.2}$$

Step I. As usual we denote by $C^1 = C^1(\bar{G})$ the family of maps $\bar{G} \to E^n$ which admit a differential $df(x;h)$ which, as a linear map in h, is continuous in x for $x \in \bar{G}$. Equivalently, if $x_1, x_2, \ldots, x_n$ and $h_1, h_2, \ldots, h_n$ are the coordinates of x and h, respectively, and if $f_1, f_2, \ldots, f_n$ are those of f, then the partial derivatives $\partial f_i/\partial x_j$ are continuous functions of $x_1, x_2, \ldots, x_n$. We recall that the linear map $df(x;h)$ is given by

$$df_i(x;h) = \sum_{j=1}^{n} \frac{\partial f_i}{\partial x_j} h_j, \qquad i = 1,2,\ldots,n,$$

and that the determinant $J(x)$ of this map is called the Jacobian of f at x.

Definition 2.1. The point $x_0 \in G$ is called a regular point of f if $df(x,h)$ as linear map in h is non-singular or, equivalently, if $J(x_0) \neq 0$, and x_0 is called a critical point of f if it is not regular. A point $y_0 \in E^n$ is called a critical value of f if it is the image under f of some critical point. The set of critical values of f will be denoted by $\Gamma(f)$.

Our next aim is to give a definition of the degree $A(f,G,y_0)$ for $f \in C^1$ and for a y_0 which, in addition to (2.2), satisfies

$$y_0 \notin \Gamma(f). \tag{2.3}$$

First we need a lemma.

Lemma 2.1. If f and y_0 satisfy the conditions just stated, then equation (2.1) has at most a finite number of solutions in G.

Proof. If the assertion were not true, there would exist a convergent sequence $x_1, x_2, \ldots$ of different roots of (2.1) in G. Its limit x_0 would be a root of (2.1) in $\bar{G}$. By (2.2), $x_0 \in G$, and by (2.3), $J(x_0) \neq 0$. This, together with the implicit function theorem, would imply the existence of a neighborhood of x_0 in which x_0 is the only root of (2.1). This contradicts the fact that x_0 is the limit of the roots x_i.

Now we can state the definition

Definition 2.2. Let $f \in C^1$, and let y_0 satisfy (2.2) and (2.3). Denote by p and q the number of roots $x \in G$ of (2.1) at which $J(x)$ is positive and negative, respectively. Then

$$A(f,G,y_0) = p - q. \tag{2.4}$$

This number is called the degree of f at y_0 with respect to G.

Theorem 2.1. i) If I is the identity map on G then $A(I,G,y) = 1$ if $y \in G$ and 0 if $y \notin \bar{G}$.

ii) Let f and y_0 be as in Definition 2.2, and suppose that $A(f,G,y_0) \neq 0$. Then equation (2.1) has at least one solution in G.

iii) Let f be as above, and let $y(t)$ be a continuous curve in E^n defined for t in the closed interval $[0,1]$. We assume that

$$y(t) \notin f(\partial G) \quad \text{for every } t \text{ in } [0,1], \tag{2.5}$$

while

$$y(t) \notin \Gamma(f) \quad \text{for some } t \text{ in } [0,1]. \tag{2.6}$$

Then $A(f,G,y(t))$, defined for the t satisfying (2.6), is independent of t.

iv) Let Δ be the product of $\bar{G}$ with the closed interval $[0,1]$. Let $f(x,t)$ be a continuous map $\Delta \to E^n$. Suppose, moreover, that for each $t \in [0,1]$ the map $f_t(x) = f(x,t)$ from $\bar{G}$ into E^n belongs to C^1, and that $y_0 \notin f_t(\partial G)$ for $t \in [0,1]$. Then $A(f_t,G,y_0)$ is independent of t for those t for which $y(t) \notin \Gamma(f)$.

v) $A(f,G,y_0)$ depends only on the restriction of f to ∂G.

Remark 1. Statements (i) and (ii) follow trivially from Definition 2.2. Nevertheless they are important. Thus (ii) contains a basic existence theorem.

Remark 2. Statement (v) is an immediate consequence of statement (iv). For the complicated proof of the latter, as well as for that of statement (iii), we refer to [10]. Here we only notice that an essential tool for the proof is the following lemma.

Lemma 2.2. Let the point $y_0 \in E^n$ satisfy (2.2) but not necessarily (2.3). Then every neighborhood of y_0 contains a point not belonging to $\Gamma(f)$.

This lemma is a special case of a well known theorem by Sard (see [10], Theorem 1).

Remark 3. We note the following applications of statements (iii) and (iv) of Theorem 2.1:

a) If $\bar{f}$ and $\bar{y}_0$ are good enough approximations of f and y_0, respecitvely, then $A(\bar{f},G,\bar{y}_0) = A(f,G,y_0)$.

b) Given f, it is often possible to find an $f(x,t)$ such that $f(x,0) = f(x)$ and such that the degree for $f_1(x) = f(x,1)$ is easier to compute than that of the original map f. In many cases $f(x,t)$ can be chosen in such a way that f_1 is the identity map. In that case statement (iv), together with statements (ii) and (i), guarantees the existence of a root of (2.1) in G.

c) Statement (iii) will be used to get rid of the restriction (2.3) (see Step II below).

d) In Step III, statement (iv) will be used to free us of the differentiability assumption on f.

Remark 4. At first view it might seem more natural to define the degree simply as the number of roots of (2.1) in G instead of using (2.4) as definition. But then statement (iv) would not be true.

Example: Let G be the open interval $(0,\pi)$ in E^1 and $y_0 = 1/2$. Let $f(x,t) = (1 - t)\sin x$ and $f_0(x) = f(x,0)$, $f_1(x) = f(x,1)$. Then (2.1) has two roots $x_1 < x_2$ in G if $f = f_0$ and no roots there for $f = f_1$. Note that $J_0(x_1) = f_0'(x_1) > 0$ and $J_0(x_2) = f_0'(x_2) < 0$ (J_0 denotes the Jacobian of f). Thus the degrees of f_0 and f_1, as defined in (2.4), are both zero. It is just as simple to see that (iv) would not be true without the restriction (2.2).

Step II. Elimination of the Restriction (2.3). Let y_0 be as in Lemma 2.2. Then the distance ρ of y_0 to $f(\partial G$ is positive and the open ball $B(\rho, y_0)$ of radius ρ and center y_0 contains no point of $f(\partial G)$. By Lemma 2.2 this ball contains points which are not in $\Gamma(f)$. Let y_1 and y_2 be two such points and let s be the line segment connecting them. Then $s \in B(\rho, y_0)$. Therefore s has no points in common with $f(\partial G)$, and we conclude from part (iii) of Theorem 2.1 that $A(f,G,y_1) = A(f,G,y_2)$. Thus $A(f,G,y)$ is constant as y varies over $B(\rho,y_0) - \Gamma(f)$. This makes the following definition possible.

Definition 2.3. Let y_0 be a point of $\Gamma(f)$ with positive distance ρ from $f(\partial G)$. We then set $A(f,G,y_0) = A(f,G,y)$ where y is a point of $B(\rho,y_0) - \Gamma(f)$.

Step III. Elimination of the Differentiability Assumption on f. Let f be a continuous map $\bar{G} \to E^n$. We first recall that for a given $\rho > 0$, there exists a map $\bar{f}:\bar{G} \to E^n$ which is in C^1 and for which

$$|f(x) - \bar{f}(x)| < \rho \quad \text{for all } x \in \bar{G}. \tag{2.7}$$

Now let y_0 satisfy (2.2) and choose ρ as the distance of y_0 from $f(\partial G)$. Then

$$y_0 \notin \bar{f}(\partial G). \tag{2.8}$$

Indeed, by definition of ρ

$$y_0 \notin B(\rho, f(x)) \quad \text{for } x \in \partial G \tag{2.9}$$

while by (2.7)

$$\bar{f}(x) \in B(\rho, f(x)) \quad \text{for } x \in \partial G. \tag{2.10}$$

Equations (2.9) and (2.10) obviously imply (2.8). Thus $A(\bar{f},G,y_0)$ is defined if $\bar{f}$ satisfies (2.7).

We claim that this number is independent of the special choice of such $\bar{f}$. Indeed, let $\bar{f}_0$ and $\bar{f}_1$ be two such maps. Then (2.7) remains true if $\bar{f}$ is replaced by $\bar{f}_t = (1 - t)f_0 + tf_1$, and the same is therefore true for (2.9) and (2.10). Using part (iv) of Theorem 2.1, we see that $A(\bar{f}_0, G, y_0) = A(\bar{f}_1, G, y_0)$.

This shows that the following definition is legitimate.

Definition 2.4. Let $f: \bar{G} \to E^n$ be continuous, let $y_0 \notin f(\partial G)$, and let ρ be the distance of y_0 from $f(\partial G)$. We then set $A(f, G, y_0) = A(\bar{f}, G, y_0)$ where $\bar{f}: \bar{G} \to E^n$ is a C^1-map satisfying (2.7).

Theorem 2.2. The assertions of Theorem 2.1 remain valid if the following modifications are made: The reference to Definition 2.2 in part (ii) is replaced by reference to Definition 2.4. Restriction (2.6) is deleted. The assertions of parts (iii) and (iv) are stated for all $t \in [0,1]$. (For the proof see [10].)

The following lemma will be of use later.

Lemma 2.3. If f and y_0 are as in Definition 2.4, then $A(f, G, y_0) = A(f - y_0, G, 0)$. (For the short proof see [11], p.505, footnote.)

Closely related to the notion of the degree of a mapping is that of the winding number.

Definition 2.5. Let G be a bounded open subset of E^n, and let f be a continuous map of ∂G into E^n. Then the winding number $u(f(\partial G), y_0)$ of $f(\partial G)$, with respect to a point $y_0 \notin f(\partial G)$, is defined as follows: Let $\bar{f}$ be a continuous extension of f to $\bar{G}$, then

$$u(f(\partial G), y_0) = A(\bar{f}, G, y_0). \tag{2.11}$$

Note, the existence of a continuous extension $\bar{f}$ is guaranteed by Tietze's extension theorem (see e.g. [1], p.76), and part (v) of Theorem 2.1 guarantees that $A(f,G,y_0)$ is independent of the particular choice of the extension $\bar{f}$.

We describe an intuitive interpretation of $u(f(\partial G),y_0)$: Let B be a ball with center y_0 whose closure has no point in common with $f(\partial G)$ and, for $y \in E^n - y_0$, let $P(y)$ denote the projection of y from y_0 on $S^{n-1} = \partial B$. Then $u(f(\partial G),y_0)$ equals the number of times S^{n-1} is covered by $P(f(\partial G))$. In particular, in the two dimensional case, $u(f(\partial G),y_0)$ tells us how often the curve $f(\partial G)$ winds around the point y_0. That the number of these windings equals the number of solutions, $x \in G$, of equation (2.1), for f a complex-valued, analytic function, was established a long time ago in classical analysis. Note, that in this case the number q in Definition 2.2 of the degree equals zero since the Jacobian equals the square of the absolute value of the complex derivative.

There is another intuitive interpretation of the winding number: it is number of properly defined intersections of $f(\partial G)$ with a half-line h issuing from y_0. Therefore the degree theory may developed as follows: First develop a theory of intersections. Then define the winding number $u(f(\partial G),y_0)$ as the number of intersections of $f(\partial G)$ with h. And finally define the degree $A(f,G,y_0)$ of an $\bar{f}$ defined on $\bar{G}$ by (2.11) where now f denotes the restriction of $\bar{f}$ to ∂G. This method of building up the degree theory is used in the topology book by Alexandroff and Hopf, [1] Chapters XI and XII.

Remark Concerning Terminology. In older literature $u(f(\partial G), y_0)$ is called the order of y_0 with respect to $f(\partial G)$, see e.g. [1], p.419. If one thinks of $f(x)$ as a vector attached to the point $x \in \partial G$ then in particular $u(f(\partial G), 0)$ is called the characteristic of the vector field f on ∂G ([1], p.478). In [8] this is called the rotation of the vector field f.

We conclude this section with the following remark: The degree theory for Banach spaces to be developed in Section 3 presupposes the theory for Banach spaces $\tilde{E}^n$ of finite dimension n. But the theory for Euclidean E^n developed in this section carries over to $\tilde{E}^n$ in an obvious manner. This is clear from the following two facts: a) The degree defined for E^n is invariant under a linear isomorphism of E^n onto itself. b) $\tilde{E}^n$ is linearly isomorphic to E^n (see e.g. [7], IV, 3.3, p.245).

3. The Generalized Mapping Degree. As mentioned in Section 1, this concept was introduced by Browder and Petryshyn [4]; the present exposition will not treat it in the full generality given by these authors.

Definition 3.1. Let E be a real Banach space with norm $\|\cdot\|$ and let E_n be a sequence of finite dimensional subspaces of E (n is not necessarily the dimension of E_n). Let P_n be a projection of E onto E_n, i.e. a linear continuous map n such that $P_n^2 = P_n$. Then $\gamma = (E, E_n, P_n)$

is called an approximation scheme if the following two conditions are satisfied:

$$\lim_{n\to\infty} P_n(x) = x \quad \text{for every } x \in E, \tag{3.1}$$

$$\|P_n(x)\| \leq k\|x\| \quad \text{for every } x \in E, \text{ where is independent of } n; \tag{3.2}$$

i.e. the P_n are uniformly bounded.

Remark. Suppose E admits a Schauder base, i.e. it contains a sequence of elements $\varphi_1, \varphi_2, \ldots$ such that every $x \in E$ admits a unique representation $x = \sum_{i=1}^{\infty} \varphi_i x_i$, where the x_i are real numbers. Let E_n be the space spanned by $\varphi_1, \varphi_2, \ldots, \varphi_n$, and let P_n be the map $x \to \sum_{i=1}^{n} \varphi_i x_i$. Then (3.1) is obviously satisfied and also (3.2), with $k = 1$ (after possibly giving E a new norm equivalent to the original one; see e.g. [2], p.111). In particular this is true, without re-norming, if E is a separable Hilbert space and $\varphi_1, \varphi_2, \ldots$ is a complete orthonormal system.

Definition 3.2. Let G be a bounded open set in E and $\bar{G}$ its closure. Let f be a continuous map $\bar{G} \to E$ and $\gamma = (E, E_n, P_n)$ an approximation scheme (Definition 3.1). Then the restriction, f_n, of $P_n f$ to the closure of $G_n = G \cap E_n$ is called the n-th approximation of f with respect to γ.

It is our aim to define, for

$$y_0 \notin f(\partial G), \tag{3.3}$$

a degree $A(f, G, y_0)$ by some limit process in terms of the degrees

$$A(f_n, G_n, P_n(y_0)). \tag{3.4}$$

But the latter are defined if, and only if,

$$P_n(y_0) \notin f_n(\partial G_n) \tag{3.5}$$

(cf. the last paragraph of Section 2). Therefore we have to restrict ourselves to mappings f for which assumption (3.3) implies the validity of (3.5), at least for large enough n. It follows from Lemma 3.1 below, that this implication is true if f has the property characterized in the following

<u>Definition 3.3</u>. The continuous map $f:\bar{G} \to E$ is called approximation proper (A-proper, for short) with respect to the approximation scheme $\Upsilon = (E, E_n, P_n)$ if it has the following property: If, for some subsequence $\{n_i\}$ of positive integers, there exist points $x_{n_i} \in \bar{G}_{n_i}$ and a point y such that

$$\lim f_{n_i}(x_{n_i}) = y, \tag{3.6}$$

then the sequence $\{n_j\}$ has a subsequence $\{n_{j(i)}\}$ such that the $x_{n_{j(i)}}$ converge and such that

$$y = f(x), \tag{3.7}$$

where

$$x = \lim x_{n_{j(i)}}. \tag{3.8}$$

The monotone maps (Definition 5.1) are an examples of A-proper maps (with respect to a suitable Υ). Other examples are the maps of the following class

<u>Definition 3.4</u>. A continuous map $f:\bar{G} \to E$ ($G \subset E$, bounded and open) is called a Leray-Schauder map if $F(x) = f(x) - x$ is completely continuous, i.e. if the closure of $F(\bar{G})$ is compact.

That a Leray-Schauder map is A-proper with respect to any Υ is easily verified by using (3.1) and (3.2).

We now state the basic lemma.

Lemma 3.1. If f is A-proper with respect to γ, then (3.3) implies the existence of a positive integer n_0 and a positive number ρ such that the distance of $P_n(y_0)$ from ∂G_n is at least ρ for $n \geqq n_0$.

Proof. Suppose the lemma were not true. Then there would exist a sequence of points

$$x_{n_i} \in \partial G_{n_i} \tag{3.9}$$

such that $f_{n_i}(x_{n_i}) - P_{n_i}(y_0) \to 0$ or, by (3.1), such that (3.6) were true with $y = y_0$. Then, by definition 3.3, there would exist a convergent subsequence $x_{n_{j(i)}}$ of the sequence x_{n_i} such that (3.7) and (3.8) were true with $y = y_0$. Now it may be verified that $\partial G_{n_i} \in E_{n_i} \cap \partial G \subset \partial G$. Therefore (3.9) shows that $x_{n_i} \in \partial G$ which by (3.8) implies that $x \in \partial G$ since ∂G is closed. Thus (3.7) (with $y = y_0$) contradicts (3.3).

By the lemma just proved the numbers (3.4) are, from a certain n on, well defined if f is A-proper and y_0 satisfies (3.3). If the sequence of these real numbers is bounded it contains a convergent subsequence. If it is not bounded it contains a subsequence convergent to $+\infty$ or to $-\infty$. Therefore, in any case, there exists a subsequence convergent to an element z of the set Z of the integers augmented by $\pm\infty$. We are now ready for the definition.

Definition 3.5. Let $f:\bar{G} \to E$ be A-proper with respect to $\gamma = (E, E_n, P_n)$, and let y_0 satisfy (3.3). Let Z be the integers augmented by $\pm\infty$. Then the subset $\{z\}$ of Z

consisting of those elements which are limits of some subsequence of the numbers (3.4) is called the generalized degree $A(f,G,y_0)$ of f at y_0 with respect to G and Υ. (This set is not empty, as shown in the preceding paragraph.)

If z_0 is a fixed element of Z, then $\{z_0\}$ will denote the subset of Z consisting of the single element z_0.

Theorem 3.1. a) If f is A-proper with respect to Υ and λ, c are constants, $\lambda \neq 0$, then λf and $f + c$ are A-proper with respect to Υ.

b) The identity map I is A-proper with respect to any Υ.

c) $A(I,G,y_0) = \{1\}$ if $y_0 \in G$; $\{0\}$ if $y_0 \notin \bar{G}$.

This theorem follows easily from the definitions and from Theorems 2.1 and 2.2.

Theorem 3.2. Let f be A-proper, let y_0 satisfy (3.3), and suppose that

$$A(f,G,y_0) \neq \{0\}. \tag{3.10}$$

Then equation (2.1) has at least one solution in G.

Proof. By Definition 3.5, the assumption (3.10) implies the existence of a nonzero element z_0 of Z and a sequence $\{n_i\}$ of integers such that the degrees $A(f_{n_i}, G_{n_i}, P_{n_i}(y_0))$ converge to z_0. These degrees are then different from zero for n_i big enough. For such n_i Theorem 2.2, together with the part (ii) of Theorem 2.1, assert the existence of an $x_{n_i} \in G_{n_i}$ such that $f_{n_i}(x_{n_i}) = P_{n_i}(y_0)$. It follows now from (3.1) that the $f_{n_i}(x_{n_i})$ converge to y_0. But by Definition 3.3, this fact implies the existence of a convergent subsequence $x_{n_{j(i)}}$ of the sequence x_{n_i} whose limit x satisfies (2.1). Obviously $x \in \bar{G}$, and (3.3) implies that $x \in G$.

Theorem 3.3. Let $f(x,t):\bar{G} \times [0,1] \to E$ be continuous and suppose that the continuity of $f(x,t)$ in t is uniform for $x \in \bar{G}$. Suppose, moreover, that $f_t(x) = f(x,t)$ is A-proper for every $t \in [0,1]$ and that

$$f_t(x) \neq y_0 \quad \text{for} \quad x \in \partial G, \quad t \in [0,1]. \tag{3.11}$$

Then $A(f_t, G, y_0)$ is independent of t.

For the proof we refer to [4], Theorem 1(c).

Theorem 3.4. Let $y = y(t)$ be a continuous curve in E defined for $t \in [0,1]$. Suppose that

$$y(t) \notin f(\partial G) \quad \text{for} \quad t \in [0,1] \tag{3.12}$$

where f is an A-proper map $\bar{G} \to E$. Then $A(f,G,y(t))$ is independent of t.

Proof. We recall that f_n is the restriction of $P_n f$ to G_n and that therefore $f_n - P_n y(t) = (f - y(t))_n$ by linearity of P_n. From this equality and Lemma 2.3 we see that $A(f_n, G_n, P_n y(t)) = A((f - y(t))_n, G_n, 0)$ for every positive integer n. Consequently, by Definition 3.5, $A(f,G,y(t)) = A(f - y(t), G, 0)$. But the right member of this equality is independent of t by Theorem 3.3 with $f(x,t) = f(x) - y(t)$ (cf. the corresponding proof in the Leray-Schauder case in [11], p.505).

We conclude this section with a remark on the definition of a generalized winding number, more properly called winding set. Assume that f is defined only on ∂G. For such f

Definition 3.3 makes sense if $\bar{G}$ is replaced by ∂G. Now $\partial G_n \in \partial G$, as was noticed in the proof of Lemma 3.1. This lemma is still valid; therefore (3.3) implies (3.5). Therefore, if (3.3) holds the winding numbers

$$u(f_n(\partial G_n), y_0) \tag{3.13}$$

are defined (for n big enough). Therefore, under this assumption, the winding set $u(f(\partial G, y_0))$, can be defined in terms of these numbers in the same way as the generalized degree was defined in terms of the numbers (3.4) (Definition 3.5). If f is defined in all of $\bar{G}$, then the generalized degree equals the winding set since then, by Definition 2.5 the numbers (3.4) are equal to the numbers (3.13) for each n.

4. The Theorem of Poincaré-Bohl and Its Consequences.

Definition 4.1. A family $\mathfrak{F}$ of A-proper mappings from $\bar{G}$ into E is said to have property M if it is convex, i.e. $f_i \in \mathfrak{F}$ $(i = 0,1)$ implies that $\mathfrak{F}$ contains the map

$$f_t = (1 - t)f_0 + tf_1 \quad \text{for} \quad t \in [0,1]. \tag{4.1}$$

And $\mathfrak{F}$ is said to have property M_0 if, for $f \in \mathfrak{F}$ and c a constant, $f + c \in \mathfrak{F}$. Finally $\mathfrak{F}$ is said to have property M_1 if it contains the identity map I. (Here, as often in the sequel, A-proper means A-proper with respect to a fixed approximation scheme Υ (cf. Definition 3.3.)).

Theorem 4.1. Let $\mathfrak{F}$ have property M, and let f_0, f_1 be two elements of $\mathfrak{F}$ for which $f_0 - f_1$ is bounded. Then, under the assumption that the f_t of (4.1) satisfies (3.11),

$$A(f_0, G, y_0) = A(f_1, G, y_0). \tag{4.2}$$

Proof. The theorem is a special case of Theorem 3.3, whose assumptions are seen to be satisfied if one notes that $f_{t_2} - f_{t_1} = (t_2 - t_1)(f_1 - f_0)$.

Theorem 4.1 will be referred to as theorem of Poincaré-Bohl since it is a generalization of the theorem of that name in the finite dimensional case. (See [1], p.459 where that theorem is stated for the winding numbers.) It also holds for Leray-Schauder maps (Definition 3.4) where, as in the present case, it is a special application of a general homotopy theorem ([9], Lemma 3).

Theorem 4.2. a) Let f_0, f_1 be as in the first sentence of Theorem 4.1. Let $y_0 \notin f_0(\partial G)$ and suppose that

$$\|f_1(x) - f_0(x)\| < \|y_0 - f_0(x)\| \quad \text{for } x \in \partial G. \tag{4.3}$$

Then (4.2) holds.

b) If $\mathfrak{F}$ also has property M_1 and if (4.3) holds with $f_1(x) = x$, then

$$A(f_0, G, y_0) = \{1\}. \tag{4.4}$$

Proof. For $x \in \partial G$, let $r(x) = \|y_0 - f_0(x)\|$. Then

$$y_0 \notin B(f_0(x), r_0(x)) \tag{4.5}$$

where, as usual, $B(y,r)$ denotes the open ball with center y and radius r. Now $B(f_0(x), r(x))$ contains its center $f_0(x)$ and, by (4.3), $f_1(x)$, Therefore it contains $f_t(x)$ (see (4.1)). But this, together with (4.5), implies (3.11), and part (a) of our assertion follows from Theorem 4.1. Part (b) follows now from Theorem 3.1, (c).

Part (a) of Theorem 4.2 is called the theorem of Rouché since it is a generalization of the classical complex variable theorem of that name. For the corresponding theorem in n-space see [1], p.459 and for its generalization to Leray-Schauder maps see [15], Satz 2a and [16], Theorem 4.

Theorem 4.3. In addition to our usual hypothese on G, we assume that this set is a stardomain with respect to the point $y_0 \in G$. Let $\mathfrak{F}$ have property M, let f_0, f_1 belong to $\mathfrak{F}$, and suppose that

$$f_i(x) \neq y_0, \quad i = 0,1 \quad \text{for} \quad x \in \partial G \tag{4.6}$$

and that

$$A(f_0, G, y_0) \neq A(f_1, G, y_0). \tag{4.7}$$

Then there exists an $x_0 \in \partial G$ and a positive λ such that

$$f_1(x_0) - y_0 = -\lambda(f_0(x_0) - y_0). \tag{4.8}$$

Proof. By the Poincaré-Bohl theorem it will be sufficient to prove that if

$$f_1(x) - y_0 \neq -\lambda(f_0(x) - y_0) \tag{4.9}$$

for all $x \in \partial G$ and all positive λ, then (3.11) is true for the f_t defined by (4.1). Now, from assumption (4.6), we see that (3.11) is true for $t = 0$ and $t = 1$. Suppose then that (3.11) were not true for some t in the open interval $(0,1)$ and some $x_0 \in \partial G$. Then $f_t(x_0) - y_0 = 0$ for such t, or, by (4.1), $f_1(x_0) - y_0 = -\lambda(f_0(x_0) - y_0)$ with $\lambda = t^{-1}(1 - t)$ which contradicts assumption (4.9).

For a finite dimensional version of the basic theorem just proved see [1], p.479, Statement (a) (where it is also called theorem of Poincaré-Bohl). We now derive some corollaries to Theorem 4.3.

Corollary 1 (cf. [4], Theorem 5). Let G be as in Theorem 4.3. Let $\mathfrak{F}$ have properties M and M_1 and let $f_0 \in \mathfrak{F}$. Suppose that

$$f_0(x) - y_0 \neq -\mu(x - y_0) \quad \text{for all } x \in \partial G \text{ and all } \mu > 0. \tag{4.10}$$

Then for $y \in G$

$$A(f_0, G, y) = \{1\} \tag{4.11}$$

and the equation

$$f(x) = y \tag{4.12}$$

has at least one solution $x \in G$.

Proof. Applying Theorem 4.3, with $f_1 = I$ and $\lambda = \mu^{-1}$, together with Theorems 3.1 and 3.2 yields the assertions for $y = y_0$. Their validity for arbitrary $y \in G$ follows then from Theorem 3.4.

For a version of Corollary 1 referring to Leray-Schauder maps see [15], Satz 7b.

Corollary 2. Let E be a Hilbert space with scalar product $(\cdot,\cdot)$ and let $G, \mathfrak{F}, f_0, f_1, y_0$ be as in Theorem 4.3. We assume that

$$(f_0(x) - y_0, \ f_1(x) - y_0) > 0 \quad \text{for all } x \in \partial G. \tag{4.13}$$

Then

$$A(f_0, G, y) = A(f_1, G, y) \quad \text{for } y \in G. \tag{4.14}$$

Proof. By Theorem 3.4 it will be sufficient to give the proof for $y = y_0$. Now if (4.14) were not true for $y = y_0$, then by Theorem 4.3, the relation (4.8) would hold for some $x_0 \in \partial G$ and some $\lambda > 0$. Scalar multiplying (4.8) by

$f_0(x_0) - y_0$ we have $(f_1(x_0) - y_0, f_0(x_0) - y_0) = -\lambda\|f_0(x_0) - y_0\|^2 < 0$ by (4.6). This contradicts (4.13).

Corollary 3. Suppose that E, G, $\mathfrak{F}$, f_0, y_0 are as in the first sentence of Corollary 2. In addition we assume that $\mathfrak{F}$ has property M_1 and that

$$(f_0(x) - y_0, x - y_0) > 0 \quad \text{for all} \quad x \in \partial G. \tag{4.15}$$

Then for $y \in G$, (4.11) holds and equation (4.12) has at least one root $x \in G$.

Proof. Applying Corollary 2, with $f_1 = I$, together with Theorems 3.1 and 3.2 yields the assertions for $y = y_0$. Now apply Theorem 3.4.

Corollary 4. Let E be a Hilbert space. Let G, $\mathfrak{F}$ and f_0 be as in the first two sentences of Corollary 1, except that G is not necessarily a star domain. Let $y_0 \in G$. We assume that there exists a positive r such that $B(y_0,r) \in G$ and

$$(f(x) - f(y_0), \frac{x - y_0}{r}) > \|f(y_0) - y_0\| \quad \text{for} \quad x \in \partial B(y_0,r) \tag{4.16}$$

Then for $y \in B(y_0,r)$

$$A(f,B(y_0,r),y) = \{1\}, \tag{4.17}$$

and equation (4.12) has at least one solution $x \in B(y_0,r)$.

Proof. For $x \in \partial B(y_0,r)$

$$\begin{aligned}(f(x) - y_0, x - y_0) &= (f(x) - f(y_0), x - y_0) \\ &\quad + (f(y_0) - y_0, x - y_0) \\ &\geqq (f(x) - f(y_0), x - y_0) \\ &\quad - \|f(y_0) - y_0\|r.\end{aligned}$$

Here the right member is positive by (4.16). Thus, if in Corollary 3 G is replaced by $B(y_0, r)$, assumption (4.15) is satisfied, and our assertions follow from that corollary.

Remark to Corollary 4. In particular, this corollary is useful under the following circumstances: Suppose that f is defined in all of E and that the assumptions made in the first sentence of the corollary are satisfied for every bounded open set G in E. In addition, assume the "coerciveness" condition, that the left member of (4.16) tends to $+\infty$ as $r = \|x - y_0\|$ tends to infinity for fixed y_0 in E. We can then choose r such that assumption (4.16) is satisfied and G such that the assumption $B(y_0, r) \in G$ is also satisfied.

As a final application of the Poincaré-Bohl theorem, we state the following fixed point theorem.

Theorem 4.4. Let G be an open, bounded, stardomain in Banach space E. Suppose that the family $\mathfrak{F}$, consisting of A-proper maps from $\bar{G}$ into E, has properties M, M_0 and M_1 (cf. Definition 4.1). Let F be an element of $\mathfrak{F}$ for which $F(\partial G) \in G$. Then $F(x_0) = x_0$ for some $x_0 \in G$.

We omit the proof which is similar to the one used in the corresponding fixed point theorem for compact F in the Leray-Schauder degree theory (see [15], p.186; [16], p.381; or [6], p.139).

5. Application to Monotone Mappings.

For the purpose of this paper we adopt the following:

Definition 5.1. (Cf. [4], p.235; Example (i).) Let E be a separable Hilbert space with scalar product $(\cdot,\cdot)$. Let $f: E \to E$ be continuous. Then f is called monotone if, for all x and y in E,

$$(f(x) - f(y), x - y) \geqq c(\|x - y\|) \tag{5.1}$$

where $c = c(r)$ is defined and continuous for $r \geqq 0$ and has the following properties: i) $c(0) = 0$; ii) $c(r) > 0$ for $r > 0$; iii) if $r_1, r_2, \ldots$ is a sequence for which $c(r_j) \to 0$ then $r_j \to 0$; iv) $c(r)/r \to \infty$ as $r \to \infty$. (Example: $c(r) = r^{1+a}$, $a > 0$.)

We recall the following.

Definition 5.2. A map f from E into E is called bounded if, for every bounded subset G of E the restriction of f to G is bounded.

Lemma 5.1. Let $\mathfrak{F}_0$ be the family of bounded, continuous, monotone maps on E, and let G be an open, bounded subset of E. Then the restriction of an $f \in \mathfrak{F}_0$ to $\bar{G}$ is A-proper with respect to $\Upsilon = (E, E_n, P_n)$ where E_n and P_n are as in the remark following Definition 3.1 (with a complete orthonormal system $\{\varphi_j\}$ used as base for E). Moreover, the family $\mathfrak{F} = \mathfrak{F}(G)$ of the restrictions obtained as f varies over $\mathfrak{F}_0$ has properties M, M_0, M_1 (see Definition 4.1).

Proof. The verification of properties M, M_0, M_1 is an obvious one. We therefore prove only that the restriction of f to $\bar{G}$, again denoted by f, is A-proper with respect to Υ (see [4], p.236; cf. also [13], Theorem 3).

We have to prove that if, for the sequence of points $x_{n_i} \in G_{n_i} = G \cap E_{n_i}$, we have

$$\lim P_{n_i} f(x_{n_i}) = y_0 \tag{5.2}$$

then there exists a subsequence $x_{n_{j(i)}}$ of the sequence x_{n_i} and an $x \in E$ such that

$$\lim x_{n_{j(i)}} = x \tag{5.3}$$

and

$$f(x) = y_0. \tag{5.4}$$

Now $x_{n_i} \in G_{n_i} \subset G$ and G is bounded. Therefore the sequence x_{n_i} contains a subsequence which converges weakly to a point x in E. For the sake of simplicity we change our notation and denote this subsequence by x_{n_i}. Then

$$x_{n_i} \rightharpoonup x \tag{5.5}$$

where $\rightharpoonup$ denotes weak convergence. We claim that x is the limit of the x_{n_i} (which in our notation proves (5.3)). (Unless indicated otherwise, the terms limit and convergence denote strong limit and strong convergence, respectively.) By properties (i) and (iii) of $c(r)$ this assertion is equivalent to asserting that

$$\lim c(\|x - x_{n_i}\|) = 0. \tag{5.6}$$

Now by (5.1)

$$\begin{aligned} c(\|x - x_{n_i}\|) &\leq (f(x) - f(x_{n_i}), x - x_{n_i}) \\ &= [(f(x),x) - (f(x), x_{n_i})] \\ &\quad + [(f(x_{n_i}),x_{n_i}) - (f(x_{n_i}),x)] \end{aligned}$$

Here the first bracket converges to 0 by (5.5). It remains to show that the second bracket converges to zero.

Now we know from (3.1) that

$$\lim P_{n_i}(x) = x. \tag{5.7}$$

This is a motivation to add and subtract $(f(x_{n_i}), P_{n_i}(x))$. Noting that this scalar product equals $(P_{n_i} f(x_{n_i}), x)$ on account of the selfadjointness of the projection, we see that the second bracket in (5.6) equals $[f(x_{n_i}), P_{n_i}(x) - x] + [(f(x_{n_i}), x_{n_i}) - (P_{n_i} f(x_{n_i}), x)]$. Here the first bracket converges to 0 by (5.7) and of the boundedness of f. In the second bracket we add and subtract (y_0, x_{n_i}). Again using the selfadjointness of P_{n_i} and the fact that $x_{n_i} = P_{n_i}(x_{n_i})$ we see that this second bracket equals

$$(P_{n_i}(f(x_{n_i}) - y_0, x_{n_i}) + (y_0, x_{n_i}) - (P_{n_i}(f(x_{n_i}), x). \tag{5.8}$$

Here the first scalar product converges to zero by (5.2) and the boundedness of the x_{n_i}, the second one converges to (y_0, x) by (5.5), and the last term to $-(y_0, x)$ by (5.2). Thus the sum (5.8) converges to zero. This completes the proof of (5.6) and, therefore, of (5.3). Assertion (5.4) follows now by the use of (3.1), (3.2), (5.2) and from the decomposition $f(x) - y_0 = [f(x) - P_{n_i} f(x)] + [P_{n_i} f(x) - P_{n_i} f(x_{n_i})] + [P_{n_i} f(x_{n_i}) - y_0]$.

This finishes the proof of Lemma 5.1. Note that assumption (iv) has not been used. This condition however plays an essential role in the following.

Theorem 5.1. Let f be monotone (Definition 5.1), and let y_0 be an arbitrary point of E. Then the equation $f(x) = y_0$ has one and only one solution in E, and this solution lies in the ball $B(y_0, r)$ if r is such that $\|f(y_0) - y_0\| < c(r)/r$. (This condition, by property (iv) of Definition 5.1, is satisfied for r big enough.)

Proof. The uniqueness follows trivially from (5.1) and property (ii) of $c(r)$. The existence follows from Corollary 4 in Section 4 since by Lemma 5.1 this corollary may be applied to an open bounded set G containing $B(y_0, r)$.

REFERENCES

1. Alexandroff, P. and H. Hopf, Topologie, Springer, Berlin 1935.

2. Banach, S. Théorie des operations linéaires, Warszawa 1932.

3. Browder, F.E., "Nonlinear functional analysis," Proc. of symposia in Pure Mathematics 18, part 2 (1975) Amer. Math. Soc.

4. Browder, F.E. and W.V. Petryshyn, "Approximation methods and the generalized topological degree for nonlinear mappings in Banach spaces," J. Functional Analysis 3 (1969), 217-245.

5. Brouwer, L.E.J., "Beweis der Invarianz der Dimensionszahl, Math. Ann. 70 (1911), 161-165.

6. Cronin, J., Fixed points and topological degree in nonlinear analysis, Mathematical Surveys 11 Amer. Math. Soc. 1964.

7. Dunford, N. and J.T. Schwartz, Linear operators part I, Interscience Publishers, New York 1958.

8. Krasnoselskii, M.A., Topological methods in the theory of nonlinear integral equations, Moscow 1956, English translation Pergamon Press 1964.

9. Leray, J. and J. Schauder, "Topologie et équations fonctionnelles," Ann. Sci. École Norm. Sup. Paris 51 (1934), 45-78.

10. Nagumo, M., "A theory of degree of mappings based on infinitesimal analysis," Amer. J. Math. 73 (1951), 485-496.

11. Nagumo, M., "Degree of mapping in convex linear topological spaces," Amer. J. Math. 73 (1951), 497-511.

12. Nirenberg, L., Topics in functional analysis, Lecture Notes 1973-1974, Courant Institute of Mathematical Sciences.

13. Petryshyn, W.V., "On linear P-compact operators in Banach spaces with applications to constructive fixed point theorems," J. Math. Anal. Appl. 15 (1966), 228-242.

14. Petryshyn, W.V., "On the approximation-solvability of equations involving A-proper and pseudo-A-proper mappings," Bull. Amer. Math. Soc. 81 (1975), 223-312.

15. Rothe, E., "Zur Theorie der topologischen Ordnung und der Vektorfelder in Banachschen Räumen," Compositio Mathematica 5 (1937), 177-197.

16. Rothe, E., "The theory of topological order in some linear topological spaces," Iowa State College Journal of Science 13 (1939), 373-390.

BOUNDARY VALUE PROBLEMS FOR $x'' = f(t,x,x')$ AND NONLINEAR FUNCTIONAL ANALYSIS

Jerry D. Schuur

Michigan State University

1. Introduction. We consider the equation

$$x'' = f(t,x,x'), \tag{1.1}$$

where $f:[0,1] \times R^n \times R^n \to R^n$ is continuous, subject to one of the boundary conditions:

$$x(0) = x_0, \quad x(1) = x_1 \tag{1.2}$$

(the Dirichlet or Picard Problem);

$$x(0) = x(1), \quad x'(0) = x'(1) \tag{1.3}$$

(the periodic problem).

Using these problems as models, we shall apply the techniques of nonlinear functional analysis to unify a number of classical results and discover new results. Extensions to higher order, more general equations will be apparent.

For background, Hartman's book [7, Chapter XII] gives a good discussion of these two problems from the standpoint of ordinary differential equations and Cesari's monograph in this volume thoroughly covers the techniques and development of the alternative method.

We have selected three cases - the Picard problem in $C^1[0,1]$, the periodic problem in $C^1[0,1]$, and the periodic problem in $L^2[0,1]$ - which should outline our theme and yet meet constraints of space and readability.

2. <u>The Picard Problem in $C^1[0,1]$.</u> For $x,y \in R^n$, let $x \cdot y = \sum_1^n x_i y_i$ and $|x| = (x \cdot x)^{1/2}$.

Let $[0,1] = I$ and let $C = C(I)$ denote the Banach space of continuous functions on $[0,1]$ with the norm $|x|_0 = \sup\{|x(t)|: 0 \le t \le 1\}$. Let $C^n = C^n(I)$ $(n = 1,2)$ denote the Banach space of functions which are continuous up to their n-th derivatives on I with the norm $|x|_n = \max(|x|_0, \ldots, |x^{(n)}|_0)$.

If L is the linear differential operator defined by

$$Lx = -x'', \quad D_L = \{x \in C^2 : x(0) = x_0,\ x(1) = x_1\} \tag{2.1}$$

and if N is the nonlinear Nemytskii operator defined by

$$Nx = f(\cdot, x(\cdot), x'(\cdot)), \quad D_N = C^1, \tag{2.2}$$

then the Picard problem (1.1), (1.2) can be written as

$$Lx + Nx = 0. \tag{2.3}$$

Since f is continuous on $I \times R^n \times R^n$, $N: C^1 \to C$ is continuous and maps bounded sets into bounded sets (i.e., N is continuous and bounded).

Since $L: D_L \subset C^2 \to C$ has a trivial null space, an inverse, H, exists which we write as

$$Hx = \int_0^t (1 - t)sx(s)ds - \int_t^1 t(s - 1)x(s)ds - (1 - t)x_0 - tx_1. \tag{2.4}$$

Further $H: C \to D_L$ is linear, continuous, and compact. (Compact means that H maps bounded sets into precompact sets. This holds whether we regard D_L as a subspace of C^2 or a subset of C^1.)

Then (2.3) is equivalent to

$$x + HNx = 0, \quad \text{where} \quad HN: C^1 \to D_L \subset C^1 \tag{2.5}$$

is continuous and compact. We treat this as a problem from C^1 into C^1, but when we find the solution x, since $x = -HNx$, it will be in $D_L \subset C^2$.

To such a problem we can immediately apply fixed point theorems or degree theory and this is summarized in the next three theorems.

<u>Theorem 2.1</u>. Let problem (1.1), (1.2) be given and assume that

$$|f(t,x_1,y_1) - f(t,x_2,y_2)| \leq k|x_1 - x_2| + m|y_1 - y_2|$$

on $I \times R^n \times R^n$, where k and m are sufficiently small. Then the problem has a unique solution.

<u>Proof</u>. Choose k, m such that $(-HN)$ is a strict contraction and the proof follows directly from the Contraction Mapping Principle. This was originally proved by Picard [22].

<u>Theorem 2.2</u>. Let problem (1.1), (1.2) be given and assume that $|f(t,x,y)| \leq M$ on $I \times R^n \times R^n$. Then the problem has at least one solution.

Bass [1] relates the work of Birkhoff and Kellogg and of Scorza-Dragoni on this problem and also gives a proof using the Schauder fixed point theorem. For if $r = 4M + |x_0| + |x_1|$ and $S_r = \{x \in C^1: |x|_1 \leq r\}$, then $(-HN): S_r \to S_r$.

<u>Theorem 2.3</u>. Let problem (1.1), (1.2) be given and assume there exists a bounded, open set $\Omega \subset C^1$ such that

i) the function $(1 - t)x_0 + tx_1 \in \Omega$; and ii) if $x_\lambda(t)$ is a solution of

$$x'' = \lambda f(t,x,x'), \quad x(0) = x_0, \quad x(1) = x_1, \tag{2.6}$$
$$(0 < \lambda < 1)$$

then $x_\lambda(\cdot) \notin \partial\Omega$. Then the problem has at least one solution in Ω.

Proof. With H given by (2.4), set $x_0 = 0 = x_1$ and call it H_0. The hypotheses state that $(1 + \lambda H_0 N)x - [(1 - t)x_0 + tx_1] \neq 0$ for $x \in \partial\Omega$, $0 < \lambda < 1$, where here I is the identity map. Hence

$$\text{Deg}(I + H, \Omega, 0) = \text{Deg}(I - [(1 - t)x_0 + tx_1], \Omega, 0) = 1$$

and the conclusion follows.

This theorem, the Leray-Schauder continuation method, can be used on our Picard problem to obtain the following sequence of results.

Corollary 2.4. Let problem (1.1), (1.2) be given and assume that there exists constants $C > 0$, $K \geq 0$ such that

$$|y|^2 + x \cdot f(t,x,y) \geq -K(1 + |x| + |x \cdot y|) + C|f(t,x,y)| \quad \text{on} \quad I \times R^n \times R^n. \tag{2.7}$$

Then the problem has at least one solution.

As stated the lemma is due to Lasota and Yorke [15]. By a careful analysis of the growth of solutions, they show there is an $r > 0$ such that any solution $x(t)$ of (2.6) satisfies $|x|_1 < r$.

They also prove that if (2.7) is replaced by

i) there exist constants $C > 0$, $K \geq 0$ such that

$|y|^2 + x \cdot f(t,x,y) \geq -K(1 + |x| + |x \cdot y|) + c|y|$

on $I \times R^n \times R^n$;

and

ii) there exists a Nagumo function for f on $I \times R^n \times R^n$;

then Corollary 2.4 is still true.

A continuous function $\varphi:[0,\infty) \to (0,\infty)$ satisfying $\int_0^\infty \frac{sds}{\varphi(s)} = \infty$ and $|f(t,x,y)| \leq \varphi(|y|)$ on $I \times R^n \times R^n$ is called a Nagumo function for f on $I \times R^n \times R^n$.

A version of Corollary 2.4 was proved earlier by Hartman [6]. Loosely speaking, Hartman proves

Corollary 2.4a. Let problem (1.1), (1.2) be given and assume

i) $|y|^2 + x \cdot f(t,x,y) > 0$ on $\{(t,x,y) \in I \times R^n \times R^n: |x| \geq r > \max(|x_0|, |x_1|),\ x \cdot y = 0\}$;

ii) there exists constants $c > 0$, $K \geq 0$ such that $|y|^2 + x \cdot f(t,x,y) \geq -K + c|f(t,x,y)|$ on $I \times R^n \times R^n$; and

iii) There exists a Nagumo function for f on $I \times R^n \times R^n$.

Then the problem has at least one solution.

If $x_\lambda(t)$ is a solution of (2.6) and $u(t) = |x_\lambda(t)|^2$, then (i) is a condition on $u''(t)$ which implies a) that $|x_\lambda(t)|$ has a maximum value less than r, and b) $|x_\lambda(t_0)| = r$ and $|x_\lambda(t)| < r$ for $t \in (t_0 - \delta, t_0 + \delta) \subset I$ does not occur.

Conditions (ii) and (iii) (the vector version of a Nagumo condition - discussed thoroughly in [6]) imply that with $|x_\lambda(\cdot)|_0 \leq r$, there is a constant $s = s(r) > 0$, independent of λ and $x_\lambda(t)$, such that $|x_\lambda'(\cdot)|_0 < s$.

To prove Corollary 2.4a we apply the theorem to $\Omega = \{x \in C^1 : |x(\cdot)|_0 < r,\ |x'(\cdot)|_0 < s\}$. No $x_\lambda \in \partial\Omega$, because no $(x_\lambda(t), x_\lambda'(t)) \in \{(x,y) : |x| \leq r,\ |y| < s\}$ for $0 \leq t \leq 1$ with $x_\lambda(t) = r$ at some $t = t_0 \in I$.

To be precise Hartman has ≥ 0 in place of > 0 and $= r$ in place of $\geq r$ in (i). He also assumes that f is defined and satisfies (ii) and (iii) only on $\{(t,x,y) : 0 \leq t \leq 1,\ |x| \leq r,\ |y| < \infty\}$. He then extends f to be defined for all x, finds the solution, and by (i) observes that the solution lies in the original domain of f. For the details see [6] or [7].

Remarks on Corollary 2.4a.

1) As Lasota and Yorke show, with (ii) holding on $I \times R^n \times R^n$ conditions (i) and (iii) are not needed for the Picard problem. Condition (i), or something more, is needed if f is defined only for $|x| \leq r$ or for the periodic problem where the end points of the solution are not "tied down."

2) Some authors replace (ii), (iii) by a "generalized Nagumo condition" of which the following is one variant: Given $r > 0$ there exists an $s(r) > 0$ such that if $x_\lambda(t)$ is a solution of (2.3) with $|x_\lambda(t)| < r$ for $t \in J \subset I$, then $|x_\lambda'(t)| < s$ for $t \in J$. See [25], for example.

3) Another variation is to replace $|x|^2$ by a Lyapunov-type function. We assume there exists a function $V:R^n \to R$ which is of class C^2 and satisfies: i) for some $r > 0$ the set $\Omega_0 = \{x:V(x) < r\}$ is bounded and $(1 - t)x_0 + tx_1 \in \Omega_0$ for $0 \leq t \leq 1$; ii) $(V_{xx}(x)y) \cdot y + \lambda V_x(x) \cdot f(t,x,y) > 0$ on $\{(\lambda,t,x,y):0 < \lambda < 1,\ 0 \leq t \leq 1,\ V(x) = r,\ y$ satisfies $V_x(x) \cdot y = 0\}$; iii) there exists a constant $s = s(V(x),r) > 0$ such that if $x_\lambda(t)$ is a solution of (2.3), then $v(x_\lambda(t)) < r$ implies $|x_\lambda'(t)| < s$, and we conclude that the problem has at least one solution.

The proof applies the theorem to $\Omega = \{x \in C^1:x(t) \in \Omega_0,\ |x'(t)| < s$ for $0 \leq t \leq 1\}$ as in Corollary 2.4a. See [5] and [17].

4) A third variation replaces the ball $|x| \leq r$ by a convex set. We assume that Ω_0 is a bounded convex set in R^n such that $(1 - t)x_0 + tx_1 \in \Omega_0$ for $0 \leq t \leq 1$ and that there exists a function $N:\partial\Omega_0 \to R^n$ which satisfies: i) $N(x) \cdot x > 0$ for all $x \in \partial\Omega_0$ and $\bar{\Omega}_0 \subseteq \{y:N(x) \cdot (y - x) \leq 0$ for each $x \in \partial\Omega_0\}$; ii) $N(x) \cdot f(t,x,y) > 0$ on $\{(t,x,y):0 \leq t \leq 1,\ x \in \partial\Omega_0,\ y$ satisfies $N(x) \cdot y = 0\}$; and we assume that there exists a Nagumo function for f on Ω_0. Then, we conclude that the problem has at least one solution. See [5].

5) Some results using upper and lower solutions can be obtained by this method. We assume that x and f are scalar valued (to simplify the statement of the conditions); we assume that there exist real valued functions $\alpha(t)$, $\beta(t)$

of class C^2 on I such that: i) $\alpha''(t) \geq f(t,\alpha(t),\alpha'(t))$, $\beta''(t) \leq f(t,\beta(t),\beta'(t))$; ii) $\alpha(0) \leq x_0 \leq \beta(0)$, $\alpha(1) \leq x_1 \leq \beta(1)$; iii) $\alpha''(t) \geq 0 \geq \beta''(t)$ all for $0 \leq t \leq 1$; and we assume that there exists a Nagumo function for f on $\Omega_0 = \{(t,x,y): 0 \leq t \leq 1,\ \alpha(t) \leq x \leq \beta(t),\ |y| < \infty\}$. Then we conclude that the problem has at least one solution.

This is not the best form of the result in that by using other methods hypothesis (iii) is not needed. But by modifying f, it should be possible to drop (iii).

The corollary was first proved by Nagumo [21], without (iii). Hukuhara [8] discusses the work of Nagumo and also Okumura, and Jackson and his collaborators have studied the problem extensively. Recent results can be found in [9] and in Schrader's papers [24], [25].

6) The use of Leray-Schauder methods, as discussed in the preceding remarks, can be found in Bebernes [2] and Mawhin [17].

3. The Periodic Problem in $C^1[0,1]$. Let $C = C(I)$, $C^n = C^n(I)$, $|x|$, $|x|_0$, and $|x|_n$ be as in Section 2 and let L and N be as in (2.1) and (2.2), respectively, except now $D_L = \{x \in C^2 : x(0) = x(1),\ x'(0) = x'(1)\}$. Then the periodic problem (1.1), (1.3) can be written as

$$Lx + Nx = 0. \tag{3.1}$$

Again, $N: C^1 \to C$ is continuous and bounded.

Let S_0 be the constant functions on $[0,1]$ and define the operator P by $Px = \int_0^1 x(t)\,dt$, $D_P = C$. Then the null space of L, $N_L = S_0$; $P:C \to S_0$; and P is a projection and it is compact.

Let I be the identity operator, this will not be confused with $I = [0,1]$, and define the operator H by

$$Hx = \int_0^t\int_0^s x(r)drds + t\int_0^1\int_0^t x(s)dsdt + x(0)$$

$$= \int_0^t (1 - t)sx(s)ds - \int_t^1 t(s - 1)x(s)ds + x(0),$$

$$D_H = (I - P)C. \tag{3.2}$$

Then $H(I - P):C \to D_L$ is linear, continuous, and compact and this holds whether we regard D_L as a subspace of C^2 or a subset of C^1.

It is easily verified that i) $H(I - P)Lx = (I - P)x$ for all $x \in D_L$; ii) $PLx = LPx$ for all $x \in D_L$; and $LH(I - P)Nx = (I - P)Nx$ for all $x \in C^1$. Hence (see Cesari, this volume), (3.1) is equivalent to the alternative problem

$$x - Px + H(I - P)Nx = 0 \tag{3.3}$$

$$PNx = 0. \tag{3.4}$$

As in Section 2, we will treat this as a problem from C^1 into C^1, but the solution, when we find it, will be in $D_L \subset C^2$.

Let $S_0 = PC^1$ (Strictly speaking this is $P|C^1$. Also, $PC^1 = PC^0$.) and $S_1 = (I - P)C^1$. If $x = x_1 + x_0 \in S_1 \oplus S_0$, let the norm of x be $|x_1|_1 + |x_0|$. Then $C^1 = S_1 \oplus S_0$ is a Banach space and we will treat (3.3) as a problem from $S_1 \oplus S_0$ into $S_1 \oplus S_0$.

We first note that neither the Contraction Mapping Principle nor the Schauder fixed point theorem can be applied to (3.3) set in $S_1 \oplus S_0$. And the example $x'' = 1$, which has no periodic solutions, shows that periodic analogs to Theorems 2.1 and 2.2 do not exist.

Define $T_\lambda : S_1 \oplus S_0 \to S_1 \oplus S_0$ by

$$T_\lambda \begin{pmatrix} x_1 \\ x_0 \end{pmatrix} = \begin{pmatrix} \lambda H(I - P)N(x_1, x_0) \\ -x_0 + PN(x_1, x_0) \end{pmatrix} \quad \text{for } 0 \le \lambda \le 1 \qquad (3.5)$$

($N(x_1, x_0) = N(x_1 + x_0)$; (x_1, x_0) will be written as either a row or column vector.)

The following hold:

a) For each λ, T_λ is compact and $T_\lambda(x)$ is continuous in (λ, x).

b) For $0 < \lambda \le 1$, $(I + T_\lambda)(x_1, x_0) = 0$ if and only if $Lx + \lambda Nx = 0$.

c) If $\Omega \subset S_1 \oplus S_0$ is open and bounded and if $0 \notin (I + T_\lambda)(\partial\Omega)$ for $0 < \lambda \le 1$ and $0 \notin (PN|S_0 \cap \partial\Omega)$, then $\deg(I + T_1, \Omega, 0) = \deg(PN|(S_0 \cap \Omega), S_0 \cap \Omega, 0)$.

To see (c), note that $\deg(I + T_1, \Omega, 0) = \deg(I + T_0, \Omega, 0)$. But $I + T_0$ is a finite dimensional perturbation of the identity so following Schwartz [26], for example, $\deg(I + T_0, \Omega, 0) = \deg((I + T_0)|(S_0 \cap \Omega), (S_0 \cap \Omega), 0) = \deg(PN|(S_0 \cap \Omega), (S_0 \cap \Omega), 0)$.

From (a)-(c) we obtain

Theorem 3.1. Let $\Omega \subset C^1$ be a bounded open set and suppose

i) $Lx + \lambda Nx \ne 0$ for $x \in \partial\Omega$, $0 < \lambda < 1$;

ii) $PNx \ne 0$ for $x \in (S_0 \cap \partial\Omega)$; and

iii) $\deg(PN|(S_0 \cap \Omega), S_0 \cap \Omega, 0) \ne 0$.

Then (3.1) has at least one solution in Ω.

(Note: For $a \in S_0$, $PNa = \int_0^1 f(t,a,0)dt$.)

Corollary 3.2. Let problem (1.1), (1.3) be given and assume

i) $|y|^2 + x \cdot f(t,x,y) > 0$ on $\{(t,x,y) \in I \times R^n \times R^n$: $|x| \geqq r$, for some $r > 0$, and $x \cdot y = 0\}$;

ii) $|y|^2 + x \cdot f(t,x,y) > -K + c|f(t,x,y)|$ on $I \times R^n \times R^n$; and

iii) there exists a Nagumo function for f on $I \times R^n \times R^n$.

Then the problem has at least one solution, $x(t)$, and $|x(\cdot)|_0 \leq r$.

Proof. Proceeding as in Corollary 2.4a we find a region Ω such that hypothesis (i) of the theorem holds.

Then (i) of the corollary with $y = 0$, implies $x \cdot f(t,x,0) > 0$ if $|x| = r$ and this implies (ii) and (iii) of the theorem.

1) The approach of Theorem 3.1 is adapted from Mawhin, who uses it in ordinary and partial differential equations (see [16]-[20]).

2) The evolution of Corollary 3.2 can be followed through Hartman [6], Knobloch [13], Schmitt [23], and Bebernes and Schmitt [3]. As in Section 2, Hartman states this corollary with weaker assumptions in (a) and with f defined only for $|x| \leq r$.

3) This idea of a direct sum decomposition, we use $C^1 = S_1 \oplus S_0$ here, can be found in Cesari [4]. Knobloch

cites Cesari and here we again see how Knobloch's result fits into this general functional analytic method. The results of Bebernes and Schmitt [3] and of Bebernes [2] also fit into the framework of Theorem 3.1.

4) Variations like those given in Remarks 2-5, Section 2 can also be made on Corollary 3.2 (see [2]). Also perturbation methods (see Kannan, this volume, and Bebernes [2]) can be used to prove such results.

4. The Periodic Problem in $L^2[0,1]$. In addition to the spaces of Section 2 let $L^p = L^p[0,1]$ $(1 \leq p \leq \infty)$ denote the Lebesgue spaces with the usual norm denoted by $\|\cdot\|_p$. On L^2, $(\cdot,\cdot)$ is the inner product and we usually write the norm $\|\cdot\|$. By $H^p = H^p[0,1]$ $(p = 1,2)$ we mean the Sobolev space of functions which have a p-th derivative that is in L^2. For $x \in H^p$, let $\|x\|_{Hp} = |x|_0 +\cdots+ |x^{(p-1)}|_0 + \|x^{(p)}\|$. This is equivalent to the norm $\|x\| +\cdots+ \|x^{(p-1)}\| + \|x^{(p)}\|$.

Let L and N be as in (2.1) and (2.2), respectively, except now $D_L = \{x \in H^2 : x(0) = x(1),\ x'(0) = x'(1)\}$ and $D_N = H^1$. Then the periodic problem (1.1), (1.3) can be written as

$$Lx + Nx = 0. \tag{4.1}$$

Define the operator P by $Px = \int_0^1 x(t)dt$, $D_p = L^2$ and define H as in (3.2) except $D_H = (I - P)L^2$. Again the null space of L is the constant functions, $N_L = S_0$; $P: L^2 \to S_0$ is a projection and it is compact; and we denote $(I - P)L^2$ by S_1.

Further, L is a positive, self-adjoint operator with the range of L, R_L, equal to S_1 and $H = (L|(D_L \cap S_1))^{-1}$, $R_H = (D_L \cap S_1)$. We shall assume $|f(t,x,y)| \leq |y|^2$ on $I \times R^n \times R^n$ (for simplicity - later we relax this), so $R_N \subset L^1 \not\subset D_H$ and this will have to be corrected.

Again (i)-(iii) of page 9 are satisfied (read L^2 in place of C^1) and (4.1) is equivalent to

$$x - Px + H(I - P)Nx = 0 \tag{4.2}$$

$$PNx = 0. \tag{4.3}$$

A solution x to (4.2), (4.3) will be in $D_L \subset H^2$. Then $Nx \in C[0,1]$, $H(I - P)Nx \in C^2$, and hence $x \in C^2$.

To correct the trouble with R_N we shall split the operator H. This device is studied in Kannan and Locker [10].

Define the operator T by

$$D_T = \{x \in H^1 : x(0) = x(1)\}, \quad Tx = x'$$

and define the operator T^* by $T^* = -T$. Then T^* is the adjoint of T; $N_T = S_0 = N_{T^*}$; and $T^*:(D_L \cap S_1) \to (D_T \cap S_1)$ and $T:(D_T \cap S_1) \to R_L = S_1$ are one-to-one and onto. Let J^* and J be the inverses of T^* and T, respectively.

If $y = x - x_0$, then (4.2), (4.3) may be written as

$$y + J^*J(I - P)N(y + x_0) = 0, \quad x_0 = Px \in S_0 \tag{4.4}$$

$$PN(y + x_0) = 0. \tag{4.5}$$

If a solution y to (4.4), (4.5) exists, then $y = J^*z$ where $z \in (D_T \cap S_1)$. Also, in (4.4), J^* is one-to-one. Thus (4.4), (4.5) is equivalent to

$$z + J(I - P)N(J^*z + x_0) = 0, \quad x_0 \in S_0 \tag{4.6}$$

$$PN(J^*z + x_0) = 0. \tag{4.7}$$

The operators J, J^* and P have integral representations: $Jx = \int_0^t x(s)ds - P(\int_0^t x(s)ds) = -J^*x$ and $Px = \int_0^t x(t)dt$, for $x \in L^2$.

Using these representations, let K be the same as J except $D_K = L^1$ and let K^* be the same as J^* except $D_{K^*} = L^2$.

We shall solve the problem

$$z + K(I - P)N(K^*z + x_0) = 0, \quad x_0 \in S_0 \tag{4.8}$$

$$PN(K^*z + x_0) = 0 \tag{4.9}$$

as a problem from L^2 into L^2. The solution z (we do not have to worry about the constant x_0 in the discussion at this point) will be in S_1; $(I - P)N(K^*z + x_0)$ will be in L^1 (because of the bound on f); $K(I - P)N(K^*z + x_0)$ will be absolutely continuous; and hence z will be absolutely continuous. Repeating this procedure we will have z in C^1 and thus z is in the domain of J^*, $(I - P)N(J^*z + x_0)$ is in the domain of J, and z is a solution of (4.6), (4.7). From there we know it is a solution of (4.1).

We shall treat (4.8), (4.9) as a problem from $L^2 = S_1 \oplus S_0$ into $S_1 \oplus S_0$. If $\beta = z + x_0 \in S_1 \oplus S_0$, let the norm of β be $\|\beta\|_+ = (\|z\|^2 + \|x_0\|^2)^{1/2}$ and for an inner product let $\langle\beta,\tilde{\beta}\rangle = (z,\tilde{z}) + (x_0,\tilde{x}_0)$. Then we still have a Hilbert space. Define $T:S_1 \oplus S_0 \to S_1 \oplus S_0$ by

$$T\begin{pmatrix} z_1 \\ x_0 \end{pmatrix} = \begin{pmatrix} K(I - P)N(K^*z_1,x_0) \\ -x_0 + PN(K^*z_1,x_0) \end{pmatrix}, \quad (N(K^*z_1,x_0) = N(K^*z_1 + x_0))$$

Then (4.8), (4.9) is equivalent to

$$(I + T)(z_1,x_0) = 0. \tag{4.10}$$

We claim that T is a continuous, compact mapping.

To show that $K^*:S_1 \to H^1 \cap S_1$ is continuous and bounded is straightforward; we shall show that $N:H^1 \to L^1$ is continuous and bounded; $(I - P)$ is a projection; P is a finite dimensional projection; and to show that $K:L^1 \to L^2$ is continuous and compact is again straightforward. From this the claim follows.

$N:H^1 \to L^1$ is bounded:

$$\|Nx\|_1 = \int_0^1 |f(t,x(t),x'(t)|dt \leq \int_0^1 |x'(t)|^2 \, dt \leq \|x\|_{H2}$$

$N:H^1 \to L^1$ is continuous:

$$\|Nx - Ny\|_1 \leqq \int_0^1 |f(t,x(t),x'(t)) - f(t,y(t),x'(t))|dt$$
$$+ \int_0^1 |f(t,y(t),x'(t)) - f(t,y(t),y'(t))|dt.$$

The first integral: With $x'(t) \in L^2$,

$$g(t,c) = f(t,c,x'(t))$$

is continuous in c and measurable in t. Hence by Krasnosel'skii's version of Lusin's theorem [14] we may divide $[0,1]$ into disjoint subsets I_1 and I_2 such that

I_1 is closed and $g|I_1 \times R^n$ is continuous;

I_2 has arbitrarily small measure.

We have just seen that the integrands are bounded so we can choose I_2 such that the integral over I_2 is less than $\epsilon/4$. On $I_1 \times \{x:|x - y(t)|_0 \leq 1\}$, f is uniformly continuous, hence if $|x(t) - y(t)|_0$ is sufficiently small, the integral over I_1 is less than $\epsilon/4$.

We now state the theorem.

<u>Theorem 4.1</u>. Let problem (1.1), (1.3) be given and assume

i) $|f(t,x,y)| \leq c|y|^2$, $c > 0$ constant, on $I \times R^n \times R^n$; and

ii) $\|x'(\cdot)\|^2 + \int_0^1 \{x(t) \cdot f(t,x(t),x'(t)\}dt \geq 0$ on ∂B_r where $B_r = \{x \in H^2: |\int_0^1 x(t)dt|^2 + \|x'(\cdot)\|^2 \leqq r^2\}$ for some $r > 0$.

Then the problem has at least one solution which lies in B_r.

Proof. We need to solve (4.10). Note that if $<(I + \lambda T)(z,x_0),(z,x_0)> > \lambda[<(I + T)(z,x_0),(z,x_0)>]$ if $0 < \lambda < 1$ and $(z,z) + (x_0,x_0) > 0$. Hence, if $[—] \geq 0$, then by usual degree theoretic arguments (4.10) will have a solution.

Now $<(I + T)(z,x_0),(z,x_0)> = (z,z) + (x_0,x_0) + (K(I - P)N(K^*z,x_0),z) + (-x_0 + PN(K^*z,x_0),x_0) = \|z\|^2 + [(I - P)N(K^*z,x_0),J^*z] + (PN(K^*z,x_0),x_0)$ (Here $K(I - P):L^1 \to (L^2 \cap S_1)$ and $z \in (L^2 \cap S_1)$. So the dual operator $K':L^2 \to L^\infty$. But on L^2, $K' = J^*$. And $[u,v]$ is the pairing between $u \in L^1$, $v \in L^\infty$.) $= \|z\|^2 + [N(K^*z,x_0),J^*z + x_0] - [PN(K^*z,x_0),J^*z] - [(I - P)N(K^*z,x_0),x_0]$. (Since $[u,v]$ is an integral we are just adding and subtracting integrals.) $= \|x'(\cdot)\|^2 + \int_0^1 \{f(t,x(t),x'(t)) \cdot x(t)\}dt$ (Since $PN \in S^0$, $J^*z \in S^1$ the second bracket is an inner product and vanishes. The third bracket vanishes by computation - it is not an inner product. Then $x = J^*z + x_0$ and $x' = z$.)

The theorem follows.

1) The advantages of this theorem are that it does not require a Nagumo condition and it may be extended to higher order problems.

2) By using Orlicz-Sobolev spaces in place of H^2, hypothesis (i) can be replaced by

i′) $|f(t,x,y)| \leq g(t,x) + \exp(|y|^9)$, on $I \times R^n \times R^n$ where $g: I \times R^n \to [0,\infty)$ takes bounded sets into bounded sets. The exponential is merely an example of a function satisfying a so-called Δ_3-condition.

For the extension to higher order problems and the use of Orlicz-Sobolev spaces see [12] by R. Kannan and this author.

BIBLIOGRAPHY

1. Bass, R.W., "On nonlinear repulsive forces," Contributions to the theory of nonlinear oscillations (edited by S. Lefschetz) Annals of Mathematics Studies No. 41, Princeton University, Princeton, 1958, 201-211.

2. Bebernes, J.W., "A simple alternative problem for finding solutions of second order ordinary differential systems," Proc. Amer. Math. Soc. 42 (1974), 121-127.

3. Bebernes, J.W. and K. Schmitt, "Periodic boundary value problems for systems of second order differential equations," J. Differential Equations 13 (1973), 32-47.

4. Cesari, L., "Functional analysis and Galerkin's method," Mich. Math. J. 11 (1964), 385-414.

5. Gustafson, G. and K. Schmitt, "A note on periodic solutions for delay-differential systems," J. Differential Equations 13 (1973), 567-487.

6. Hartman, P., "On boundary value problems for systems of ordinary, nonlinear, second order differential equations," Trans. Amer. Math. Soc. 96 (1960), 493-509.

7. Hartman, P., <u>Ordinary Differential Equations</u>, John Wiley, New York, 1964.

8. Hukuhara, M., "Kneser family and boundary value problems," Proc. U.S. - Japan seminar on differential and functional equations (edited by W.A. Harris and Y. Sibuya), W.A. Benjamin, New York, 1967, 99-111.

9. Jackson, L.K. and K.W. Schrader, "Comparison theorems for nonlinear differential equations," J. Differential Equations 3 (1967), 248-255.

10. Kannan, R. and J. Locker, "Nonlinear boundary value problems and operators TT*," to appear.

11. Kannan, R. and J.D. Schuur, "Boundary value problems for even order ordinary differential equations," Bull. Amer. Math. Soc. 82 (1976), 80-82.

12. Kannan, R. and J.D. Schuur, "Nonlinear boundary value problems and Orlicz spaces," to appear.

13. Knoblock, H.-W., "On the existence of periodic solutions of second order vector differential equations," J. Differential Equations 9 (1971), 67-85.

14. Krasnosel'skii, M.A., <u>Topological Methods in the Theory of Nonlinear Integral Equations</u>, Macmillan, New York, 1964.

15. Lasota, A. and J.A. Yorke, "Existence of solutions of two-point boundary value problems for nonlinear systems," J. Differential Equations 11 (1972), 509-518.

16. Mawhin, J., "Equivalence theorems for nonlinear operator equations and coincidence degree theory for some mappings in locally convex topological vector spaces," J. Differential Equations 12 (1972), 610-636.

17. Mawhin, J., "Boundary value problems for nonlinear second-order vector differential equations," J. Differential Equations 16 (1974), 257-269.

18. Mawhin, J., "Nonlinear perturbations of Fredholm mappings in normed spaces and applications to differential equations," Univ. de Brasilia Trabalho de Mat. No. 61, May 1974.

19. Mawhin, J., "Topology and nonlinear boundary value problems," International Symposium on Dynamical Systems, Brown Univ., August 1974.

20. Mawhin, J., "Recent results on periodic solutions of differential equations," International Conference on Differential Equations (edited by H. Antosiewicz), Academic Press, New York, 1975.

21. Nagumo, M., "Über die Differentialgleichung $y'' = f(x,y,y')$," Proc. Phys. Math. Soc. Japan (3) 19 (1937), 861-866.

22. Picard, É., "Sur l'application des méthodes d'approximation successive a l'etude de certaines équations differentielles ordinaires," J. Math. pures appl. (4) 9 (1893), 217-271.

23. Schmitt, K., "Periodic solutions for systems of second order differential equations," J. Differential Equations 11 (1972), 180-192.

24. Schrader, K.W., "Boundary-value problems for second-order differential equations," J. Differential Equations 3 (1967), 403-413.

25. Schrader, I.W., "Existence theorems for second order boundary value problems," J. Differential Equations 5 (1969), 572-584.

26. Schwartz, J., Nonlinear Functional Analysis, Gordon and Breach, New York, 1969.

A NONLINEAR ELLIPTIC BOUNDARY VALUE PROBLEM

Howard Shaw

University of Massachusetts

1. Introduction and Preliminaries. We are concerned here with existence theorems for nonlinear elliptic boundary value problems of the form

$$Lu = h(x) - g(u) \tag{1}$$

for $x \in \Omega$, where $u = u(x)$ satisfies homogeneous Dirichlet conditions on $\partial\Omega$. Here L is a uniformly elliptic, linear partial differential operator of order $2m$ with smooth, real coefficients on $\bar{\Omega}$, and Ω is a bounded, connected, open set in $\mathbb{R}^n$ with smooth boundary $\partial\Omega$. We assume that L is "simply resonant", i.e. there exist non-zero functions θ and ψ in $\overset{\circ}{H}{}^m(\Omega)$ such that for $u \in \overset{\circ}{H}{}^m(\Omega)$, $Lu = 0$ if and only if $u = c\theta$ for a scalar c and such that for a given $f \in L^2(\Omega)$, $Lu = f$ is solvable in $\overset{\circ}{H}{}^m(\Omega)$ if and only if f is orthogonal to ψ (in the sense of $L^2(\Omega)$). Here g is a continuous real-valued function on $\mathbb{R}$ for which the limits $g(\infty) = \lim_{s\to\infty} g(s)$ and $g(-\infty) = \lim_{s\to-\infty} g(s)$ both exist and are finite. We define $\Omega_+ = \{x \in \Omega | \psi(x) > 0\}$ and $\Omega_- = \{x \in \Omega | \psi(x) < 0\}$. Let h be an element of $L^2(\Omega)$ satisfying

$$g(-\infty)\int_{\Omega_+}|\psi| - g(\infty)\int_{\Omega_-}|\psi| < \int_\Omega h\psi < \\ < g(\infty)\int_{\Omega_+}|\psi| - g(-\infty)\int_{\Omega_-}|\psi| \tag{*}$$

We also assume that θ and ψ share regions of positivity and negativity, i.e., that $\theta(x) > 0$ for $x \in \Omega_+$ and $\theta(x) < 0$ for $x \in \Omega_-$.

<u>Theorem</u>. Under the preceding assumptions, there exists a (weak) solution $u \in \overset{\circ}{H}{}^m(\Omega)$ to the nonlinear Dirichlet problem (1).

This theorem is proved in a most noteworthy paper by Landesman and Lazer [6] for L a self-adjoint operator. In this case $\theta \equiv \psi$, and so the hypothesis of "shared regions of positivity" is trivially satisfied. This hypothesis can be verified, however, in some important non-self-adjoint cases, as we will see later. The proof in [6] appears to be quite complicated; we will give here a simpler proof that permits generalizations of the theorem in several directions and also gives some geometric insight into the meaning of the condition (*). Here, as in [6], use is made of the alternative method of Cesari [2], [3].

2. <u>Proof of the Theorem</u>. We first define two linear operators P and Q on $L^2(\Omega)$ by orthogonal projection onto θ and ψ, respectively. We may assume that $\|\theta\|_{L^2} = \|\psi\|_{L^2} = 1$, and so $Pu = \langle u, \theta\rangle\theta$, $Qu = \langle u, \psi\rangle\psi$. We wish to solve

$$Lu = h - g(u), \quad u \in \overset{\circ}{H}{}^m(\Omega). \tag{1}$$

Write $u = v + Pu = v + c\theta$ where v is orthogonal to θ, and c is a scalar. (1) now takes the form

$$Lv = h - g(v + c\theta), \quad v \in \overset{\circ}{H}{}^m(\Omega). \tag{1'}$$

We now construct Γ, a partial inverse of L. By the linear theory for elliptic operators, L is a continuous map from

$H^{2m} \cap \overset{\circ}{H}{}^m$ into L^2. Our assumptions on L imply that as a map from $\{v \in H^{2m} \cap \overset{\circ}{H}{}^m | v \perp \theta\}$ to $\{f \in L^2 | f \perp \psi\}$, L is a bijection. We define Γ from $\{f \in L^2 | f \perp \psi\}$ to $\{v \in H^{2m} \cap \overset{\circ}{H}{}^m | v \perp \theta\}$ as the inverse of L; Γ is continuous by the open mapping theorem. We regard the range of Γ as embedded in $\overset{\circ}{H}{}^m$; as such, Γ is a compact map by the Rellich compactness theorem. Consider the equations:

$$v = \Gamma(I - Q)\{h - g(v + c\theta)\} \tag{2}$$

$$Q\{h - g(v + c\theta)\} = 0. \tag{3}$$

It is easy to see that the solvability of (1′) is equivalent to the simultaneous solution of (2) and (3), where $v \in \overset{\circ}{H}{}^m$ and c is a scalar; see [2], [3].

For any fixed $c \in \mathbb{R}$, the (auxiliary) equation (2) has a solution $v \in \overset{\circ}{H}{}^m$ by the usual Schauder fixed point theorem. The map $v \to \Gamma(I - Q)\{h - g(v + c\theta)\}$ is a compact continuous map from $\overset{\circ}{H}{}^m$ into $\overset{\circ}{H}{}^m$, with range contained in a fixed ball in $\overset{\circ}{H}{}^m$. This last statement follows from an easily derived, a priori bound on $\|\Gamma(I - Q)\{h - g(v + c\theta)\}\|_{H^m}$ that is independent of v and c. We will use a form of the Schauder fixed point theorem that is somewhat stronger than the usual one; it concerns a one-parameter family of Schauder-type maps.

<u>Fixed Point Theorem</u>. Let B be a closed bounded convex subset of a Banach space, and let $I = [p,q]$ be a closed finite interval in $\mathbb{R}$. Assume that $F: B \times I \to B$ is a compact continuous map. (The usual Schauder theorem asserts that for each $c \in I$, $F(v,c) = v$ for some $v \in B$). Then there exists a connected set $S \subset B \times I$ of fixed points of F, i.e.,

$F(v,c) = v$ for $(v,c) \in S$. Moreover, this set S meets both $B \times \{p\}$ and $B \times \{q\}$ (the "end-discs" of $B \times I$).

This theorem is a consequence of the Leray-Schauder fixed point theory; for a proof, see Browder, [1], or Leray and Schauder, [7]. In our application, $F(v,c) = \Gamma(I - Q)\{h - g(v + c\theta)\}$. We conclude that for any compact interval $[p,q] \subset \mathbb{R}$, there is a connected set (in $\overset{\circ}{H}{}^m \times [p,q]$) of solutions to (2) that "stretches" from $\overset{\circ}{H}{}^m \times \{p\}$ to $\overset{\circ}{H}{}^m \times \{q\}$. If we pick $[p,q]$ properly, we will be able to satisfy the (bifurcation) equation (3) on this connected set, thus proving the main theorem.

Equation (3) may be written as

$$\int_\Omega (h - g(v + c\theta))\psi = 0. \qquad (3')$$

We may expand this as

$$\int_\Omega h\psi - \int_{\Omega_+} g(v + c\theta)|\psi| + \int_{\Omega_-} g(v + c\theta)|\psi| = 0. \qquad (3'')$$

If we compare the left side of (3″) with

$$\int_\Omega h\psi - g(\infty)\int_{\Omega_+} |\psi| + g(-\infty)\int_{\Omega_-} |\psi|, \qquad (4)$$

it is quite plausible that the left side of (3″) approaches (4) for sufficiently large positive c. (This requires the hypothesis of "shared regions of positivity and negativity" for θ and ψ.) It is not difficult to prove this rigorously (using our a priori bound on $\|v\|_{L^2}$); we omit the details. The expression (4) is negative by hypothesis (*); it follows that for some large positive q, any solution (v,q) to (2) $F(v,q) = v$ gives us

$$\int_\Omega (h - g(v + q\theta))\psi < 0.$$

Similarly, for some large negative p, any solution (v,p) to (2) $F(v,p) = v$ gives us

$$\int_\Omega (h - g(v + p\theta))\psi > 0.$$

By the fixed point theorem quoted earlier, we can find a connected set S which "stretches" across $[p,q]$. Considering $(v,c) \to \int_\Omega (h - g(v + c\theta))\psi$ as a continuous real-valued function on the connected set S, this function assumes values both positive and negative, and hence assume the value zero somewhere on S. There (3) is satisfied; we are on S, so (2) is satisfied, and the proof is complete.

3. <u>Remarks</u>. (a) A simple example from [6] is given by

$$\Delta u + 2u = K - \text{Arctan } u, \tag{5}$$

where $\Omega = [0,\pi] \times [0,\pi] \subset \mathbb{R}^2$, $u = 0$ on $\partial\Omega$, and $|K| < \frac{\pi}{2}$. It is clear that the theorem remains valid if both inequalities in (*) are reversed, hence we can also solve

$$\Delta u + 2u = \text{Arctan } u - K, \quad \text{with } \Omega \text{ and } K \text{ as before,} \quad u = 0 \text{ on } \partial\Omega. \tag{5'}$$

(b) Some nonlinearities involving derivatives of u can be handled by this method of proof. An example is

$$\Delta u + 2u = \text{Arctan } u + \text{Exp}(-u_x^2 - u_y^2) \tag{6}$$

on $\Omega = [0,\pi] \times [0,\pi]$, $u = 0$ on $\partial\Omega$.

(c) We give several examples of non-self-adjoint operators L that satisfy the hypothesis of "shared regions of positivity." We note first that a second order ordinary differential operator may be made self-adjoint with respect to a (topologically) equivalent inner product on L^2 (see [5]), and in this case $\psi = \rho\theta$ for some positive function ρ.

"Shared regions" therefore occur in these cases:

(i) $L = a_0(x)D_x^2 + a_1(x)D_x + a_2(x)I$ is a second order ordinary differential operator $(n = 1)$. In this case, $\rho = a_0(x)^{-1} \operatorname{Exp}(\int^x \frac{a_1(t)}{a_0(t)} dt)$. Uniform ellipticity here means that $a_0(x) \geq \mu > 0$.

(ii) L is second order and admits a "separation of variables"; specifically, L (for $n = 2$) is of the form $L = a_0(x)D_x^2 + b_0(y)D_y^2 + a_1(x)D_x + b_1(y)D_y + c_2(x,y)I$.

(iii) L is second order and has constant coefficients (except possibly for the zero order term). We can pick a new basis for $\mathbb{R}^n$ with respect to which L is the Laplacian (no mixed partials appear) plus lower order terms (see [4]). Case (ii) now applies.

(iv) L is of the form $Lu = \Delta u + (\nabla q(x)) \cdot \nabla u + c(x)u$, where q is a smooth function on $\bar{\Omega}$. In this case, $\rho = \operatorname{Exp} q(x)$.

(v) "Shared regions" always occurs at the lowest eigenvalue. Specifically, if L is second order and uniformly elliptic, one can prove (using the maximum principle; see [8]) that L has a unique eigenfunction θ that is positive in Ω, and the corresponding eigenvalue is real, simple, and is strictly smaller than the real parts of any other eigenvalues. The same holds true for L^*, whose lowest eigenfunction is ψ, and hence θ and ψ share regions of positivity (all of Ω).

(d) This method of proof can be generalized to handle the case of Fredholm indices (m,m) where $m > 1$. See [9] for the proper generalization of (*); we must assume in addition that $g' \to 0$ at ∞. As in (c), (ii)-(iv), L need not be self-adjoint. The details of this (and other extensions) will appear in [10].

(e) This technique can be applied to other boundary conditions as well. Two such examples are

$$\Delta^2 u - 4u = \text{Arctan } u + \cos(\Delta u) \tag{7}$$

where $\Omega = [0,\pi] \times [0,\pi]$, and $u = \Delta u = 0$ on $\partial\Omega$, and

$$\Delta u = K - \text{Arctan } u, \quad \frac{\partial u}{\partial n} = 0 \text{ on } \partial\Omega,$$

where Ω is a smooth bounded region and $|K| < \frac{\pi}{2}$.

BIBLIOGRAPHY

1. Browder, F., "On continuity of fixed points under deformations of continuous maps," Summa Brasil. Math. 4 (1960), Fasc. 5.

2. Cesari, L., "Functional analysis and periodic solutions of nonlinear differential equations," Contributions to Differential Equations 1, Wiley, 1963, 149-187.

3. Cesari, L., "Functional analysis and Galerkin's method," Michigan Math. J. 11 (1964), 385-414.

4. Epstein, B., Partial Differential Equations, McGraw-Hill, New York, 1962, 42-47.

5. Hildebrand, F.G., Advanced Calculus for Applications, Prentice-Hall, Englewood Cliffs, N.J., 1962, 207.

6. Landesman, E.M. and A.C. Lazer, "Nonlinear perturbations of linear elliptic boundary value problems at resonance," J. Math. Mech. 19 (1970), 609-623.

7. Leray, J. and J. Schauder, "Topologie et equations fonctionelles," Ann. Sci. Ecole Norm. Sup. #3 51 (1934), 45-78.

8. Sattinger, D., Topics in Stability and Bifurcation Theory, Lecture Notes #309, Springer-Verlag, New York, 1973, 20-22.

9. Williams, S.A., "A sharp sufficient condition for solution of a nonlinear elliptic boundary value problem," J. Differential Equations 8 (1970), 580-586.

10. Shaw, H.C., "A nonlinear elliptic boundary value problem at resonance," to appear.

CHOOSING PROJECTION OPERATORS FOR THE ALTERNATIVE METHOD

Daniel Sweet

University of Maryland and
University of California, Los Angeles

1. Introduction. In the standard functional analysis setting we let X, Z be Banach spaces and $L:D(L) \subset X \to Z$ a linear closed operator with properties that $\mathcal{N}(L)$ and $\mathcal{R}(L)$ admit projections U, E respectively. That is, $U:X \to X_U = \mathcal{N}(L)$ and $E:Z \to Z_E = \mathcal{R}(L)$. Under these assumptions there exists a unique bounded linear operator $M:\mathcal{R}(L) \subset Z \to D(L) \subset X$ where $LMz = z$ for all $z \in \mathcal{R}(L)$ and $UM = 0$ so that $MLx = (I - U)x$ for all $x \in D(L)$.

Letting $N:X \to Z$ be another operator, the standard alternative formalism tells us that solving the problem

$$Lx = Nx \tag{1}$$

is equivalent to determining an $x \in D(L)$ for which

$$x = Ux + MENx \tag{2}$$

and

$$(I - E)(Lx - Nx) = 0. \tag{3}$$

The advantages of these techniques are documented ([1], [2], [3], [5], [7]).

The alternative formalism can be generalized by searching for projections $S:X \to X_S \subset X_{I-U} \cap D(L)$ and $J:Z_E \to Z_J \subset Z_E$

whereby (1) can be equivalently expressed by the pair of equations

$$x = (U + S)x + M(I - J)ENx \tag{4}$$

and

$$(I - (I - J)E)(Lx - Nx) = 0. \tag{5}$$

That is $\mathcal{N}(L) \rightarrow$ is "expanded" by S with a corresponding "shrinking" of $\mathcal{R}(L)$ by J so that the alternative method is preserved. This is advantageous if $\|M(I - J)E\| < \|ME\|$ since we can allow for "larger" mappings N.

Defining $\mathcal{J} = \{(S,J) \mid (1)$ is equivalent to (4), (5)$\}$, it can easily be shown that for a given S, $J = LSM$ will suffice. We can see, however, the more realistic and difficult problem is that given a particular J, to determine the corresponding S where $(S,J) \in \mathcal{J}$.

2. The General Theory. From here on we discuss the important case ([4]) where Z_J is finite dimensional.

We say that J is admissible if there exists an S whereby $(S,J) \in \mathcal{J}$. A necessary condition that this happen is that $S = \overline{MJL}$ becomes a bounded operator.

Theorem 2.1. J is admissible if and only if JL is bounded.

The proof relies strongly on the compactness of Z_J ([8]).

This theorem is advantageous since the analysis of M, which often is an extremely difficult task, is avoided.

If the corresponding adjoint operator of L, L^* is employed, a constructive and natural means of determining helpful admissible projections J can be established. Letting X^*, Z^* be the corresponding dual spaces for X, Z respectively, we know that since Z_J is finite dimensional, there exists $\{z_1,\ldots,z_n\} \subset Z_E$, $\{z_1^*,\ldots,z_n^*\} \subset Z^*$ where $\langle z_i, z_j^*\rangle = \delta_{ij}$ and J can be expressed by $Jz = \sum_{j=1}^{n} \langle z, z_j^*\rangle z_j$. We then have the following results.

<u>Theorem 2.2.</u> J is admissible if and only if $\{z_1^*,\ldots,z_n^*\} \subset D(L^*)$. Furthermore $Sx = \sum_{j=1}^{n} \langle x, x_j^*\rangle x_j$ where $L^* z_j^* = x_j^*$ and $Mz_j = x_j$, $j = 1,\ldots,n$.

Using adjoint theory to determine projections for alternative problems is not novel ([6]) and this technique aids in choosing the optimum projections for determining periodic solutions for ordinary differential equations.

3. Periodic Solutions for Ordinary Differential Equations.

Consider the n-dimensional system

$$\dot{x} = A(t)x + f(t,x) \tag{1}$$

where A, f are continuous and 2π-periodic in t and also the related homogeneous and corresponding adjoint systems given by

$$\dot{x} = A(t)x \tag{2}$$

and

$$\dot{y}^T = -y^T A(t), \tag{3}$$

respectively.

We place this problem in functional form by letting $X = Z = C_{2\pi} = \{x \in C([0,2\pi],R^n) \mid x(0) = x(2\pi)\}$ with uniform norm and $L: C^1_{2\pi} \subset C_{2\pi} \to C_{2\pi}$, where $C^1_{2\pi}$ are the continuously differentiable elements of $C_{2\pi}$, by $(Lx) = \dot{x}(t) - A(t)x(t)$. Letting $X(t)$ represent the principal matrix solution of (2) at $t = 0$ $(X(0) = I)$, we can employ the Variation of Parameters formula to determine M which will have the form: for any $y \in \mathcal{R}(L)$, $(My)(t) = X(t)[x_0(y) + \int_0^t X^{-1}(\lambda)y(\lambda)d\lambda]$ where $x_0(y) \in R^n$ is chosen $UMy = 0$.

We then search for a sequence of paired finite projections $(S_\ell, J_\ell) \in \mathcal{J}$ where $\|M(I - J_\ell)\| \to 0$ as $\ell \to \infty$.

Case A. Total Resonance

Here we assume $X(t + 2\pi) = X(t)$, that is, all solutions of (2) are 2π-periodic. We can then show that the set $\{\Phi_k(t) = X(t)e^{ikt}\}$, $-\infty < k < \infty$, will accomplish the desired task. Defining, for any $h \in C_{2\pi}$, $m[h(\cdot)] = \frac{1}{2\pi}\int_0^{2\pi} h(\lambda)d\lambda$ we have that any $x \in C_{2\pi}$ has the expansion $x(t) = \sum_{-\infty}^{\infty} \Phi_k(t)m[\Phi_k^{-1}(\cdot)x(\cdot)]$. Defining for a given integer ℓ, $(J_\ell x)(t) = \sum_{|k|\leq\ell} \Phi_k(y)m[\Phi_k^{-1}(\cdot)x(\cdot)]$, it can be shown that $\|M(I - J_\ell)\| \to 0$ as $\ell \to \infty$. Also $(S_\ell x)(t) = \sum_{|k|\leq\ell} ik\psi_k(t)m[\Phi_k^{-1}(\cdot)x(\cdot)]$ where $M\Phi_k = \psi_k$. In the case (2) is self-adjoint $(A^T(t) = -A(t))$ then $M\Phi_k = (1/ik)\Phi_k$ so that $S_\ell = J_\ell$ for all ℓ.

Case B. Non-resonance

We assume (2) has no 2π-periodic solutions. In this case, letting $\{\nu_1, \ldots, \nu_n\}$ represent the characteristic exponents of (2) with corresponding eigenvectors $\{v_1, \ldots, v_n\}$, we have $X(2\pi)v_i = e^{2\pi\nu_i} v_i$, $i = 1, \ldots, n$, so that defining the non-singular matrices $V_0 = (v_1, \ldots, v_n)$, $\Lambda = \operatorname{diag}(\nu_1, \ldots, \nu_n)$ we then have $X(2\pi)V_0 = V_0 e^{\Lambda 2\pi}$. The modified functions needed to express the desired projections are then expressed by $\Phi_\ell(t) = X(t)V_0 e^{-\Lambda t} e^{i\ell t}$. In fact, by direct computation, we see $M\Phi_\ell = \Phi_\ell(i\ell I - \Lambda)^{-1}$.

By employing Floquet theory, we can greatly simplify our characterizations. We know there exists a non-singular 2π-periodic matrix $P(t)$, $P(0) = I$, and a constant matrix K whose eigenvalues are $\{\nu_1, \ldots, \nu_n\}$ where $X(t) = P(t)e^{Kt}$.

Then defining $\Gamma_k(t) = P(t)e^{ikt}$, we have

$$(J_\ell x)(t) = \sum_{|k| \le \ell} \Gamma_\ell(t) m[\Gamma_\ell^{-1}(\cdot)x(\cdot)] = (S_\ell x)(t).$$

Furthermore, $\|M(I - J_\ell)\| \le \|P\|\|P^{-1}\| \sqrt{2\pi}\, \beta_\ell$ where $\beta_\ell = (\sum_{|k|>\ell} |(ikI - K)^{-1}|^2)^{1/2} \to 0$ as $\ell \to \infty$.

The case of partial-resonance can easily be seen to be a combination of the previous two cases.

Example. $\ddot{x} + m^2 x = f(t,x)$ where $m \ne 0$, an integer. We then have the corresponding vector system $\dot{x}_1 = mx_2$, $\dot{x}_2 = -mx_1 + (1/m)f(t,x_1)$. Using the development presented for total resonance, we have $\|M(I - J_\ell)\| \le ((2\sqrt{2})/m)(\sum_{|k|>\ell} (1/k^2))^{1/2} \to 0$ as $\ell \to \infty$ independent of the choice of m. This is completely a different situation than if we used the standard Fourier series "chop-off" for which the estimate of M on the "tail-end" of the series is a function of $1/(\ell - m)$.

BIBLIOGRAPHY

1. Bancroft, S., J. Hale and D. Sweet, "Alternative problems for nonlinear functional equations," J. Differential Equations 4 (1968), 40-55.

2. Cesari, L., "Functional analysis and periodic solutions of nonlinear differential equations," Contributions to Differential Equations 1 (1963), 149-187.

3. Cesari, L., "Functional analysis and Galerkin's method," Mich. Math. J. 11 (1964), 385-414.

4. Hale, J., "Applications of alternative problems," Lecture Notes 71-1, Center for Dynamical Systems, Brown University, 1971.

5. Hale, J., Ordinary Differential Equations, Wiley-Interscience, 1969.

6. Locker, J., "An existence analysis for nonlinear equations in Hilbert space," Trans. Amer. Math. Soc. 128 (1967), 403-413.

7. Osborn, J. and D. Sather, "Alternative problems for nonlinear equations," J. Differential Equations 1 (1975), 12-31.

8. Sweet, D., "Projection operators for the alternative problem," J. of Appl. Anal., to appear.